Farming in the Dark
A Discussion About the Future of Sustainable Agriculture

Rhonda R. Janke

To Mary & Gene,
For a sustainable
future!
Rhonda Janke

University Readers™
San Diego, CA

Dr. Rhonda R. Janke has over 25 years experience in sustainable agriculture research and extension work in sustainable cropping systems. A graduate of Cornell University, she worked at the Rodale Institute for eight years overseeing cropping experiments and later as Research Director. Dr. Janke is currently an Associate Professor at Kansas State University teaching sustainable and organic agriculture. She conducts research and extension in the areas of soil and water quality, alternative crops, whole farm planning, and local foods. She continues to farm her own ten acres with her husband, where they raise vegetables, sheep, and chickens.

First published in the United States of America in 2008 by University Readers

Printed in the United States of America
12 11 10 09 08   1 2 3 4 5

ISBN (paper): 978-1-934269-18-3

Cover design: Monica Hui Hekman
Cover photo credit: Pete Garfinkel

The website for the book
can be found at
http://www.farminginthedark.net

University Readers
800.200.3908 I www.universityreaders.com

Dedication: in memory of the women and men who make sustainable agriculture possible, including all those who have gone before.

****

The Alliance & Labor Songster
A Collection of
Labor and Comic Songs
For the use of
Alliances, Grange Debating Clubs and Political Gatherings,

Compiled by Leopold Vincent, Winfield, Kansas.
1891

NO. 24 THE FARMER'S DAUGHTER
(to the tune of Yankee Doodle Dandee)

Oh, here we are as thus you see,
Each one a farmer's daughter
We know just when to legislate
And when we had not oughter

Chorus
So we won't have any of your banker's sons
To kneel to us and bow, sir;
For we can do without a man,
If he can't follow the plow, sir.

So we won't have any like Geo. A. Peck,
At Washington to loiter;
He schemes to work the farmer hard,
And swindle the farmer's daughter.

So when we're hunting a candidate,
Oh, never do you doubt, sir;
We're sure to find a man that's read
How Peffer planned The Way Out, Sir.

When brokers are freed from all their harm,
And lobbyists are dead, sir
The banker'll bow unto the farm
And *come to us* for bread, sir.

****

Contents

Preface

Creating a compass to guide sustainable agriculture in the future

After working in sustainable agriculture research and extension for nearly 25 years, I realized that I had been at this for a while, and was wondering if we had really accomplished anything? I was looking for some new direction, so that I wouldn't be guilty of doing research that had outlived its usefulness. I decided to consult the farmers themselves, and also sustainable agriculture activists in the directions of the four winds. We are scattered far and wide, so it also seems appropriate to use a compass approach to find our way into the future. My journeys to collect these interviews took me North, East, South, and West. This was not a comprehensive tour, but a search for role models and leaders in the movement.

One doesn't often hear from farmers. They are less than one percent of the U.S. population, and are usually portrayed as simple minded or as victims in the media, depending on the story of the hour. You will see from these interviews that these individuals are neither. Public perception of farmers needs to change if they are ever to be treated as the professionals that they are.

Sustainable agriculture has accomplished a lot. For one thing, it has managed to stick around through the controversial years. Two important sustainable agriculture advocacy organizations in Kansas have celebrated the first 25 years of their existence; the Kansas Rural Center and the Land Institute. The national Sustainable Agriculture Research and Education (USDA-SARE) program, which provided the first meaningful (but still not adequate) funding for sustainable research has reached the 20 year mark. The Rodale Institute has existed more than 30 years, and Rodale Press, one of America's first champions for organic gardening and farming, has been around for 60.

But what have we missed? What could be done better if we were to do it over again? What should be done better if we keep at it now? Why haven't we made more progress? Why is the funding base so small for sustainable agriculture compared to where we should be? Why aren't there more farmers practicing sustainable agriculture? What slowed us down, blocked our way, and how can we overcome those barriers? What are new barriers to sustainable agriculture, emerging, and in the future? Do our overall goals need to be revisited? These are the questions I took to my colleagues; farmers, researchers, activists.

This book sets the personal stories within a broader policy analysis of the environment within they/we work and live. It's like studying a fish in a bowl. If you've ever had a sick goldfish, you may have realized at some point that you need to change the water! If the environment in which the fish is trying to live (the water) is polluted and unhealthy, how can we expect the fish to thrive or even survive? I think a similar situation may be happening with sustainable agriculture. We need to look at the context within which we are doing our work, and maybe we need to change that, or work towards that, at the same time that we are doing our own little piece of field or laboratory work.

We need to do a self-critique to move forward. Maybe there are things we can do in the next 25 years that we couldn't do before? Or maybe we just need to change the water in that fishbowl.

Who else cares about food besides farmers? Consumers. Even children know what is good for them. They are bombarded with advertisements, but they know the score. I asked my four nieces and nephews to tell me about their favorite foods, and where they go to find them. Elizabeth, age 13 likes fruit, especially cherries. She specifically mentioned the farmers' market as the place to go to get them, because "they were fresher, they weren't shipped all over." Michael, age 10, likes meat, especially if it is "cooked enough," but also specifically mentioned a man selling peaches, "really, really good peaches," at the farmers' market. Nicholas, age 5, likes the grapes at the farmers' market, because "they are really great, better than at the store." And Rebecca, age 3, said she "likes vegetables." When I asked which ones, she said "watermelon!" When I asked her where to get it, expecting her to echo her siblings farmers' markets comments, she replied emphatically, "at home!"

Interestingly, none of them knew any farmers personally, or had visited a farm except for my 10 acre small farm, even though they live in a semi-rural part of Missouri, and my brother is a veterinarian. I was surprised at how "healthy" their responses to my questions were, and so I asked if they liked pizza? "Of course! they replied. We thought you just wanted to know about foods on the food pyramid!" Their cousin, Ben from upstate New York, age 12, mentioned that some of the kids at his school only bring chocolate pudding some cookies, and Dr. Pepper for lunch. When I asked why he thought they did that, he said, "Maybe their parents don't care." He also said that people shouldn't waste food, "because it takes time to grow." His favorite foods come from "Adams market," a farm stand near Poughkeepsie, NY.

Yes, food is important, and even young children have opinions on what they like, and they know where to find good food. I hope that this book will help everyone understand the food system a little better, and create those connections between the farmers that raise our food, and the place that we eat it, our homes. I also hope that after reading this book, we won't be "farming in the dark," anymore, but "dancing in the dawn," as my friend Nancy suggests. I'm writing this from Kansas, the center of everything in the Universe. I figure it's a pretty good place to start, if you are going to draw a map of the future.

1

A Critical Look at the Sustainable Agriculture Movement

"We aren't a movement. A movement is basically identified as a wide-spread consistent ability to mobilize people over a long period of time, quite focused to act.
This is a mish-mash of 30 different things. It is really hard to say that everybody is working from the same framework, the same value system, and they are all wanting the same thing in the end"[1]

Why I wrote this book

To my knowledge, no one has really done a critique of sustainable agriculture[2]. There are many critiques of conventional agriculture, and many enthusiastic and positive books about sustainable agriculture, but to really move forward, we need to take a critical look at ourselves, and the movement we've created. This book does that. Some would even debate whether we've even created what could be technically defined as a "movement." Others would argue that it doesn't matter—we've spent enough time already defining sustainable agriculture, lets not spend another 20 years defining what is, or is not a movement! This discussion will be revisited in the last chapter of this book. The primary objectives of the book are:

–To offer a critique of sustainable agriculture from an insider's point of view and ideas for the future.

–Give a voice to farmers and others who usually aren't heard from, or only viewed in media "sound bites" or farm narratives. What do they really think?

–Provide students, consumers, and other players in the food system an inside look at what is going on in agriculture.

–Point out future directions for research and activism for all of us interested in a sustainable future.

My Methods/Structure of this book:

Most of the book is made up of case studies or interviews. I think this will be a fresh way to look at a topic that hasn't been treated this way before. In the interviews I asked three primary questions:

1. What have we accomplished in the sustainable agriculture movement throughout the past 25 years?

2. What have we not accomplished? Where are the gaps? Where did we fail? What are the barriers?

3) Where do we go from here? What do we do for the next 25 years?

Occasionally other questions would slip in, such "has your agricultural university been of help to you," and also "do you consider this a movement?" Always, the conversation also included specifics about their farm, and I also asked them how they happened to come into farming and sustainable agriculture.

Most of these interviews were conducted from July through December 2005 while I was on sabbatical leave to do this project. Most were conducted in person; one was by telephone. Most were with people that I've known for many years, some as long as 20 or more years. The original intent of the sabbatical leave was to conduct these interviews with farmers, and also with university faculty, agricultural agency employees, and non-governmental organization employees and activists. I did this, making trips across the country to Washington D.C. and Pennsylvania, to California, to Texas, and to Minnesota and many points in-between.

During this time I collected approximately 70 interviews. The unfortunate thing about having a short sabbatical is that I used the entire time to conduct interviews, and didn't leave enough time at the end for the writing part. Here are 20 interviews, in 18 chapters, but there are another 50 interviews waiting to be summarized. I have chosen to publish the farmer interviews first, along with three interviews with consumers, because they seemed to cover the broadest spectrum of issues. I don't know how else to describe this sense except that they came across as three-dimensional. The other interviews with university and other employees were also quite informative, but more focused on their particular piece of the puzzle. Farmers have to deal with the whole puzzle.

Though I conducted interviews throughout the country, these interviews have a Midwest/Great plains emphasis or focus and are heavily weighted toward Kansas. Eight of the farms and two of the consumer interviews were conducted in Kansas, but six other states are also represented here. Someone writing this in another region of the country might capture a different flavor of farming, but I think this cross-section is a good place to start. This has been a valuable learning experience for me, and I encourage my colleagues and their students in other regions of the country to replicate this exercise.

These are personal stories, by real people, doing real jobs. They don't give a blow by blow historical account of sustainable agriculture throughout the past 25 years, but these are a summary of the accomplishments and failures of sustainable agriculture, from the point of view of people who have been players for at least 20 years or more. I'll highlight some of the historical trends and major events of the past 25 years later in this chapter, but students of sustainability are encouraged to sample the books on the reading list in Appendix A, and in the Notes of each case study/interview.

Some of my motivation to do this work was my own background of growing up on a farm in Kansas. After getting my B.S. in Agronomy at Kansas State University, I attended graduate school at Cornell, doing research on weed ecology, or non-chemical weed control in forage crops. I spent the first eight years of my career at the Rodale Institute, conducting research on organic cropping systems. I've spent the past 14 years at Kansas State University (K-State), doing research and extension in the area of sustainable and organic cropping systems. I have maintained ties with colleagues from my graduate student days in the "Ecological Agriculture Research Collective," at Cornell. Though most of us have dispersed far and wide, many are still involved in ecological agriculture in one way or another. I have links to many non-profit organizations involved in sustainable agriculture from my days at Rodale, and had some interaction with various people and organizations in Washington D.C. during those years. At that time, Rodale was lobbying heavily for more sustainable agriculture funding from Washington. In Kansas, I've been collaborating with the Kansas Rural Center and have served as an advisory board member, and served on the board of directors for the Land Institute for six years. I continue to farm part-time[3], and have developed other linkages to the farming community in Kansas through my extension work, as a consumer, and also through friends and relatives.

The people that I've selected to interview were not a random sample. A sociologist might call this purposeful sampling, but they were chosen for a couple of reasons. First, most have been working in sustainable agriculture for 20-25 years, and second, they are also in a category I call thinking people. As it turns out, three of the farmers and one of the consumers interviewed have degrees in philosophy, two of these are doctorates. These aren't your average farmers—or are they?

These are farmers who have pioneered farming methods, marketing methods, and are also leaders in their communities. I would hesitate to call this information "data," but instead, these are stories of their lives, and their thoughtful answers to my probing questions. It takes courage to criticize something you've put your heart and soul into for 20 years, but these people were brave enough to do that. Some of that has to do with trust. These interviews were conducted colleague-to-colleague, sometimes farmer-to-farmer and also friend-to-friend. I've asked each person who was interviewed to approve of their portion of the manuscript prior to publication. Throughout this book, I discuss both sustainable agriculture and organic agriculture. These are not exactly the same thing, though there is quite a bit of overlap. The farmers in this book might all identify themselves as sustainable agriculture farmers, but only some of them practice organic methods, and a smaller group is certified organic. For more precise definitions, see the 1990 farm bill[4].

Why Stories?

There are several jokes that continue to make the rounds in farming communities. In the first, a farmer is buying hay at $4 per bale. He has a market in the next county over, and loads it on his truck and sells it for $3 a bale. At the end of the year, his accountant points out that he's not making any money, and in fact, is losing money. The farmer says, "Well, what should I do then? Maybe buy a bigger truck?" You'll see various versions of this played out in real life, as the farmers interviewed struggle to make ends meet in an economy where wholesale prices for grains and meat often don't cover the cost of production. Solutions sometimes involve value-added and direct marketing to get a better price, marketing cooperatives, community supported agriculture (CSA) marketing systems, and even farm agritainment. These are not perfect solutions either, as you'll soon see. The farm bill subsidies often make up that extra $1 per bale for the commodity grain farmers.

Another joke involves a farmer and a county agent. The farmer wants to raise chickens. He buys a batch of day-old chicks from the hatchery, and plants them about six inches deep. They all die. He decides that maybe that's too deep, so he orders another batch of chicks, and plants them three inches deep. These all die as well. He decides to consult his county agricultural agent. He tells him about the experience with the first two batches of chicks. The county agent has an idea, smiles, and says, "Well, have you done a soil test?" This little bit of humor points out the disconnect some farmers have with their extension staff, and the lack of useful information they get. I can offer my own criticisms now, as I'm part of that system. One thing that surprised me in the interviews was the honesty with which everyone seemed to offer critique to the agricultural universities and their staff, even knowing I was part of that. Perhaps they would have offered even more if I wasn't coming from a land grant university? Maybe they were just being polite? I'd hate to see what was on their list if they weren't holding back!

In the third joke, two county agents meet for lunch. One is really down in the dumps. In fact, he is nearly in tears. The other agent asks, "What has you so down today?" The first one replies. "My farmer died." This joke is only funny if you're familiar with the on-going downward trend in farm numbers and the loss of the middle-sized farms in the past 50 years. Total farm numbers have declined from 2.3 million in the 1978 Census of Agriculture[5] to 2.1 in 2002, for a total of 128,793 farms lost, or six percent. If this was divided equally among all 50 states, this is 2,576 farms per state, but in reality, these would be more heavily weighted in the farming states.

All three of these jokes are funny, mainly because they contain some truth. The interviews were also conducted in a spirit of light-heartedness at times, even though we were discussing serious issues. You'll see several places where laughter occurred, and these are noted, to give the reader a sense that fun, and laughing at ourselves can be a part of sustainability too.

Another reason that this book is written as a series of interviews, which are sort of like life stories, is that is a time-tested communication and teaching tool. My colleagues in the education department tell me that in their field this is called "narrative inquiry," and there are books and even a journal dedicated to this methodology. Many of our most revered aspects of culture include our stories, including books, opera, plays, even songs, etc. We have lost much of our oral tradition, but for many cultures it is alive and well. These interviews are preserved largely as they were conducted, with relatively minor editing for ease of reading, putting topics in a consecutive order, also to change some grammatical style from normal "speaking style" to something more like we are used to reading. As much as possible, however, the spirit of the language and figures of speech are preserved.

In attempting to write this book, I tried several times to distill the interviews down to a few "pithy quotes," or summarize the main points in more academic language. Each time I did that, it seemed like I was taking the spirit out of it, and what I don't want to do is contribute an academic treatise on sustainable agriculture, describing farms, but not really listening to farmers. There are a lot of wonderful books out there already on other aspects of sustainable agriculture (see Appendix A.)

Another thing that I've noticed is that there are certain phrases from each of the interviews that stay with me after I read them, like echoes, or ghosts, haunting refrains. Even after doing the interview, typing the rough transcript, and editing several times, new things jump out each time I read them. I hope the reader will find this same sort of quality to the words.

Chapter 1—A Critical Look at the Sustainable Agriculture Movement

Time-line of Major Milestones

The 1980 USDA *Report and Recommendations on Organic Farming*[6] makes a good starting point for a review of the past 25 years. There are many wonderful books[7] published before that (see Appendix A), and a whole time-line of human history going back at least 10,000 years of sustainable ag practices. These will not be reviewed here.

To understand agricultural policy within the context of U.S. history, 1862 was a key year. Both the USDA (United State Department of Agriculture) and the University Land Grant System (the Morrill Land Grant College Act) were established that year. As Hildreth[8] (1982) points out, these did not come out of the blue. Agricultural committees were established by the House of Representatives in 1820 and in the Senate in 1825. Professor Jonathan Turner of Illinois began to campaign for "industrial universities" as early as 1850. In 1855, both Michigan and Pennsylvania passed legislation establishing their agricultural schools. In 1887, Congress passed the Hatch Act, which provided research funds to these agricultural universities. In 1890, Land Grant Universities were established at 18 sites for African Americans and in 1994 for Native Americans. The Cooperative Extension service was created as the outreach branch of these universities in 1914 with the Smith-Lever Act.

Little has changed in this structure since this time with the exception of levels of funding coming from the federal government to the states, and the slow shift from exclusively "formula funds" to more and more allocation to a system of competitive grants. The pros and cons of this will be discussed in some of the interviews. Although extension has federal dollars supporting it, most states match those dollars with state and local funds for positions at the state and county levels. As you'll see from the interviews, the ratio of these various income streams may be changing, with less support from the local level than before, especially as less comes in from the other two sources too, and as farmers are getting their information needs met by other sources.

When we begin looking for sustainable practices, our time-line might begin in the mid to late-1800s, with the Louisiana Purchase under President Lincoln, and the westward expansion to the Midwest and prairie states. Much of this expansion was made possible by the railroads, and also the switch to steel mold-board plows, which allowed for the now infamous sod-busting. Though old fashioned, agriculture during this period could not be considered sustainable by today's standards. Crop rotation may have been practiced in the east on George Washington and Thomas Jefferson's farms, but the plowed ground in the western prairie states was in continuous wheat, until the dust bowl years of the 1930s destroyed farms and also lives. Fruit orchards were regularly sprayed with lead-arsenic and other toxins. Fertilizers were just being invented, and were not widely available yet, so farmers used manures from their livestock, or used the fertility that had been stored in the soil organic matter from the plowed prairie as long as they could and then abandoned the land. These fields are still identifiable today with partial covers of grass, and are called "go-back" lands. Legume cover crops and rotation crops were in use in some places however, and some of K-State's earliest bulletins were on cowpeas as a nitrogen fixing crop for Kansas. Cover crop research continued to be an area of activity in most agricultural colleges until the late 1940s, when nitrogen fertilizer became cheap and plentiful.

Farming in the Dark

The post WWII era is generally agreed to be the chemical era or industrial era of farming in the United States. Tractors and fuel became available and affordable, though many farmers continued to farm with both horses and tractors for many years during the transition. Nitrogen fertilizers and concentrated forms of potassium and phosphorus, the other two nutrients most needed by plants became commercially available. Nerve toxins developed during the war years became the new insecticides, and herbicides like 2,4-D also were now in use. Much of what we think of now as the alternative agriculture movement was a reaction to these post-WWII chemicals, and organic farming standards prohibit their use. However, you'll see from the reading list that Lady Eve Balfour in England and J.I. Rodale, and Edward Faulkner in the United States were writing in the 1940s before the chemical era of farming really took off, in reaction to other un-sustainable practices they were observing in agriculture.

Some of the political movements that we see today can trace their roots to the early Populists of the late 1800s. As the railroads and other industrialists were encouraged to develop the west, the farmers were seeing lean years in the 1880s, and argued that it was up to the government to create a balance of power, as the railroads were charging exorbitant prices for shipping. Some say the populist movement eventually died out, but others point to many of their platforms that were later adopted by the other major parties and eventually made into legislation. Recently, a news program commentator noted that the Democratic contenders in this year's election have added a populist tone to their stump speeches, suggesting that some of that corporate power in the United States today is in need of balance again, and that we should become a nation of citizens, not consumers[9]. Important legislation in the late 1800 and early 1900s in support of that balance included the Sherman Anti-Trust Act of 1890[10] and subsequently the Federal Trade Commission Act of 1914. The Capper-Volstead Act of 1922 encouraged the creation of rural cooperatives.

The 1930s dust bowl era brought much of farming to its knees, and the first federal direct aid to farmers was passed in 1933 as the Agricultural Adjustment Act. It was an attempt to raise farm prices and improve rural farm income. That same year, the Soil Erosion Service was established, later changing its name to the Soil Conservation Service in 1936, and becoming part of USDA. Various farm bills have been passed at approximately five year intervals since 1933, and each has unique attributes and the potential to change the face of farming. Many had supply control provisions in them, as a way to raise farm prices without subsidies, but all that changed if we fast forward to 1973, under President Richard Nixon and Secretary of Agriculture Earl Butz. His fence-row-to-fence-row policies led to over-supply once again and the potential for soil erosion and water pollution from these intensive farming systems. He also promoted the "get big or get out" philosophy of agriculture. By the 1970s, the EPA had been created and the Clean Water Act was passed in 1972[11]; but some of its provisions, including the control of non-point source pollution from agriculture were not enforced until recently in response to lawsuits brought by environmental groups.

Conservation programs, and cost share funds, have been available since the creation of the Soil Conservation Service, but the 1985 farm bill was the first time that they became mandatory for farmers before they could receive any other program payments or benefits. The 1985 farm bill was also the first one to authorize spending specifically on sustainable agriculture research, for what is now the SARE

(Sustainable Agriculture Research and Education[12]) program. These were both watershed events in terms of policy that promote sustainable agriculture practices and research.

After the publication of *Silent Spring* in 1962, the public became aware of the risks, and not just the benefits of the use of pesticides, and an agricultural system based on their use. By the mid-1970s, many of the organizations that are active today in sustainable agriculture were founded. Though J.I. Rodale had been publishing on the topic for decades, the Rodale Institute as a research center was founded in 1976. Back in Kansas, the Kansas Organic Producer organization was founded in 1974 as an educational organization, and later became a marketing cooperative; in 1976. The Land Institute was created to give form to Wes Jackson's vision of a perennial agriculture and training to hundreds of idealistic interns, and the Kansas Rural Center was founded in 1979. On the coasts, similar things were happening. On the west coast, the California Certified Organic Farmers (CCOF) began certifying in 1973, and on the east coast, the Natural Organic Farmers Association (NOFA) was forming chapters in several states.

After the 1985 farm bill authorized funding specifically for sustainable agriculture research, appropriation of that funding followed in 1987 after some heavy lobbying by a coalition of sustainable agriculture organizations. The 1990 farm bill gave a more precise definition of sustainable agriculture, still widely used today. It also authorized the creation of the National Organic Standards Board (NOSB), and the creation of uniform national organic standards. Certification would be performed by some of the same organizations providing this service before, but they would have to be accredited by USDA, and all use the same minimum set of standards. The USDA standards were supposed to be ready in three years, but were hashed over until December of 1997 when they were released for public comment. An overwhelming and record breaking 275,000 comments were received, mostly negative, in response to USDA watering down the standards that had been created by the NOSB. The USDA then back-peddled, and put back in the stricter standards requested by the farmers, consumers and organic food companies. Revised standards were re-released in 2000 and came into effect in 2002.

During these years the commodity crop portion of the farm bill still received the lion's share of the funding, dwarfing the handful of dollars spent on sustainable and organic agriculture, and even the small but growing portion of the farm bill allocated to conservation practices. Subsequent farm bills removed the base acres requirement that limited crop rotation in 1996. A new conservation program in the 2002 farm bill[13] was designed to reward farmers for "doing the right thing" rather than reward them with funding to correct problems, as previous programs were likely to do. The program is called Conservation Security Program (CSP), and included three levels of participation and reward, based on practices that farmers could adopt such as soil testing, crop rotation, as well as soil conservation. The sustainable agriculture community takes, and deserves credit for this piece of legislation, and they also saw it as a opening for expanding what are often called green payments, now becoming more common in Europe. Unfortunately, the CSP has never been fully funded, is only available in certain limited areas, and at the local or county level, some farmers find that their Natural Resource Conservation (formerly Soil Conservation Service) staff are not familiar with the program, or how to help them sign up.

A Critique of Sustainable Agriculture

With all these successes, why would I want to write a book critical of sustainable agriculture? In addition to these legislative successes, the number of farmers in the United States identified as organic continues to grow (12,000 in the 2002 Census of Agriculture), although their numbers still only makes up less than one percent of farmers. Consumer demand for organic has been growing at 20 percent per year for more than 10 years. Research on organic and sustainable agriculture practices has expanded from almost non-existent (with the exception of a few programs on biological nitrogen fixation, and land application of organic wastes) in 1980, to several USDA/ARS labs with significant scientific effort in this area (Beltsville Sustainable Systems Lab, and the Tilth Lab in Iowa, to name a few) in addition to the increased effort within universities. University presence has grown from a handful of somewhat related courses in 1980 in 19 institutions (USDA 1980 Report), to courses in 157 colleges and universities, and degree programs in at least five universities (National Agriculture Library 2005). [14]

Other milestones signifying official recognition at the national level include the American Society of Agronomy publication in 1984, *Organic Farming*[15], based on a symposium held in 1981. The National Research Council Board on Agriculture published two reports, "Alternative Agriculture" in 1989[16], followed by "Sustainable Agriculture Research and Education in the Field: A Proceedings"[17] in 1991. These, combined with the trickle of funding from the USDA/SARE program helped give credibility to the somewhat isolated faculty and fledgling sustainable agriculture programs getting started at universities during those years, including myself.

My inspiration was initially all the farmers who also had to have a day job to pay the bills, obtain health insurance, etc. and are literally farming with headlights on the tractors, and using flashlights to feed the livestock. What is wrong with sustainable agriculture? Why is everyone I know, even the sustainable agriculture and organic farmers, farming at night, and having to support their farm and their family with "the day job?" Why hasn't sustainable agriculture addressed this problem? It seems that direct marketing, niche marketing and all the latest innovations make longer hours for farmers, but they still have trouble making a living. What should we have done differently? What can we do now? It seems there are some issues that have not yet been addressed, some things that have not been said, and some topics that haven't been explored. Thus, the other meaning in the title; what don't we yet understand? What are we still "in the dark" about?

We also haven't seen huge environmental gains from sustainable agriculture, or at least not ones that we can document. Some even claim the environmental movement, of which sustainable agriculture it could be argued is a part, is dead[18]!

More personal questions also nag at me; why does my friend Nancy have to work in an envelope factory to support her farm? Why am I selling cucumbers at the farmers market for the same price today that I sold them for 10 years ago, and the price of gas, movies, etc. has gone up by 20 to 50 percent during that same period of time?

Farmers often tend to blame themselves for their financial problems. "If I was only smarter, more clever, I would be able to make it." They see their neighbors suffering too, and only a few realize that there is a systemic problem. It's not just them. Consumers too, often take the view that the way to be healthy is to "make wise food choices," not realizing that the whole food system[19] might need to be changed. I hope that consumers, students, as well as farmers and activists will find something useful here.

I also wanted to explore several issues, some obvious, and some underlying or hidden. We need to shed light on them, to "put them on the map" of our consciousness. We need to address some things that haven't been put out there before. How has the U.S. farm bill shaped the rules of the game? Yes, there are farm support payments, but are they truly helping farmers? What about international trade? Is "free trade" what we really want, or is "fair trade" a better model? In the present, farmers are still losing their farms, or choosing to take the job in town, and we (the sustainable agriculture community) manage to keep ourselves busy, but are we really helping? Are we just rearranging the chairs on the Titanic? Serving as hospice care for the dying family farm? Do we need to "raise more hell and less corn"[20] to borrow a phrase from George Pyle's book (originally from Mary Elizabeth Lease) and the populist tradition? How do we truly understand the politics of the food system, and come to a point where we can do something about it? How many of us are "eating in the dark," and not knowing important details about what is in our food? Do consumers even care about where their food comes from, and if they do, how can they help?

Hypotheses about what people might say:

Before conducting the interviews, I wrote down some initial thoughts about what we did right, what we didn't do so well, and some of my ideas about the future. I tried not to look at these lists during the six months that I did the interviews. As it turned out, many of the items on my "accomplishments" list were listed by farmers as negatives, and vice versa. To be honest, most of the things we can point to as milestones, even the passage of the new organic standards, turn out to have very mixed effects.

In Chapter 20 I identified certain topic areas, stated where things were in 1980, what had happened between then and 2005, and responses to those changes. Space only allows for a certain number of responses to be listed, so I suggest reading this whole book, and not skipping to chapter 20 to see the Cliff Notes version. You are also welcome to add your own insights, either privately, or publicly. Another thing that has changed in the last 25 years has been our ability to communicate, effectively, at long distances, even with strangers. Websites and Wikis have changed the way we teach, the way we learn, and now the way we collaborate. Join us at www.farminginthedark.net to continue the discussion.

Organization of the book

Chapters 2-4 are interviews with people who primarily raise meat for a living. Chapters 5-9 are crop farmers, or mixed crop and livestock farmers. Three of these have degrees in philosophy, thus the title of the section. Chapters 10-12 are interviews with vegetable growers; two in Kansas and one from Illinois. Chapter 12 also includes an interview with a soil scientist, who slipped in as the co-owner of Blue Moon Farm. Other researcher's interviews will be featured in Volume II of Farming in the Dark, tentatively called "Dancing with the Devil: Perfecting Hell." Chapters 13-16 are people that I worked with during my years at Rodale, but hadn't seen for more than 15 years. George DeVault was editor of New Farm Magazine, and the other three were on-farm research cooperators in Illinois, Ohio, and Minnesota, giving some diversity to my other Kansas-focused chapters. Finally, chapters 17-19 are with consumers, but not just any consumers. These are people interested in the Slow Food movement, organic food, and they talk about the difficulty of finding good food in a conventional world. At the end of each chapter, five or more questions have been added relating back to the content of the chapter. These are called discussion questions, but they could be used in a classroom setting for post-reading discussion, for term paper ideas, or for class debates, another new learning style/model.

I think this information will be useful to all of us in the sustainable agriculture community. In recent years, sustainable farming groups have been joined by consumer and chef's organizations including the Slow Food Movement, the Chef's Collaborative, and numerous Buy Local campaigns, that often include sustainable agriculture within their circle of priorities. Many non-agricultural colleges are also looking at buy local campaigns for their dininghalls[21], often at the request of students. I think this manuscript will be useful to all involved in this work. It will also be useful to students who are looking for a way to get involved, and wanting to understand what is going on before they decide. This will give them an insiders' look at some of the important work, directly from the mouths of those doing the work

Notes

[1]Anonymous quote, but for a more thorough discussion of this issue, see "Warrior, Builder, and Weaver Work; Strategies for Changing the Food System," by S. Stevenson, K. Ruhf, S. Lezberg and K. Clancy in *Remaking the North American Food System,* University of Nebraska Press, 2008.

[2]Actually, there have been plenty of critiques from people opposed to sustainable agriculture, just not critiques from those working to promote it. There are now starting to be critiques of organic agriculture which are quite interesting, for example *Agrarian Dreams—the Paradox of Organic Farming in California*, by Julie Guthman,2004, University of California Press.

[3]We raise sheep, chickens, vegetables and fruit on ten acres in Wamego, KS. For more details see "Pedagogical Farming: One Experience Combining Horticultural Research, Extension Education, and Organic Farming." Chapter 17 in *Community and the World, Participating in Social Change.* Ed. By Torry D. Dickinson. Nova Science, Publisher.

[4]The 1990 Farm Bill, officially called "Public Law 101-624, Nov. 28, 1990, Food, Agriculture, Conservation, and Trade Act of 1990."

[5]USDA Census of Agriculture. Note that the census is conducted only about every five years, so the closest estimates for 1980 and 2005 were the 1978 and 2002 census data. http://www.agcensus.usda.gov/

[6]Report and Recommendations on Organic Farming. 1980. USDA Study Team on Organic Farming, USDA.

[7]Gold, Mary V. and Jane Potter Gates. 1988 (updated and expanded May 2007) Tracing the Evolution of Organic/Sustainable Agriculture: A Selected and Annotated Bibliography. USDA National Agricultural Library, Beltsville, MD, Bibliographies and Literature of Agriculture No. 72.

[8]R.J. Hildreth, "The Agricultural Research Establishment in Transition." 1982. in Food Policy and Farm Programs. Ed. By Don F. Hadwiger and Ross B. Talbot. Proceedings of the Academy of Political Science, New York. Vol 34, No. 3.

[9]See Bill Moyers interview with Nell Irwin Painter, Feb. 29, 2008 at www.pbs.org/moyers/journal

[10]West's Encyclopedia of American Law. http://iris.nyit.edu/~shartman/mba0101/trust.htm

[11]Clean Water Act—EPA http://www.epa.gov/r5water/cwa.htm

[12]SARE (Sustainable Agriculture Research and Extension) Funds research, extension, and training programs throughout the U.S. See www.usda.sare.org.

[13]NRCS Fact Sheet, Farm Bill 2002—Conservation Security Program. March 2005. http://agriculture.senate.gov/ag/cspfacts05.pdf

[14]National Agriculture Library Publication—"Educational and Training Opportunities in Sustainable Agriculture," www.nal.usda.gov/afsic

[15]ASA Special Publication Number 49. Organic Farming: Current Technology and Its Role in a Sustainable Agriculture. 1984 American Society of Agronomy, Madison, WI (Proceedings of a symposium December 1981) Ed. D.F. Bezdicek, J.F. Power, D.R. Keeney, and M.J. Wright.

[16]Alternative Agriculture. 1989. Board on Agriculture, National Research Council. National Academy Press, Washington, D.C.

[17]Sustainable Agriculture Research and Education in the Field; A Proceedings. 1991. Board on Agriculture, National Research Council. National Academy Press, Washington, D.C.

[18]The Death of Environmentalism—Global Warming Politics in a Post-Environmental World. By Michael Shellenberger and Ted Nordhaus. 36 pp. Web

published 2004. Available at http://www.thebreakthrough.org
/PDF/Death_of_Environmentalism.pdf

[19]Perceptions of the U.S. Food System: What and How Americans Think about their Food. 2005, 88 pp. A series of 4 papers written by The Frameworks Institute (frameworksinstitute.org), commissioned by the W.K. Kellogg Foundation Food and Society Initiative. For more information or a copy of the report see www.foodandsociety.org.

[20]Raising Less Corn, More Hell, by George Pyle, 2005. Public Affairs.

[21]For example, see New York Times article, "Fresh Gets Invited to the Cool Table," Aug. 25, 2005.

2

Dan and Mary Howell
Yonder Road, Frankfort, Kansas

"Selling land for me is like selling one of my kids. You just never know if the next person will take care of it properly, the way I would."

Dan and Mary Howell were a major inspiration for this book and especially for the title. Mary was collaborating with me on testing some soil and water test kits on their farm. "Rhonda, I just thought I'd let you know we finished our soil tests last night at about 3 a.m. Dan got home from work, we had dinner, and around 11 o'clock he said it was going to rain, and we'd better get those tests done, so we did it!" *As we continue our phone call, I told Mary that the data wasn't more important than the sleep they may have missed, but she reassured me that they do this (work late) all the time, especially during calving season.*

As will be evident in the interview, "farming in the dark," or at least at night, is something that Dan and Mary have been doing for quite a few years, as Dan had worked full time at the Jeffrey Energy Center in St. Marys Kansas for nearly 25 years. Dan recently was able to qualify for early retirement from Jeffrey but less than two years after his first chance to farm full time he became seriously ill from West Nile virus. His time now is limited by the need to get his strength back. This doesn't limit his enthusiasm for agriculture, or his activism. He recently served a one-year term as the chair of the advisory board for the Kansas Center for Sustainable Agriculture and Alternative Crops at Kansas State University (K-State), and prior to that served on the board for several years. His critique of K-State in the following dialog is from a position of first-hand experience working with K-State, and attempts to improve their standing with farmers.

Mary's interest in sustainable agriculture gained more momentum a few years ago, when she began working part time at the Kansas Rural Center as a field staff person for the Clean Water Farms Project. She and Dan were early participants in the program, and had hosted numerous visits from Research and Extension, farmer field days and Kansas Biological Survey staff putting in water quality monitoring equipment on their farm. She also plays a major role as a contact person for the Kansas Graziers Association. Just to make sure she doesn't run out of things to do, Mary also works part-time for the Marshall County Fair, which is actually more than part time during the weeks leading up to the fair in mid-July. Mary has invited me to come and judge the 4-H and open class vegetables, which takes the better part of a morning. Dan then invited my husband Raad and me out for lunch at the local diner. Mary stays back at the Fair office to handle the never-ending questions that will come her way.

As we started the interview and after I explained the three questions I had about sustainable agriculture, Dan asked, "When you are talking about the sustainable agriculture movement—are you talking about the sustainability of the family farm and staying on the farm? Or are you asking more about the organic side and where we need to go with the environment? *Both, I answered.*

Dan then proceeded to answer all my questions, with very little additional prompting from me. I contentedly ate my burger and fries in air-conditioned comfort, and tried to soak in all that he was saying.

The sustainable deal, for me, got started in '93 due to the flood, when we lost so many crops and everything was under water. The incubation of the thought process started with reading *New Farm* magazine.[1] I've never been an organic farmer, but I joined the Kansas Organic Producers[2] (KOP), to hear about what people are doing. I still haven't figured out how I can get organic to work for me. There's the difficulty of getting there, and I don't find that there are many answers to the problems of how to get there, to stay away from the commercial system. For example, how do I control spots of bindweed in scattered places? I haven't been through a really good course on how to build your soil. I've worked with hairy vetch and different things. It's a challenge to make the whole system work and put it together. You almost need somebody to work with you to work a whole plan. I get these little bits and pieces. Even the people who are organic now, the ones I've been visiting with, they have the same problems I do conceptually. I have a great deal of respect for the organic farmer, but I haven't figure out how I can get there.

The other thing is, I don't consider myself a visionary, and I have seen for a long time that we are on an energy table that is unsustainable. I haven't figured out how we can sell grain at prices we are now, even organic prices. Another consideration is the price of fuel, the way it is now . . . I actually expected it would happen quicker. That's why our plan is to plant grass, legumes and forbs to complement our cow-calf enterprise, and get rid of machinery, get rid of depreciating assets, capital costs. I'm talking about depreciating assets at 1975 prices. Now at 1990, 2000, and 2005 prices—they were hard to pay for back when prices were a tenth or what they are now, and I'm still getting the same prices for commodity grain.

I've seen this train wreck coming. Now that it is here, I'm not sure what the answer is to it. For me, I am not going to get rich doing cow calf, but if I can get everything seeded back to grass, I have some control over the inputs and expenses. I'm not losing any soil, I can inter-seed legumes and grow my own nitrogen, and I'm probably pulling up micronutrients from down below. The biggest thing for my area, the soil tests call for some sulfur, some phosphate and my pH needs to come up some. I can do these pretty much organically.

We're still on the same plan we were six or eight years ago. Four years of dry weather, from 2000 to 2004, slowed the process down. We had to back off of our plan; we couldn't plant new stands when it was so dry. But we're getting moisture now; we should be able to get back on track. I don't see my future in growing grain, even organically. Even with organic grain I still have a terrible time with markets, and I have to store the grain. I would have to spend a bunch of capital for good grain holding facilities, keep aeration on it all the time to keep the bugs down without chemicals. To have that system work, there's a lot of capital cost to that. Most of those guys have to hold their grain at least six months to a year. We're talking a buck and a half to three bucks per bushel in facility costs. At my age, I don't see getting over the capital cost to where I can generate enough income to pay the bills along with the interest payments, to pay for all this.

Chapter 2 – Dan and Mary Howell

The faster approach to sustainability for us is growing grass, legumes and forbs to feed our cow-calf operation.

Do you still make or use hay?
Some but we are moving away from it all the time. I can feed less hay if I can get more warm season grass mixes, and stockpile grass for our dry cows. My goal is to get to feeding hay for 30 to 60 days at the most. I don't know in Kansas if you can get to 100 percent grass, but I think we can get down to a lot less than where we are now. I can also go that way by keeping my cow-calf numbers a little lower, by bringing in stockers during the big flush in the summer, and move cattle numbers up and down all the time.

We are more cow-calf at the moment. I'm not against the other system, and I'll look at it, but I'm not ready for that wrinkle. We reduced our cow numbers from 270 to about 100 because of the four years of drought. We started doing that early to preserve our grass. One can destroy grass a lot faster than it can be resurrected.

How did your gamma grass do in the drought?
Good. We should take a drive and go look at it. Both years we planted it, were kind of dry and we didn't get as good a stand as we should have. The years we got moisture, we went ahead and let it seed, and only ran cattle on it in the winter. Now it is filling in, it's looking good. The first couple of years were tough (grimace) . . . and the seed is so high-priced to start with. If I could get a year with the right moisture at seeding time, I could cut two or three years off the establishment side, but I just can't predict. Mother Nature is in control!

Did you put cattle on it yet?
The first planting, we hayed it. I'll never do that again. Then it got wet. The weeds in it got big, then it got brittle, the hay wasn't very good. Now the second one, we planted and flash grazed it hard for short durations, to eat the weeds, rather than the mechanical process. That worked a lot better. Besides that, it had less inputs and less work. For rotational grazing, we just put in a mile and a half of water line. There are some capital costs to this—some water line and some portable tanks, but it's not like buying and depreciating new paint (tractors). Usually there are some cost-share programs out there to get help with improving water quality issues, which will partially help for paying for these.

Do you think farmers in general have figured out this capital cost pinch?
The average commercial farmer, the big guys, their answer to the whole thing is to farm their neighbor's ground, and everything that joins it. The answer for the need to farm more ground is always the same; to cover equipment costs. A tractor used to be $20,000 and now is $100,000. But, it's not just the tractor, it's the combine, more trucks, help; capital costs and nitrogen prices have gone up. There is just a whole boat-load of things wrong with this. The capital costs and operating costs always stay ahead. Some are making it work and doing all right, but John Deere and the bank own them. That's just not what I consider more friendly, more sustainable or fun! That's not the trip I want to take, not the train I want to ride. Once you are on it, I just think you never get off it.

So what is your definition of sustainable agriculture?
My idea of sustainable is actually somebody can stay on the farm and not work two or three other jobs, to be environmentally friendly, to have soil erosion to a minimum, have decent water quality and to be family friendly.

It's not just agriculture that makes that difficult, it's like health insurance and all the other things that used to be a minute cost of the family budget, but now is the equivalent to a really, really nice house payment for a month. Many wives work off the farm to pay for health insurance and groceries. But, how do we do without these things?

Part of sustainability is not to push Ashley (their daughter) to come back to the farm, but if she wants to, she can. My Dad wanted me to come back, but we didn't have the vision and the plan to have the income level without borrowing a ton more money to expand. We looked at the normal plan of bigger this and bigger that. We were already as far as we wanted to go on that trip, I wasn't exposed to *New Farm* magazine, or *Acres*[3], or Jim Gerrish's[4] rotational grazing, Kit Pharo's[5] philosophy. If I had been exposed to some of these people when I was 20, I think I could have maybe made it work and started farming and never worked off the farm. But I wasn't exposed to that thought process and wasn't a visionary, so I didn't come up with those ideas myself.

Do you have other sustainability goals? How close are you?

We were hoping by now we'd have most of our farm ground in grass. We'd like to put the rest of our farmland back into grass mixtures and legumes; we were trying to do about 50 to 100 acres per year, but we were slowed down by four years of drought, and are now behind. So we are looking at about 2010 before we'll be finished. I've been working with Gary Kilgore (a K-State agronomist) on different ideas for different legume mixes for the different grass. That's the next thing I need to work on more, not to buy inputs. Warm season grasses don't need as much nitrogen as the cool season ones, but we want to grow both, and start out with legumes in the mix. I need to expand grazing in the spring and fall with cool season grass, I have no intention of buying lots of nitrogen for the grass. I would rather grow my nitrogen than buy it. I'm willing to give up a little production to save on nitrogen expenses.

How are you doing with respect to your water quality goals?

Most of that works really well with rotational grazing, not leaving them on the same grass too long. With the rotation, I need water in each paddock, so it works out well to take water to the paddocks. Those two things complement each other. There isn't anything that is antagonistic about water quality and rotational grazing when managed properly; there's no conflict in principle.

What are things about sustainable agriculture that you say, "I wish someone was doing this, or someone was doing that?"

Two or three years ago, I heard a speaker talk about working with soaps, trying to develop natural chemical sprays for weeds. She had taken this to K-State, and wanted to work with them, and they weren't interested. She's now at the University of Missouri or Oklahoma. This is more to the organic side, which pertains to health issues, and not having to handle pesticides. There are spot treatments in an old grain farm that I want to take care of, like the persistent spot of bindweed that I don't really want to spread all over the place. It's hard to handle in an organic system. I have yet to find anyone with any remedy except disking once a week. If I have little spots scattered all around that is kind of tough. If we had that herbicide made out of soaps . . . I don't know where she is at with that research now.

Being that I'm so compromised from West Nile, I would love to figure out how not to use any pesticide chemicals at all. I'm not a radical about it, but I just don't like to be around it. I'm not saying there isn't a place for them, but if I could get away from using them, I'd be tickled pink. The more I learn, the more I'm concerned. I think we have no comprehension the problems we are causing ourselves. I'm pretty liberal-minded about this, but I think there is a lot we don't know. The half-lives (breakdown rate of the pesticides in the environment) of this stuff are more than we comprehend.

I can get to the organic side faster with a grass system than grain system. I don't think I have the knowledge to make an organic grain system work and I don't know if I have the heart to want to. It's still intensive in terms of production equipment and capital investment. If I could convert everything to grass and legumes in one year I'd be delighted, but Mother Nature doesn't work that way. I can seed it, but it may be four years before I get good production. It's a cash flow deal and it only allows me to move so fast. We'd like to sell every piece of equipment we have except a swather and bailer (for hay). We aren't feeding any grain now, just forages and alfalfa. We sell feeder cattle, not fattened cattle, except for a few for personal meat. We aren't in that finishing deal. If we were, there is a more sustainable future with grass finishing as compared to grain anyway. How reliable those markets are down the road, who knows. Maybe we'd have too many cattle to make that work. We could sell stockers into a grass finishing system to someone down the line if we don't do it ourselves. We aren't hung up on having to do the whole enchilada. Our deal is to stay out here and make a living without driving to places like Jeffrey (Energy Center). I want to make farming more like I thought it was when I was a kid.

What is your background, how did you get into farming?
I was raised on a farm. When I was little we sold cream, and then milk in 10-gallon cans, and then we went to grade B milk in a bulk tank. We milked about 20 to 30 cows with a Surge bucket system. We milked cows until I graduated from high school and went off to college. That's when Dad transitioned into more beef cows and grain farming.

When I got out of school, I always wanted to farm. Dad wasn't against it, but we couldn't see how there was enough income for two families. I went off to school, but ended up coming back home after college. Dad had a new set of bifocals, missed a step coming off the combine and broke his leg just above the ankle. I had already decided I didn't want to farm, because I didn't want to borrow the money. I was planning a different future. My degree was in business administration. I had worked while I was in school at Washburn at Wentz Equipment in the parts department.

Then I came back and helped Dad. He got better. During that time, I bought an 80-acre farm with a house on it. What I had put out of my mind was brought back. I wanted to farm, even though it was what I decided not to do. I still needed money, so I worked for Haven's construction company in Centralia, which was close by. Then I drove a 10-wheel truck for Suther Feed for a while.

That was about the time we had trouble with chinch bug and green bugs in milo. That milo would get about 20 inches tall, and then would just bleed to death on the ground from green bugs and chinch bugs. We had a couple of years of terrible crops. I decided I needed another source of money. I was looking for better money to speed this process up. I put in an application at Jeffrey in the parts department and Boiler Makers Apprenticeship in Kansas City. Jeffrey was the first to call, so I went there. I thought I'd work there two or three years, and get my feet on the ground. One thing led to another, equipment costs, other expenses. As the years went by, health insurance became a bigger deal, and before I knew it, I'd been there nearly 25 years. Then I got an opportunity for early retirement.

Dad was in poor health and mom had passed away. I needed to be back here, so I took the early retirement. I'm tickled to death that I took early retirement. It's still no walk in the park, but we are doing all right, partly because we aren't buying new equipment. After I get these new grasses in, there is a lot of work. The water systems need to be finished; and the trees need to be taken out of fence lines. I like trees, but the fence line is not the place for them. Once I get this system working well, I'd be interested in a five- or 10-acre market garden, something for the farmers' market, a restaurant or something else. I could do something else on the side once other stuff is completely up and running well. I am getting older, so maybe not!

Mary loves working at the Kansas Rural Center[6]. That job came open just as I was getting ready to retire from Jeffrey. It was a stabilizing force, as we moved from one ship to another ship. It worked out well. Now we need more weather cooperation, and I need to get my health back. The West Nile virus threw a wrinkle in this whole thing, just two years after I retired from Jeffrey Energy Center.

Most of the farm magazines; most of the articles and a lot of stuff is geared towards somebody selling something. A lot of the research—if I have something I wanted research on, I don't have a million dollars to get someone to do research to see if it will work. They are doing the research mostly for some companies that have something to sell. I'm not saying it won't help production, but in the long-run, is it helping enough with production to help agriculture in the total system, or does it put more that we don't need on the market, and the farmer breaks even? The research is saying, "I have to do this to stay ahead of the curve, to be the high-tech guy." Look at where we are going with GMO seed and chemicals.

I don't see much work on how to grow different crops without herbicides and insecticides. You don't see a lot of work on systems where you can grow all your nitrogen and amendments with cover crops. That isn't a big push in research. For example, some of the natural minerals one might need, instead of commercial fertilizers, like soft rock phosphate. I don't see any extension research or county research on foliar feeding, using like a few pounds instead of a hundred pounds. Farmers did many things before World War II when all these things weren't available. I'm not anti-progress or anti-changes, but change for the better; and not for a few multi-national companies owning all plant life and all input supplies. Research that is truly for keeping the family farm on the farm, and true to the farm, is very little.

In the last 10 to 20 years, can you think of any land grant research that helped keep people stay on the farm?

Several growers groups have begged the land grant universities to figure out a better marketing, economic system for rural America instead of just selling commodity grain, and exporting it over-seas. The wheat commission and several groups have just *begged* them to research changing the marketing system for wheat, how to develop a better system. They didn't do it; they weren't interested in that. The big grain companies and milling places, where the money is to support the research, weren't interested in seeing it happen. So it's follow the dollar.

Chapter 2 – Dan and Mary Howell

If I had $5 million to walk in, or maybe now it'd be $50 million, whatever that magical number is, to get somebody to listen to us; I could get some research done that I wanted done if I walked in with a pot of gold. But us family farmers out here, we don't . . . I'm on the board of the sustainable center at K-State (KCSAAC)[7], and I've been to several meetings with the upper guys at K-State. I've said to several of them, and I didn't mean this as a wise-crack, and I believe they didn't' take it that way, but they didn't have a lot of comment. They looked at me, and they didn't want to say they agreed, but I could see the wheels turning. I told these guys, the Deans at K-State, "The thing that you're missing is that if you think your university is sustainable, with less and less farmers in the country, and you will have multinational companies contracting all the grain and livestock, they'll end up providing all the inputs, you'll have serf farmers, just a contract grower. Once that chain is completed, there won't be any voting farmers out here voting to pay K-State, to keep your county offices, or keep the land grant universities going. This may be a selfish interest, but you should be more concerned about keeping more family farmers on the farm, not that they need to farm the whole county to support your structure. But we need you, and we need you to be doing research for us."

I truly believe that's the road we're headed. We're going to get to the stage where there will be just a few farmers per county. They're going to be tied in like these contract hog and chicken growers, where they don't have anything to say about anything. They bring the hogs or chickens, once they are finished they come and get them, and they provide the feed. The growers have no decisions about anything. Once at that stage in agriculture, there isn't anybody going to care about the university and their high-paid jobs. You didn't help me when I needed help. I'm done for, and gone now, so I can't help them even if I wanted to. I think they understood what I was saying, but I don't think anybody wants to step out of the chain of power and money flow even to bring those deep-seated issues to the forefront. There may be something to this. Short term, we may give up a few dollars to do some of this other stuff, but long term, we may be a hell of a lot better off.

What are some changes you've seen in past 25 years, especially in sustainable agriculture?

I didn't know back then what to look for or ask for. I didn't know what questions to ask. I didn't know what thought process to even change. For us, the thing that got us started was in '93 when Tuttle Creek backed up and we had so much stuff under water, and we lost most of our crops—we had that year's expenses, and no crops, and we weren't carrying federal crop insurance because it was more of a hassle. Back then with the federal insurance program, you had to lose everything to get anything. We were always raising decent crops. That year it just backed up, and we did lose everything, and didn't have any insurance for it. Some of the crops we put in three times. I had oats in, water backed over it, washed it out. While there was still time, I put in beans, and then beans a second time. Then it flooded over that. I carried all those expenses into 1994. It would have been very easy to get really depressed! I think I probably was in depression, but I wasn't smart enough to know it (laughter).

That was probably good!

The years following '93, were the worst, in a way. We knew we had to change the system, but weren't sure where we wanted to go. That's when we went to Missouri, to Jim Gerrish's grazing school. We went to the beginners, and then we went to the second course, the "non-beginners." That's when we got into the idea of going into more rotational grazing, more grass. Especially on flood ground, if you had eastern gamma grass, the flood won't kill it. It can stand being underwater for a while; it has a sponge core in the root where it can store some oxygen. I might lose production, but when it all settles out, it'll come back up, and I don't have a bunch of seed to pay for, a bunch of chemicals, and several thousand gallons of diesel fuel. And I don't need a combine to harvest grass; I get the cow to do the work.

If we want to talk about farm policy issues—if a person wanted to switch ground to grass, rather than a 10-year CRP[8] contract, what about a two-three year CRP agreement to get grass started to the point it is grazable, the program could get three to four times the acres with the same amount of federal dollars, and I could utilize the grass for my operation. That's what takes so long to convert to a grass system, is the time it takes to get grass established without income from those acres during those years. You have to baby it along and you can't guess what the timetable is. Mother Nature and the weather, heat and water are in control of that whole situation. Now that we have some moisture, we're in a stage to speed this grazing deal up now.

Are there new information needs, or do you have what you need to do what you want to do?
I've pretty well been to enough places, and done enough research. I didn't have the knowledge in '93, or in the '70s. Now I have the knowledge I need to move toward a grazing system; it's just a matter of implementing it, and balancing the cash flow to get it done.

Has enough been done on grazing systems? Are there unanswered questions?
The only thing I don't have totally worked out is the mix of legumes with eastern gamma grass. I've worked with Gary Kilgore, and he doesn't know for sure. I want to grow some more nitrogen for the grasses, especially the cool season grasses, and I don't want the legume to compete with the grasses. He's given me a sheet on that for warm season grasses. Eastern gamma grass is a native; I don't have to fertilize it. I've asked Kilgore about hairy vetch, it comes on early, and then dies back, and should release nitrogen. He said, that's not a bad idea, but he hasn't done any work on it. He said, if I do it, he'd like to hear about it; s so, I'm going to work with that. I have the seed bought, and I'm going to work with that this fall, if the moisture cooperates.

We then talked for a while about all the different possible weather possibilities, and how that might affect a grass-legume mix, and concluded that a person could do the same thing 10 years in a row and get 10 different answers.
I have this dream of 20 or 30 like-minded producers with a small portion of their operation in vegetables and fruit and vineyard, and salad bar stuff. We are paying taxes in the county and state, why aren't we supplying the vegetables and fruit for the nursing homes, the high school, and the grade schools. Why aren't we developing a food system instead of paying all this freight to ship food across the nation? It is raised overseas and we don't know what it's been sprayed with. Why aren't we doing research on developing local food systems that are truly local? We could have 30, 40 or 50 farmers growing different things; not everyone has to grow the same thing. This person wants a vineyard, this one wants an orchard and another guy wants an acre or two of salad bar stuff. This person wants tomatoes, and as a group work out in the fall what is going to be grown together, and market it to-

gether, through a local farmers market, a cooperative or a school district. We need to be working on those things. I'm not saying it's going to be the whole enchilada for the farm, but it could be part of that pie to make all those producers more sustainable.

It would be energy friendly, school friendly, vehicle friendly, health friendly, and local kids may learn where different things are coming from. The local FFA (Future Farmers of America youth club) and science classes come out and see these systems work. They could do papers on it, and participate in the operation of the system. Why isn't FFA working on this idea with kids? The kids are supposed to be doing so many hours, why aren't we teaching the next guy different ideas about how a farm might be sustainable? This could make an opportunity for our local youth, a little different style farming, instead of buying up all our neighbors. We could have fewer acres and be better off. Make the whole system more sustainable, instead of worrying about producing another bushel of wheat, acquiring another section of land, or another combine with a 40-foot header, instead of trying to survive by buying my five farmer neighbors out. It's going to be the meanest and fittest is the one that is going to survive. Why do we think that that's the only way agriculture has a future? It's really sad and sickening as far as I'm concerned.

What do you think the most important thing this book should communicate to the public?
They are part of this large voting block, this group that tells the state legislature, the state senators, heads of universities—what we need. They ought to be even more worried about it than us here on the farm are. Our school system; it's not really teaching anybody the trades. There's nothing wrong with being a college graduate, but not everyone is designed that way. Somebody has to do the manual work, be the mechanic. Many kids would be better off going to vocational-technical or a trade school. There is no sense in telling everybody that you don't have a future if they don't go to college. I went to college, but if had to do it over, I probably wouldn't choose the same route I did. I could have started farming if I knew then what I know now. Probably the best thing would have been for me to go to the Missouri Forages Research Center, or the Noble Foundation[9], and just worked for nothing on some sort of internship.

For someone who really wants to learn the nuts and bolts of how to make a farm work, I don't think K-State is the place to go. I'm not anti-K-State. They are opening the door a little bit with the new Sustainable Ag Center. Is their true university heart in it? They're tiptoeing into that arena, but that isn't where the big bucks are. Not to be too critical, a place like that, money is an issue. This theory of what I think is best, it's easy to think of and speak of, but how to make it happen? . . . I understand that. The big mistake is too much bricks and mortar, and the number of people on the payroll, and the number of grants and things they are doing. Instead, they could be pulling back to basics, and not trying to do everything. Maybe there are some things we need to do well. Don't worry about being bigger, and maybe be a little smaller, more efficient, and do things to keep people on the land.

We need to get rid of this damn mentality in the United States—that bigger is better, and that we need a few multi-national companies doing everything. This mentality is taking us down the wrong road. This same mentality has gotten into economics, farms, corporations and into universities. K-State ought to be more concerned about being better, and more sustainable. There should be concern about filling the needs of family farms rather than being bigger with more money coming in all the time. If I were the new dean, man—there'd be a lot of soul searching. Doing all this research to support Monsanto, Syngenta, NCBA (National Cattlemen's Beef Association), the Meat Institute and corporate America . . . We don't want to alienate all of them, but we have to be diplomatic, and change the focus.

What should this book say to sustainable agriculture farmers?
It should be inspiring like the Heartland Sustainable Ag Conference at K-State. I have to go to something like that every three to five months to regenerate my thought process. I can get discouraged with things going wrong. I'm the lone duck out here on some things. Sometimes I need to be with good, like-minded people to hobnob with them, to re-light the fire and come back home to fight the fight. This change in thought process, we aren't going to get from *Kansas Farmer* or *Successful Farmer* (two common mainstream farm magazines). I have to look for publications like *New Farm, Small Farm*; and the *Grass Farmer*[10] magazine for articles to meet my needs. *Acres* magazine is good for a wider variety, it deals more with crops and other things. I've got to look for that stuff.

Another thing we ought to do more of, people in a county, we ought to get together once a month over a cup of coffee and piece of pie. Farmers could meet, the way investment groups meet. It needs to be within a 10- or 15-mile radius, not a big deal, a ritual like church, except that it is once a month. It's kind of like the Kansas Graziers[11] Association but that's a yearly schedule of meetings and tours. To keep the mission and the drive, you need support and camaraderie a little more often and each other's support.

I'm not anti-big guy, who thinks the answer is that bigger tractor and next section of ground. For many people that's not their mission. There are a small number of people that can do that. For many people, there are other ways to make their farm sustainable, to make it survive, with probably a lot less risk. Not that there can't be some people like that, but it's not the train I want to ride.

If you were to be critical of sustainable agriculture, what are things you would mention?
For some commercial industrial farms, there are some words tied to things that just automatically shut them off. Organic is one of them. I'm not organic, may never be organic, but I like to pick up knowledge from them. Sustainable isn't quite as intrusive. Organic isn't a turn-off to me, but I'm not the typical, not normal, maybe I'm too open. I'm not anti the person that does it, but I haven't figure out how I can never use a chemical. I'd be tickled to death to figure that out but I don't have that vision, of how to get away from them 100 percent. I'd get away from them tomorrow if someone came out to hold my hand—how do I solve that problem, to get rid of these spots of bindweed. I know if I try to do it by tilling, I'd have to disk that little spot every week for two or three years, which would totally deplete its nutrient source. Most of the time it can't be done, it's not practical.

Are there other words that are turn-offs?
I'm not sure. No-till, I'm doing some no-till. I use Roundup® and some chemicals. I'm concerned about the chemicals I'm exposed to, and if some of these practices that a few years ago I thought were sustainable—if they really were sustainable, such as no-till and Roundup® spray, and stuff like that. I would like a system where I could do no-till and not use herbicides or insecticides, use cover crops and something like a soap spray. Now THAT is the thing a university ought to work on—

insecticides and pesticides as natural products, that will hold stuff back, so the crops can take over and compete. I need to look for that person who was working on the soap spray. I just thirst for that kind of knowledge. I haven't found it on the shelf, who has that information to make it work?

The organic producers that I know are still struggling with water issues, erosion issues, and cover crops. They don't have it all figured out, it's not a cookie cutter approach. I know how difficult some of that stuff is. For me to get a system like that to work I'd need one or two hired men. There are time sensitive operations, narrow windows to cultivate, rotary hoe, with only a few days and if you miss it— for example with two weeks of wet weather, I could miss getting the weeds under control. As a kid, I used to cultivate a lot, but on curvy terraces, it's not as simple as it sounds.

The whole marketing deal, the ag-economics and those departments, they don't even want to mention or touch that whole program. The board of trade, the big grain companies, they don't want a different marketing system screwing up the present system.

In the policy arena, what do you think sustainable agriculture should be doing?
With agriculture, there are some really good programs. With the federal budget deficit, I understand the need to cut budgets, but their answer is to shut down some really good programs, when maybe the answer is to bring the caps down (what a producer can receive) on all of them. That is the fairest across the board. Keep all the good programs, and even the ones I consider not-so-good programs, but put a cap on the total deal. Then they don't have to monkey with everything.

It's back to the big guy, the connections, the money deal and who has it—the dollars and connections. There are a few good guys, Senators and Representatives working on this, but they seem to be in a minority. We have the best political system in the world, but at the same time, it has many faults, and I don't have the resources and time to get my word back there and be heard. I'm talking about that elite group, that five percent that are getting 80 percent of the dollars . . . The government shouldn't pay me a ridiculous amount either, I shouldn't be getting a half million or $250,000 dollars, but the other guy shouldn't get that either. The other bad thing is they don't watch that. There gets to be enough pork built into the whole farm bill and they don't do the things that need to be done. We're a little bitty voting block. If you get the public to understand, and if they get mad enough, they'll just wipe the whole thing out. We need to be reasonable, and support something we can defend, logically, morally, ethically and financially. That's what we don't seem to do, anywhere, with anything. It's the same story I said the university should be thinking about; the same thing needs to happen with the national farm policy.

What do you think about CSP[12] (Conservation Security Program) and other types of green payments?
I would rather not draw a dollar on overproduction of grain that's not needed, and get all my dollars from doing the right environmental things. I'd rather not put on that one more pound of nitrogen to get that one more bushel of grain, which also isn't environmentally friendly, but that's the way the program was geared. NRCS (Natural Resource Conservation Service) and the environmental side of farm policy says one thing, but what they actually support (commodity payments) is in total contradiction. It's as if the twine never meets (laughter).

Farming in the Dark

It's like they are writing policy in the dark, we're farming in the dark, the universities are in the dark, everybody's in the dark—and nobody is talking to anybody! [laughter] And you *(speaking to me)* are caught in the middle! I see people like you in the university, kind of caught, where your heart is, and what you think ought to be done, and what you can get done policy-wise, you can't really change . . . . I get the impression that most of the university faculty are being told, "Here is your base, your department, you need to go find a grant." Are you going to come and get a grant from me for $50,000, $100,000, or a half million? No, but some big corporation, they can say, "We need you to do this research . . ." And they have the money to play with.

The family farm is who they should be supporting. I swear to God, if they do away with us, at some point in time . . . it won't be a gradual deal. Things always go in jerks. They'll support you until the question comes up, then "Gosh, there aren't many of us left, and Syngenta or Monsanto or Cargill will tell me what I can grow, they furnish it, I grow it, they haul it out. I don't need Research and Extension; and you want me to pay this much tax to support that? Hey representative, let's do away with them. We don't need 'em." It will be just like an overnight shift.

My advice to the universities is to really figure out who you want to support, and who you want to keep out here, and are you going to have enough people to make enough votes to make a difference to keep you when the battle comes? They might become the next dinosaur, and I truly believe they are moving toward extinction.

I envision that in 10 years, most people won't even spend much time at a university getting a degree. Half of the classes will be on the internet, on-line. They'll come to school once a week for labs, and to do other things. The way the cost of education is going, I truly believe that bricks and mortar institutions in 10 or 15 years will become obsolete. I think they'll become as obsolete as a hay barn with a hayloft that you have to put hay in with an elevator (a device for unloading hay). Energy cost, and institution cost, heating cost, faculty cost . . .

We're going to see more changes in the financial market in the next 10 years than in the past 20. It's going to be unbelievable. Every new trade and farm policy makes it more competitive to it grow crops overseas. CAFTA[13] is going to be a disaster. Most tax law promotes overseas production, and overseas manufacturing. The United States is the largest debtor nation in the world. If foreign countries would sell our treasury bills and not buy them, overnight we would have a financial collapse. We are so dependent on everything else. Our balance of trade—we don't even have a plus balance of trade in agriculture products any more. I hope to hell I'm wrong, but I think we'll see some form of 1929 in the next 10 years.

Dan then tells the story of his mother's parents experience with farm crisis and financial crash and one can see why he would be apprehensive. They were fairly prosperous farmers in the early 1900s in Kansas. Her father saw a national financial crisis looming, and felt that cash would probably be a better way to preserve farm assets rather than cattle or grain. He sold some cattle, sold the grain on hand and deposited the cash in the bank at 11 o'clock in the morning. By one o'clock that day, all his cash was gone—the bank failed. According to Dan, he had two things right, to sell farm assets, and to save cash, but the one wrong decision, to put it in the bank, canceled out the other two. The end result was a cascading series of events that lead to selling the farm, renting another farm and his grandfather work-

ing for another farmer. Then his grandfather had a farm accident with horses, which resulted in a head injury that eventually lead to his permanent hospitalization. His grandmother was left to raise five children without a husband and no money during the drought of the 1930s. According to Dan, his mother carried the emotional scars of that time with her to the grave.

That's part of being sustainable. I'm really fighting with the issue in our operation now. How can I get situated to survive something like that? To survive something like that, I just about have to be 100 percent out of debt.

He was debating whether they sell some land, in order to pay off the debt owed on other land, just in case.

Selling land for me is like selling one of my kids. I just never know if the next person will take care of it properly, the way I would. This was just in my lifetime—in the 1980s farm crisis, just a few years before that, land was bringing $800-$900 an acre. In the belly of the farm crisis, some of that same ground was selling for $150 to $200 per acre. It was just incomprehensible that in a couple of years it could go down to that. In a window of a couple of years, when they were foreclosing on a lot of places, there was a ton of ground (land) on the market, people couldn't make payments, the Federal Land Bank and FHA[14] were taking it back no matter what. In the belly of the farm crisis, I went to the bank, where I had always banked. There was a place for sale at $150 per acre. The bank wouldn't loan me the money. I told the banker, "If I can't make that work, you're going to get everything else back anyway. If you're scared of that . . . " This is part of the deal, I wasn't planning to come back to farm, I'd started a retirement insurance account in Topeka, I cashed it in, I had enough money for the down payment to buy it, set up on payments. The banker just had a fit. A couple of years later, it worked out, I was doing all right. He's not in that bank anymore.

That is just how that mindset could change, in just 24 months. At that time, implement dealers were going broke. You could buy a tractor for 50 cents on the dollar, compared to a few years back. We don't realize how perilous the times are now, with our economic system. People are buying all these new houses on floating interest rates with no down payment. The speculative deal in this whole system—and I hope I'm wrong, but we may see something that a lot of this generation has only read about. Even my age group and younger think the government can control things, and that this absolutely can't happen again. This deal is bigger than our government. If the other nations decide they're not going to float our debt. We don't understand…80 percent of what Wal-Mart sells is made in China. We are so dependent, and I don't know if that is wise. And we are so dependent on foreign oil at any price. There is a ripple effect of all that through the system after a few years. There are a lot of things we're doing in agriculture with $2 farm diesel, that's $150 per tank full per tractor per day!

But on the marketing side we are price takers, not price makers. We just accept whatever the market is. A few national grain companies control that whole deal. The wheat growers begged the economics department at K-State to work on a better wheat marketing system. They didn't want to touch that. That's been several years back.

The worst thing that's happened to agriculture is that we've become industrialized growing a few commodities instead of growing many different crops. We don't have the diversity in agriculture we used to have. That's a mistake. I can remember back when I was just starting to farm, the big deal of K-State's economics department was that you have to get big and specialize. Like you need a sow confinement, and get big, or you need to be a big feeder. Their whole deal was to move away from diversity. You have to be big and specialized, be a low-cost producer, with big volume. I disagreed with it at the time, and it's been proven that it wasn't a good plan, but their thought process still is big, specialized, low-cost production. It's a race to the bottom. My opinion is, you'd be better off growing 20 different things and have a close connection to the consumer on 5 or 10 of those items. I'm not saying everything has to go that way, but there ought to be a blend or a mix of that to make you more stable. If one product crashes and the other three or four things are holding their own, you still have something to make some money. Think of all the hog producers there were, and when hogs got to $8 (per cwt) (note: the break-even price is $50 per cwt or higher), how many of them vanished? The National Pork Producers council pushed and pushed people to get bigger. Now there is hardly any place to sell fat hogs. It's like chickens. If I want to grow chickens for the processors, I need a contract, and hogs have gone that same way. If you don't have a contract, you can't go to the sale barn and sell fat hogs. In 10 or 15 years, the cattle market might be the same. I don't know what the future holds for it. Kansas Cattlemen's Association[15] and the Ranchers-Cattlemen Action Legal Fund[16] are trying to fight that battle. The Meat Institute, and Kansas Livestock Association[17] and National Cattlemen's Beef Association[18] are in the hip pockets of the packers in the meat deal. It's just a race to the same formula in everything.

You mean specialized, vertically integrated?
Once you get that way, nobody has any say about anything.
Is sustainable agriculture going to pull us out?
I don't know. I'm optimistic, and I hope. But in the political climate today, I'm not very hopeful. The only good thing I can think of out of 9-11, is that maybe we will think more about food security, want more of our stuff grown in our own counties and states and country. I don't see that being much more than an afterthought in national security, and food security . . . I don't know.

Have you done any direct marketing?
I used to do sweet corn, but after I started work at Jeffrey, that fell away. I'd like to grow for farmers' market, where it's not just us, but a group of people helps each other, and do it together. When people come to farmers' market, instead of just sweet corn, there would be a smorgasbord of everything. Since we are paying taxes, the school ought to be required to buy as much locally as possible, and actually go out, and tell producers, a year or two ahead of time, if you produce it, and if it's locally grown, we'll buy it. I think that out to be part of state school policy. It's as if they have policy on federal projects, where you have to buy a certain percentage of your stuff from minorities. I think on state school policy, they ought to have to buy as much as 50 percent of their food products from within the state that they're in, and maybe 10 percent from within the county.
He then asked me, "What is your vision of how to change this apple cart?"
I told him that most of my work has been at the cropping systems level, for example looking at legumes and alternative crops. Then I find out we can't sell the alternative crops and make a profit because of programs like NAFTA (North American Free Trade Agreement), and the loosening of trade barriers by the WTO (World Trade Organization), so what is the point? On our farm in Wamego, selling

vegetables at the farmers market, we get nearly the same price we were getting when I started selling 10 years ago, and the prices of everything else have gone up. How can we do sustainable agriculture and farm if it isn't profitable to grow food?

I'm not convinced the farmers' market deal would work. It might work in some places, but out here? I'm not convinced we have 10 or 15 like-minded producers that could provide a total menu, own a restaurant or two and actually run it themselves. In most of these households, somebody is working an off-farm job. Why couldn't 10 or 12 of us, run it as teams of maybe five people, stagger the workload. Our mistake in agriculture is that we aren't taking our product far enough through the chain. We're selling it, and everyone else down the line and in-between is making the money. We need to quit selling it to the middleman, to the processor. I'm thinking we need to have 10, 15 or 20 people actually own the restaurant, maybe as part of a motel combination, sell the end product to get the whole dollars worth, and then divide it up among the group.

There are still people who go out and eat; and everybody's working these jobs, and the wives don't have time. The current marketing mentality is that we want to get rid of it, get the money quick and go on to the next job. That's part of how we've given away control of the whole food system. I'm not for an individual doing it by him or herself. I think it has to be a team, because you have to have the diversity and the backup. That's like an insurance policy, if somebody gets sick the rest of the team can fill in. If we're growing 20 products, and one person has a major farm problem, the others in the group can pick up that one product, and grow it on their farms. It's a team effort. Then it's however you want to divide the money, you work that out, by the hours you put in.

We've only talked about production and volume and how do we shift it off to the next person. I'm convinced we need to look at it all the way through the food chain. I'm not talking a food chain down the road; I'm talking about that local food chain, within a 20-40 mile radius. Part of that local food chain could be nursing homes, care homes, hospitals, and school districts, besides our motel/restaurant combination. Sysco and all these other places are hauling this in from God knows where. I mean, it's being bought, and it's being paid for and sold. They are paying full dollar for it.

We finished lunch and stepped out into the July heat to Dan's truck, a flatbed with a water tank and several other essential farm tools. We continued our discussion of farm policy.

"Here, read these"—and he handed me several newsletters from his office/library in the cab of the truck. "These will explain a lot."

I read the newsletters, [see notes below] and found that farmers can be well informed about the issues that affect their farming operations due to farm policy generated in Washington. Many organizations have representatives there specifically to represent farmers, to monitor legislation, and to try to influence policy through member phone calls, letters, testimony and "fly-ins." But like other policy moving its way through the legislative channels, farm programs have a visible portion; for example press releases and statements from the Secretary of Agriculture, and the votes and bills sponsored by senators and representatives. There is also the invisible part…the money part, where committees and decisions are influenced by agricultural sectors with more money than farmers.

Notes

[1]*The New Farm* magazine. Published by the Rodale Institute from 1979 until 1995. Archived articles and current news at www.newfarm.org

[2] Kansas Organic Producers (KOP)—is a marketing/bargaining cooperative for about 60 organic grain and livestock farmers located primarily in Kansas, with some members also in bordering states. KOP's purpose is to help build markets for organic grain and livestock and to represent its members in negotiating sales and coordinating deliveries of organic products. KOP was first organized in 1974 as an education association for organic farmers to promote organic agriculture, develop organic certification standards and establish organic markets. http://www.kansasruralcenter.org/kop.htm

[3]*Acres USA*. http://www.acresusa.com/magazines/magazine.html A monthly magazine on Eco-Agriculture

[4]Jim Gerrish—worked for 20 years at the University of Missouri, and was co-founder of the 3-day grazing management workshop program at the Missouri Forages Systems Research Center, in Linneaus, http://aes.missouri.edu/fsrc/ He and his wife Dawn currently provide consulting services from their home in Idaho Falls, Idaho http://www.americangrazinglands.com/.

[5]Pharo Cattle Company—founded by Kit and Deanna Pharo in Cheyenne Wells, Colorado. They publish a newsletter and sell cross-bred cattle breeding stock based on the principle that cattle genetics should fit their environment, including adaptation to grazing systems. http://www.pharocattle.com/

[6]The Kansas Rural Center is a non-profit organization that promotes the long-term health of the land and its people through research, education, and advocacy. The River Friendly Farm Program is a whole farm planning and assessment tool developed jointly between KSU and the Kansas Rural Center, promoted through the Clean Water Farms program by the KRC. See http://www.kansasruralcenter.org.

[7]The Kansas Center for Sustainable Agriculture at K-State, (KCSAAC) was established by Senate Bill 534, passed by the 2000 State Legislature, out of concern for the survival of small farms in Kansas. The Center works in partnership with state and federal agencies, nonprofit organizations, environmental groups and producer organizations to assist family farmers and ranchers to boost farm profitability, protect natural resources, and enhance rural communities. http://www.kansassustainableag.org.

[8]Conservation Reserve Program (CRP) was begun in 1985, allows farmers to enroll highly erodible acres in the program. In exchange for seeding acres to perennial grasses and/or trees, and not producing a saleable crop or graze livestock for 10

years, land owners receive annual "rent" which is determined by a bidding process. http://www.nrcs.usda.gov/programs/crp/

[9]Noble Foundation—conducts forages and grazing research in Ardmore, Oklahoma. www.noble.org

[10]The Stockman Grass Farmer magazine, since 1947, publishes articles from around the world on making a profit from grassland agriculture. Widely read by those in the U.S. interested in rotational grazing, management intensive grazing, year-round grazing and other systems. http://stockmangrassfarmer.net/

[11]Kansas Graziers Association, began in 2000 as a grassroots organization administered by farmers and ranchers and dedicated to the continuing improvement of the profitability and quality of life of livestock producers. It links the grazing clusters of the Heartland Network together by coordinating and promoting educational activities about year-round grazing using advanced techniques and forages. The association holds a one-day winter grazing conference and summer grazing tours. http://www.kansasruralcenter.org/kga.htm

[12]Conservation Security Program (CSP) is a voluntary program that provides financial and technical assistance to promote the conservation and improvement of soil, water, air, energy, plant and animal life, and other conservation purposes on Tribal and private working lands. Working lands include cropland, grassland, prairie land, improved pasture, and range land, as well as forested land that is an incidental part of an agriculture operation. The program is available (on a limited basis) in all 50 States, the Caribbean Area and the Pacific Basin area. The Farm Security and Rural Investment Act of 2002 (2002 Farm Bill) (Pub. L. 107-171) amended the Food Security Act of 1985 to authorize the program. CSP is administered by USDA's Natural Resources Conservation Service (NRCS). http://www.nrcs.usda.gov/programs/csp/

[13]CAFTA (Central America Free Trade Agreement). Passed in 2005, follows the model of its predecessor, NAFTA (North America Free Trade Agreement—passed in 1993) which linked the U.S., Canada, and Mexico. Removes trade barriers for U.S. companies selling products to the countries in the agreement, but also removes some trade barriers (tariffs) that up until now had provided some protection to farmers in the United States and farm prices. http://www.ustr.gov/Trade_Agreements/Bilateral/CAFTA/Briefing_Book/Section_Index.html

[14]FHA (Farmers Home Administration) Provides government backed loans to farmers to whom bank loans are not available. http://www.fsa.usda.gov/

[15]KCA (Kansas Cattlemen's Association)—An independent Kansas grassroots association working to restore profits in our industry to the beef and agriculture industries. http://www.kansascattlemen.com/

[16]R-CALF USA (The Ranchers-Cattlemen Action Legal Fund)—represents the U.S. cattle industry in trade and marketing issues to ensure the continued profitability and viability of independent U.S. cattle producers. http://www.r-calfusa.com/

[17]KLA (Kansas Livestock Association) Organized since, 1894 the association continues to serve members by fulfilling its mission to: Advance members' common business interests and enhance their ability to meet consumer demand. *www.kla.org*

[18]NCBA (National Cattlemen's Beef Association) http://www.beefusa.org/

Discussion Questions:

1. If the agricultural universities loose their farmer constituents, who will vote to continue to support them?
2. Can no-till farming and organic farming systems ever be compatible? What would a system like that look like? How would it work?
3. Dan and Mary are moving from grain-based farming to grass-based farming. This allows them to achieve several of their sustainability goals. What do you see as the pros and cons of a strictly grass-based system vs. a mixed farming system? After reading chapters 3, 4, 5 and 6 in this book, come back to this question and see if you would give the same answer.
4. Dan suggests that the purchase of local food by tax-supported institutions be mandatory. Do you think this is a good idea? Why or why not? If you think it is a good idea, at what percent of food would you require them to purchase locally? Would that percentage change depending on the type of food, e.g. meat vs. grains, fruits or vegetables? What institutional changes would be required for them to purchase and utilize a greater percentage of local food? Who would "lose out" in this deal?
5. Dan points out an incongruence in the federal farm support program, where, in his words, "the twine doesn't meet." What is this incongruence, and how could it be resolved?

3

Jim and Kathy Scharplaz
Prairie Natural Beef, Minneapolis, Kansas

"I'm not sure the link between food and sustainable farming has been made. Maybe that is the thing we haven't accomplished, the link."

I first met Jim and Kathy through the Kansas Rural Center[1], and I see them at the Land Institute[2] Prairie Festival every year. We drove up to their farm nestled in a valley on the Salt Creek, north of Salina, in the Smoky Hills of Kansas. Their farmhouse and outbuildings are constructed of local sandstone, built by Jim's father, an immigrant from Switzerland. Jim's Swiss roots have also led to friendships with other Swiss-Kansans such as Dr. Ernst Horber[3], K-State Entomologist, and Dr. Isidor Wallimann[4], who came from a farm in Switzerland to get his master's degree at K-State. Kathy's roots are more urban, as you'll see in the interview, but their partnership has helped their farm prosper, and both have contributed to the local food council and activities in nearby Salina. It was a beautiful fall day, the colors of the native prairie grasses were at their peak and the drive out to their farm was relaxing. We were seated at their dining room table for the interview, and afterwards were treated to a meal of homegrown hamburgers and appropriate side dishes.

Jim has a master's degree in Agricultural Engineering from Kansas State University. He worked at the Scandia Extension Experiment Station before returning home to manage the Scharplaz Ranch. He also has been a frequent contributor to the Prairie Writers Circle, a network of writers on alternative agricultural topics. Kathy's background includes a B.A. in chemistry from Rice University, medical research, freelance journalism and producing farm-safety publications for KSU Extension.

How did you each first get interested in sustainable agriculture?

Kathy begins. What got me started was that back in the '80s I got very involved in the environmental movement. I don't know why but for some reason, I got the connection that many environmentalists don't get, which is how important agriculture is. It's not just "save the world, save the rain forest." The biggest use of the land in the country is agriculture. Part of it at that time was that Jim Hightower[5] was running for Texas Agriculture commissioner, and I got involved in his campaign. From there on out, I was interested in anything that had to do with sustainable agriculture. That led to my coming and being an intern at the Land Institute. Several months into my internship, I met Jim (Scharplaz). We ended up getting together, and I just felt bad when I thought about what happens to our cows going off to the feedlot. [laughter] I was so against that.

We had talked before the tape started about her years as a vegetarian, so I asked if she was still a vegetarian when she met Jim.

No, that was long before, but what I was trying to work against was the entire industrial food system.

I suggested to my husband that maybe we should try to reach the niche market and sell some of our beef right off the farm to people, so the cattle wouldn't have to go out to the feedlots and go through the big packing plants and stuff. Jim was, "sure!" Jim's always the one that says, "Give the customer what they want, the customer is always right." We started with selling sides (halves) and quarters, and then a couple years into it we found out how easy it was to get a license to sell retail, and to sell it straight out of our freezer. That brought in a huge pool of customers, because the market for sides and quarters is limited. First, people are somewhat intimidated, second, people that can afford that much cash at one time, don't have that much freezer space. The retail is how we sell a lot of it.

Are you from Texas originally?

Yes, but from the big cities, Dallas and Houston [more laughter].

Jim: It was Kathy's idea to sell it off the farm. The first few we fed, I just did it to humor her. I said I'd feed out some beef if she could sell it, and she sure enough did!

Kathy: Well, more than that, he changed his practices. The first time he invited me to come see him and his cattle working buddies work cattle, I was so horrified by the de-horning. I was amazed to discover that they didn't seem to mind the castration very much, they didn't seem to mind putting stuff up their noses very much but they really hurt with that de-horning. It happened at that time I was working for somebody that had cattle and had mentioned to me about the dehorning paste. I asked Jim, and he said, "Yeah, I've always hated that dehorning. Sure, we'll give it a try." I really admired him for that, because it is a lot more work. You have to get out there to the calf in the first 24 hours after it is born, no matter what the weather, and get that dehorning paste on it. Sometimes the mama cow doesn't want you coming around, so you have to be quick.

Jim: It doesn't always work. If it is raining, it washes off. It depends on the weather.

I turned to Jim; how did you get involved?

Well, my folks never farmed. They just had the grass and the cattle. My dad was always very particular about not over grazing. When I was little, about five, the USDA had the soil bank[6] program. The purpose of it was to get cropland out of production. He planted a lot of this farm back to grass, old fields that were worn out. He bought them with the intention of putting them into the soil bank program, and making a pasture out of them when they came out.

Part of it was that, and part was my friend Pat Dreese who was an intern at the Land Institute. At that time, I was an agriculture engineer, and I thinking about things like robotic tractors, and Pat was just always asking questions about things like that.

Kathy: He also introduced you to Wendell Berry.[7]

Jim: Oh yeah! He was the one that said I should read *The Unsettling of America* and all those radical things.

Kathy: And that was a seminal book for me too, I read that in '85. You read it even before that. That was a seminal book for both of us.

Jim: We should all now praise Saint Wendell! [laughter]

That was one of the books we passed around at Cornell; a lot of us read in the mid-1980s, it shaped our thinking too.

Chapter 3 – Jim and Kathy Scharplaz

Kathy: It is life changing.

Jim: Part of it was with the farm crisis in the early 80's. We said, "Man, this just isn't working! It's crazy."

Kathy suggests Jim tell the story about working with extension.

Jim: You mean my farm management person? When I first started farming, I joined the Farm Management Association (an extension program offered at KSU) because I needed to learn how to keep books, do taxes and all that stuff, and they did, excellent. One day my field man said to me, "You are just wasting my time with your little operation. You should borrow some money and expand!" That was just six months before that big farm crash here in the early '80s. I started thinking, "Here are these agricultural economists, telling us how the world is going to work on into the future. They are out there making predictions about the world economy for the next 20 years. But if they can't even predict the biggest agricultural crash in the second half of the twentieth century six months before it happens right here in Kansas, how come he thinks he can predict the course of the economy of the western hemisphere for the next 20 years?" [laughter]

So you didn't take his advice?

Jim: No, but not because I'm smarter, but because I was too scared to do it.

Kathy: It would be a big financial risk.

Jim: My folks were always real risk averse; I'm risk averse. I'm what you'd call a late adopter, the people who are the last to take on new technology. But sometimes you get lucky; you just walk away.

Kathy: or like Nancy Vogelsberg says, "We didn't have to go organic, we just keep farming the way we did for the last 100 years. This is the way my great, great grandfather did it."

Jim: My old college roommate, he farms down by Sterling. He's a big operator; he's way into precision agriculture, with the satellites and self-guided tractors. Even he says, "I'm not doing anything that my grandfather didn't do. Except his farm was small enough he could do it by eye. I have to have a computer and all this stuff to keep track. My grandfather knew where in the field he needed to throw an extra load of chicken manure, and I have a computerized fertilizer spreader behind my satellite-guided tractor. But it's the same thing."

So as an engineer by training, you haven't wanted to get into gizmos for the farm?

Not any more. I really used to like gadgets. When I first started to work up at the experiment field up at Scandia, boy, I liked all the gadgets. It takes so much work to get along with them. By the time I left there, any research I couldn't do with a bucket, a stopwatch and ruler just didn't need to be done! [laughter]

Did you have automatic weather stations and data loggers and all that?

They were just trying to do that. It was just before PCs (personal computers) became readily available. When I started work up there, we were still doing our statistics with mechanical adding machines. A couple years after I left, they had half a dozen PCs out there at the field. It was a wonderful thing. We had data that we'd taken for 20 years, and nobody had time to analyze. We'd been taking soil moisture samples at every foot down to six feet, for years and years, and all we ever did was use about the top couple of feet to schedule irrigation. They learned a lot about how corn uses water during the growing season, and you should irrigate smaller, earlier in the season and less late in the season, and all those things.

Farming in the Dark

Is your friend with the precision agriculture making money, or is it just another money user?

He is either making money, or has a very generous banker. [laughter]. He is really, really smart, I don't know if he is making money or not, but he knows. There is no question in my mind about that. When they do precision agriculture things (field days, meetings), he is on the program, and all that kind of stuff. The people from John Deere that build the machinery, they talk directly to him, and he talks directly to them, and the people from K-state that do that, it's a tight little circle; they come out there, call him on the phone.

Kathy: He is like one of their poster boys I guess. [laughter]

Jim: Well, yes he is, because he's just real, real smart. It's easy for him, the concept, how the stuff works; it's just easy, easy, easy for him. He just goes right through it.

Kathy: What makes Lee interesting is that he is always willing to try new things. I find the contrast interesting between him and Don. We have another friend, Don, who is also willing to try new things but he goes a completely different route. Instead of mechanizing, he gets old machinery, and he has hired hands that help fix up the old machinery. Tell them some of the stuff Don does.

Jim: He's into grazing. He plants everything you can imagine to try to get year-round grazing.

Kathy: turnips, kale . . .

Jim: He has three people working for him, and they just move cows. His idea is to let the machinery sit as much as possible. They just put up electric fence, drive the cows down the road into this field, they get that eaten up, he's got something else ready to go. They move them over there. That's his way of doing it. Lee on the other hand has no livestock anymore, it's all mechanized, it's all machinery.

Looking back over 20 years, what do you consider some of the successes of sustainable Ag?

Kathy: Well, one is that, there is still a long, long, long way to go, but there is more awareness on the part of the consumer, I think, than there was 20 years ago, about different kinds of production. Some of it is probably simple-minded, you can put the word organic on anything and people will think it is good without looking into it. There is certainly more awareness.

Jim: I'm not sure people make the connection between organic and sustainable, like it's not necessarily the same thing.

Kathy: Organic at the consumer end is we don't want pesticide residues on our food, like Alar on the apples, and especially not on our children's food. But you are right, that doesn't necessarily connect back to the larger environmental picture. I think there is still a lot of ignorance about agriculture in general.

Have you seen an increasing sophistication of ideas and awareness from your consumers, people that buy your beef?

Kathy: Oh yes, because that is part of my job. When a first time prospect, who's seen our little one-inch ad in the directory calls up—what they see in the ad is "no hormones, no antibiotics." They call, and I describe to them what else we do. We try to butcher while it is still lean, they don't go to a feedlot, they're out in the pasture 365 days a year. We don't feed them any animal by-products. Many people that raise cattle don't realize they are feeding byproducts. On the feedbag, you have to read the ingredients, somewhere way down you might see, "hydrolyzed feather protein, chicken waste." They think they are feeding their cattle a grain mix, and don't realize they are feeding animal byproducts in there.

I do my best to educate people as to what makes our beef different, and the differences between our beef and strictly grass–fed beef. Jim does give them grain, for two or three months before we take them to the butcher, just local grown milo mostly. We explain what's the difference between our beef, and organic. When somebody calls and is interested organic, I send them to a guy over in Abilene who raises organic beef. One of the things that is nice about being in sustainable agriculture, particularly in the Salina area is the way we all try to work together, it's like the "Miracle on 34th street," with Macy's and Gimbels. We want the customer to be happy, and if we don't have what they want, we send them to someone who does. All of us that are involved in the Farmers' Market, the Prairieland Market, the Land to Hand Alliance[8] we try to educate. We're all involved in educating the consumers. I think that's why the awareness of sustainable agriculture is coming along more slowly, than organic - it takes an intensive process of education. The only way I've ever been able to do it is one-on-one conversations with these new customers. That's a slow way to get the word out. We have had the good fortune to have write-ups in the Salina Journal a couple of times. They seem to like us. They've done profiles on us two or three times.

Back to your question and the topic of education. Our customers are very aware, they call up with an open mind, they call up with questions and I answer them. I don't know how we make that happen on the big massive scale that needs to happen for there to be big scale changes.

Do your customers ask you to be organic?
Kathy continues: On a rare occasion. Usually I just explain why we are not, which is we don't farm any cropland. We don't have a source of organic grain or hay, for that matter. I tell them no, our beef is not certified organic, but on the other hand, it is lean, and since pesticide residues concentrate in the fat tissue, there is probably less residue. Plus, we feed them mostly milo as opposed to corn, and around here at least, the milo doesn't get quite the pesticides that the corn does.
Jim: Not quite as much.
Kathy: So I figure, well, it's not perfect, but it comes close, for the price.
How many do you raise?
Jim: we usually have about 200 cows, and then we keep the calves over the winter, and then graze them part of the next summer. Some times of the year, we have the cows, this year's calves and last years calves. That is approaching 600, if you count baby calves. Right now we don't have as many since it has been dry the past few years, we have been cutting back, cutting back to save the grass.
I noticed while driving out that the ponds seem low. Has it been dry this year?
Jim: This year has been closer to normal, but the subsoil is very dry from previous years. We used the rain up as it fell. It didn't recharge the ponds. There are many aquifers we tap into for stock water springs; some of them are pretty weak.

What do you see as successes in the sustainable agriculture movement?

At this point, Kathy turns to me and says, Well, you're here! [laughter]
Jim continues: That's a sea change.
Kathy: That's a huge change!

Farming in the Dark

Jim: Remember that time we were at St. Mary's, with the Rural Center, and an administrator from K-State was there. They were negative, hostile. It got contentious.

Were they asking for more research on sustainable ag?

They were telling us that we were a bunch of nuts! "There is nothing wrong with pesticides and chemical fertilizer." Many of that generation were the ones that developed and first promoted fertilizer. Many of the old farmers remembered when they first started promoting ammonium phosphate and those kinds of things. That's what those guys built their careers on. They believe in it, just like we believe in this. It's hard to change. It's hard to say, "Oh my gosh, that was a mistake."

Kathy: K-State, such big changes that we've seen, even just since I've moved here, all the difference in the world. There was no support for sustainable ag.

Jim continues: And the relationship with K-State and the Kansas Rural Center. Then there was that meeting in Salina, I don't remember the name of the dean at the time. All of a sudden, things got a lot friendlier. I wonder why that happened? All along there were professors who saw both sides. Like our friend, Ernst Horber, from entomology. He said, "Well this is my job, this is what I do." But a long, long time ago, he also gave me a copy of the Rural Papers from the Kansas Rural Center, and he said," Here, you should send them money and get on their mailing list." Dr. Horber worked on insect—host plant relationships, importing natural predators of various pest plants. Those people aren't done. They are smart, smart people; they can see ahead, hopefully that will continue.

Kathy goes on to list other achievements or milestones: And the farmer-consumer networks; they are still in the infant stages, but at least getting that started. The community supported agriculture, the CSAs, definitely on the east and west coast. That is a good change.

Jim: Even things like reduced tillage, where a person can say, "Hey, I don't have to buy near as much machinery and diesel fuel. Now if I can try to figure out a way to cut out all these chemicals I have to buy," it gets the wheels turning in that direction, though I'm not sure how much good it does to exchange atrazine for diesel fuel.

Kathy: The USDA, for all its flaws, the fact that it has an organic label . . . Neither one of us is very enthusiastic about that. At least it shows they are feeling the pressure.

The phone rings, and as Kathy takes the call, I turn to Jim. What is your take on the organic label?

Jim: I don't trust the USDA too much. I'm afraid they'll water it down under pressure from the industry. Didn't I hear that for chickens, if organic feed gets to be a certain percentage higher than conventional feed, you could feed them just regular feed and still call them organic?

I think they tried but it didn't pass.

Things like that. They are just so hand in glove with the industry, that I don't trust them a whole lot. The only way I can think of, if people are worried about the food they eat, they have to get involved personally. They just have to, but there will be more effort involved. Either they go to the farm so they can see what they eat, or be involved in a co-op they can trust.

Chapter 3 – Jim and Kathy Scharplaz

What do you see as the short falls of the sustainable agriculture movement? Do you think there is a movement?

Yes, I think there is. So far, it has been pretty much a producer movement. I think what needs to happen, much more, is it needs a bigger involvement with people that do the eating. Farmers are about two percent of the population, so for every farmer involved in sustainable ag, we need 50 non-farmers! [laughter]

That sound like a good ratio.

Because there needs to be a certain political element. I don't know how you make that step. We always looked at it as a way to preserve farms, farm families, a rural way of life, and so on. That has always been the why of the movement. So why should someone in Kansas City be interested? Maybe food safety, but I'm not sure the link between food and sustainable farming has been made. Maybe that is the thing we haven't accomplished, the link.

I always think of sustainability in terms of many small farms, but I'm sure that Cargill will be happy to tell you that they can be just as sustainable as I am!

They think they can!

But can they? That is a real hard argument. I always fall back on the idea that stability comes from diversity. My mother always told me not to put all your eggs in one basket. Having all our food raised by one company is a very bad idea.

I compare it to the Titanic. We have the world's best food system, the biggest, the shiniest, and we are rushing full speed ahead into this uncharted territory with biotechnology. If we hit something and it sinks, we don't have enough life-boats.

Kathy returns to the room. That was another beef call right there! A big order, they just ordered a whole side.

We've moved on to a critique of the movement.

Kathy: Hmmm, have you offered anything on that topic yet?

Jim: I just said we need more connections with urban people.

While you were on the phone with your customer, he was saying we need more connections with consumers . . . [laughter]

Kathy: That's the big challenge, making non-rural, non-farm background people aware of and concerned about agriculture. That's always been the big challenge. I've been on this a long time. That is what Jim Hightower was all about 23 years ago!

His book was influential to me when I first read it, Hard Tomatoes, Hard Times. I'm somewhat amazed, when I think back on it, that he was elected, and re-elected. In Texas! He got many organic operations started in Texas.

Kathy continues: It's a big state, there are many different kinds of people there.

Jim: Of which, I'm often reminded. In Kathy's hometown, there are more people than the entire state of Kansas.

Kathy: The thing about Hightower being elected, he wasn't elected with the rural vote, he was elected with the urban vote.

Jim: How did he do that?

Kathy: Somehow, enough urban people, before he ran, I can't ever remember being the least bit interested in the race for agriculture commissioner. Something like this is an elected position in Texas, not like Kansas where it is an appointment. Enough urban people cared about that. It is our shortcoming, and I'm not sure what to offer to do about it. I think that's our biggest challenge, like George Pyle[12] says, "Ag will change when the urban people who have the vote do."

Jim: Part of the problem is that some factions of agriculture work very hard to keep urban people out. They don't want people horning in on their business. "Well," they say, "It's none of their business." They are going to eat the stuff, they are going to swallow, it. What do you mean it is not their business? It is more their business than anyone else!

Kathy: Because everybody has to eat.

Jim: So we get more and more marginalized, farms get smaller and smaller, we stay behind the barricades, don't want to have any thing to do with the urban people. They keep getting pushed back, less and less, smaller and smaller. Not smart. Not smart at all.

Kathy: On the other hand, you have things like u-pick and like "The Depot Market." Our son is going on a field trip with his school later this week to pick apples and pumpkins and make apple cider, in Cortland, Kansas.

Jim: That is one hopeful thing. People come here, people from Dallas. It's a big deal. You just put a hay wagon behind a little tractor, put some hay bales on, take them for a tour, and they go nuts!

Kathy: Or like our friend the electrician. For fun the other day, he helped Jim and our ranch hand Aaron work cattle! [laughter]

Jim: I don't think it would be that hard if we did it right. I think it is a good connection.

You mean something like agro-tourism?

Kathy: Not all of that stuff has anything to do with food production. A lot of the agro-tourism is stuff like glorified pony-rides.

Jim: But that is ok. At least they have a good feeling about a farmer somewhere. "I like old so-and-so." I remember that rails to trails thing (referring to an attempt a few years earlier to turn old railroad beds into hiking and biking trails, which was opposed by certain farm groups in Kansas).

Kathy: Let's get them out here, have a good time, get a good impression, making connections.

Jim: I don't know why there is that part of it, the "stay away" kind of thing (referring to the opposition by some farm groups). That's dumb, hurtful I think.

Do you think that attitude is starting to change?

Jim: I think the new attitude is becoming more common all the time. I think people are becoming more accepting.

What do your neighbors think about sustainable ag?

Kathy: Well, our vet is a case in point. When we first started selling drug-free beef, he just thought we were loony. "You don't need to worry about those anti-biotic residues. They are all out of the cow's system by the time it goes to market. That's just silly." We said, "Well that is what people want." Now he raises pastured chickens! [laughter] And he's gotten to be an expert on sustainable pasture poultry! He helps run a web forum, he has totally changed his tune, he is way into all this sustainable stuff.

Jim: His goats are across the road here eating brush. I've noticed more interest in things like that in the agriculture press too. They quickly saw the handwriting on the wall. The declining number of farmers, means declining subscribers. "We better find somebody to sell our magazine to." There is a lot more interest. That makes it more credible . . . look, it's right here in print . . . right here in the *Farm Journal.*

Chapter 3 – Jim and Kathy Scharplaz

Kathy: And farmers are more open to it, because it is so hard financially, you are always looking for more ways to be more profitable. If you can cut down on inputs, or you can find that niche market where you eliminate that middleman and get that money, anything to make it more profitable. Conventional farmers aren't as skeptical as they once were, it seems to me, or maybe it is just our neck of the woods.

Jim: From K-State, the sustainable agriculture program there, that gives it a lot of credibility.

Kathy: That's true.

Jim: For a long time, conventional was almost like a religion. To question it was heretical. That is no longer the case. It is more "They do it their way, we do it our way," and people watch all the time.

So what do we do next, the next 25 years? Where do we put our efforts?

Jim: What I worry about, this is on my list of things to worry about right now, sustainable and organic agriculture is a small segment. It's really no threat to the conventional system. If it becomes big enough to become a threat, then I think there will be a violent backlash.

Kathy: And witness, for example, a few years ago Monsanto suing Swiss Valley farms or something about BST (hormone) labeling on milk.

Jim: Or Monsanto whipping up on Percy Schmeizer[10] in Canada.

Kathy: It can get nasty. They have the bottomless pockets for lawyers.

Jim: If you take sustainable agriculture all the way, it not only means a change in agriculture, it changes society, it changes the economic system. What we have now, you are either an extractor, that digs up resources and turns them into money, or you are a resource. If you don't want to play that game anymore, you are a threat. That's how the economy works. You dig things up, use them up, burn them and send them down the river. If you stop doing that, it's a threat to the whole system. I don't know how to get around that yet. I'm sure we can't have a sustainable agriculture on a large scale inside this other system, the economic system we have, because it becomes such an attractive resource for the rest of the system to mine and exploit. If you think about it one way, we go to Mexico and mine people, and bring them here, and use them in our industry.

Kathy: Use them up.

Jim: We've kind of gone through the easy stuff, the oil, and so on, but everything, gets mined and used.

Kathy: I think the thing; the single governmental kind of thing that would make the biggest difference is the anti-trust laws. If they were really enforced, if the Justice Department had the chutzpa to go after these biggies; the laws are being violated, and no one is prosecuting anybody. I think that could make the biggest difference if you could bust up the Tysons and IBPs and Cargils. If that were de-centralized, it would make a difference over night, I think. I'm not very optimistic about it happening, but if I try to think of a magic bullet that would solve many problems in one fell swoop. That would do it! Things are just getting bigger and bigger, more and more consolidated, and there are fewer and fewer players.

We talk a bit then about how the sustainable agriculture movement sometimes tends to focus on creating an alternative system along side the dominant, unsustainable system, assuming that the unsustainable system will collapse on its own eventually.

Jim: Or when it collapses, will the bricks fall on us? How big a crash is it going to be?

Kathy continues: Maybe the whole food security thing is finally going to be the thing that gets the so-called realists thinking about these things. It is just like, because of Saudi Arabia, and the possible oil shortages, now there are all these people on the right, so-called conservatives saying we need energy alternatives. They are coming around to it by a different path, but coming up with the same thing. I wonder if that is what is going to happen with agriculture. "You know what, for our own food security, we can't just rely on three companies for our food. We can't just rely on four varieties of corn," or what ever. You can't have all your eggs in one basket. It's not secure. Maybe that will be the way for people to start coming around. It's lousy, you wish they would come around for the health of the thing.

Jim: I just hope the transition is peaceful, and people don't suffer.

Kathy: Peaceful and sane. Like our friend Isidor Wallimann, and that apocalyptic book that he wrote.

Sure, I know Isidor. He comes to K-State every year to give seminars about his latest research.

Kathy: I call him the "Jolly Marxist" [laughter] Every other Marxist I've met has been a real downer, and reading his book, I couldn't get past the introduction, it was so depressing. Then you meet the guy and he is like Santa Clause! He's so jolly! He's so fun to be around, and he jokes with the kids, and plays. And the cover is so macabre, but god bless him, the reason he does this is that he cares about people, and the specter of mass starvation is always in the back of his mind.

Kathy continues and she raises a question related to the recent destruction of the hurricanes during the 2005 season, including Katrina.

Do you think these hurricanes are going to wake people up at all, to how vulnerable we are? It's waking people up on the energy thing.

Well, they aren't making fun of global warming anymore. But at this point, it's too late, and the trends seem to be irreversible, and also accelerating.

Jim: I used to be on the Rural Center board, and we would have a discussion, and then somebody would always say, "And now for the most cynical possible view of the whole thing, what do you think, Jim? " [laughter]

Kathy: He's the Jolly Cynic.

Jim: You don't want to hear my worries.

So what do you think?

I don't know. I think change will be coming, but I don't know when and I don't know how disruptive it will be.

Kathy: For our part, we just keep plugging away at what we are doing, trying to educate people in our little corner of the world.

Jim: People are so vulnerable. We have a crazy system, people move all the time, they don't stay close to where they grew up. Things go wrong and there is nobody to rely on. We are lucky here. My sister lives down the road 10 miles away, her husband's folks live across the road, all somewhat together. It is such a contrast to the larger society, where such a crappy system encourages people to work, pays them, then convinces them immediately to spend all their money. It doesn't matter how much money you *pay* somebody, if you can convince him or her immediately to give it back to you. Most folks have nothing. They don't have a farm.

Chapter 3 – Jim and Kathy Scharplaz

Kathy: I was going to say one other thing ... The one thing that if it would happen, would make a huge difference, would be the churches. A lot of them are sympathetic on environmental stuff. If they would take the concrete steps and network, I think the churches could be a huge part of that networking between farmers and consumers. The churches make a great distribution point. I'm guilty. I'm a member of a church in Minneapolis and I haven't really done much to try to get that network to happen. If the churches decided they wanted to take the initiative to change the food system they could. Probably more than half of American households belong to a church or a synagogue or something. If they really took the initiative, there could be big changes. Churches already have the networks there, the denominations have big national hierarchies, which many times is bad, but it can be good, in terms of setting out initiatives and policy and stuff. And you already have the inter-community networks there. If that happened, we could make big changes.

Jim: And they (the churches) also see the downside, when people are hungry, they help feed.

Kathy: Churches run food banks and stuff like that. There you have already made community networks in place. All you have to do is plug them into the food. Everybody has to eat. Many churches around the country are buying Fair Trade[11] coffee now. Equal Exchange has a big outreach program to churches. Take that model, of Equal Exchange, and just extend it to more than just coffee.

I've been thinking we should apply Fair Trade to domestic products, not just coffee and chocolate. We need Fair Trade everything!

Kathy: Support your local farmer!

Postscript: About two years after this interview, September 27, 2007, Jim passed away from cancer, which was not diagnosed at the time of this interview. From his obituary: "Jim will be remembered for his strength of character, keenness of mind, lightness of spirit, and largeness of heart." I believe he will also be remembered for his wisdom and contributions to sustainable agriculture.

Notes

[1]Kansas Rural Center is a non-profit organization that promotes the long-term health of the land and its people through research, education and advocacy. http://www.kansasruralcenter.org/

[2]Land Institute, founded in 1976 in Salina, Kansas. Conducts research on Natural Systems Agriculture, has an active perennial grain breeding program, and conducted the Sunshine Farm project to look at energy trade-offs in modern agriculture. Hosted an intern-training program from 1978 through the 1990's. www.landinstitute.org.

[3]Ernst Horber—Entomologist at Kansas State University for many years, with a program focused on biological control of weeds, including the successful introduction of several insect pest of the musk thistle, a noxious perennial weed in Kansas. In a follow-up conversation, Kathy pointed out to me that one of the things they especially enjoyed about their friendship with Dr. Horber was that he was also a Renaissance man, spoke several languages, and had many talents.

[4]Isidor Wallimann, Sociologist and Professor at the University of Applied Sciences, Basel, Switzerland and author of numerous books including *The Coming Age of Scarcity—Preventing Mass Death and Genocide in the Twenty-first Century*, (Syracuse University Press, 1998).

[5]Jim Hightower, former commissioner of Agriculture in Texas, and author of several books and a newsletter. One of his most influential books in the early sustainable agriculture movement was Hard Tomatoes, Hard Times, Schenkman Pub. Co. 1971. http://www.jimhightower.com/

[6]Soil Bank Program—established by the USDA in 1956 to remove vulnerable agricultural land from production as a conservation measure, but also functioned to reduce surplus commodity production .http://www.ers.usda.gov/Publications/aib485/aib485fm.pdf

[7]Wendell Berry—prolific author and poet. One of his most influential books was *The Unsettling of America—Culture and Agriculture*. Sierra Club Books, 1978.

[8]Salina Food Project and the Land to Hand Alliance were local food projects sponsored and organized by the Kansas Rural Center to promote local food and sound local food policies in this central Kansas community.

[9]George Pyle-author of *Raise more Hell and Less Corn, The Case for the Independent Farm and Against Industrial Food* (Public Affairs 2005). He also writes frequent newspaper columns on farm policy.

[10]Percy Schmeizer—the farmer in Canada sued by Monsanto for allegedly saving his own round-up resistant canola seed. See CBS article at http://www.cbc.ca/news/viewpoint/vp_omalley/20040521.html for more information, or the video, "The Future of Food." http://www.thefutureoffood.com

[11]Fair Trade Coffee—certification available to assure consumers that certain economic, social and environmental criteria were met during the growing and procurement. Often indicates that coffee was purchased above a certain floor price, usually from cooperatives of small growers. http://transfairusa.org/

Chapter 3 – Jim and Kathy Scharplaz

Discussion Questions:

1. Jim and Kathy mention the disconnect between consumers and their food. They suggest that there is "still a lot of ignorance about agriculture in general." How could that gap be bridged? As long as food is plentiful, and doesn't make them ill, do you think people will continue to choose to remain ignorant/uninformed?
2. Their beef is not organic, but is raised humanely, mostly on grass, and without the drugs and additives found in feedlot beef. Should there be a label for this type of beef, the way there is for organic, to distinguish it from conventionally raised beef? Why or why not? Do you think it would work in the marketplace, or is the continued one-on-one education necessary, the way Kathy is going about it now?
3. They discuss the contentious interactions between K-State administrators and sustainable agriculture farmers at a meeting in St. Mary's, then the calming of that relationship following a meeting in Salina. Why was there that antagonism in the first place? What could have been done by the university to avoid such negative interactions with one of their constituent groups?
4. As a way to create a more sustainable agriculture, Jim and Kathy suggest that since farmers are only 2 percent of the population, each should get at least 50 consumers involved. How would urban consumer involvement in sustainable agriculture, especially in the political realm, change things? Would this be a good thing?
5. Towards the end of the interview, the conversation turns to worst case scenarios, oil shortages, and some sort of collapse of the current system. How could sustainable agriculture prepare for such a future? Are sustainable agriculture farmers and consumers prepared now? What more could they do?

4

Nancy Vogelsberg-Busch
Bossie's Best Beef—Home, Kansas

"Your book, Farming in the Dark, *needs a sequel. The sequel will be called* Dancing at Dawn, *because that is what I want to do."*

I met Nancy more than 15 years ago, at a meeting sponsored by the Kansas Organic Producers, in a church basement, in Frankfort, Kansas. Since then, we have become good friends, through good times and bad. I've become one of her regular beef customers, and worked with Nancy through one of my graduate student's projects, where we sampled soils on six organic, and six conventional farms, to determine if we could detect soil quality difference due to legumes in the rotation. This interview with Nancy took place over the course of a day spent at her farm, relaxing on the porch, walking the farm, looking at the fields, and enjoying a delicious dinner (lunch to you easterners) of vegan (non-meat or dairy) spaghetti, made with home-grown tomato sauce.

The Farm

Bossie was my first milk cow. My Dad gave her to me as my dowry when I married. Probably my brothers said "heck, we don't want to milk, give her to Nancy." That cow outlasted my dishtowels, pots and pans, even the marriage! Bossie's offspring are the basis for my farm today. By the way, did you know that the word Bossie is originally from the Greek word Bos, so when you say "come boss" to call the cows, you are simply saying, "come cow"?

These 25 cows have 25 calves in the late spring. They go to pasture in the summer, along with the 25 calves from last year, the yearlings. Then there are the two-year old steers and heifers. This is about 100 animals to manage, feed and keep track of at any one time. She saves back the best 5 or so heifers as replacements for older cows that become "culls," so she has 20 to fatten and market each year. That doesn't sound like much, and even with 160 acres of cropland, and another 160 acres of pastureland, she is still considered a "small farmer" in these parts. The 20 per year are fattened are sold as quarters, which mean four customers per animal, for 80 customers per year. Assuming that they are married or part of a 2-person household, this beef feeds 160 people at least, and if it is a four-person household, she is probably feeding 320 people from this one small farm. She also markets "hot dogs" from the older cull cows, and sells some beef for special dinners or events like the Prairie Festival.[1]

Nancy is giving me a tour of her farm, in transition from organic row crops (corn and soybean) to mostly alfalfa hay and prairie grasses. We start by climbing over two electric fences, to reach her only stand of native, virgin prairie. (Walking out to her little virgin prairie patch)

Farming in the Dark

I love this; this is my little oasis, my little heaven on earth. This is all native virgin; it's never been broke out. This little knob, the lay of the land, no big farmer is going to break it out because their equipment is too big for these small fields. My grandpa always said that prairie hay is medicine. At first I thought that was bad, because I didn't like medicine, but the hay, it smells so sweet, it has all the medicinal herbs in it, I depend on it to keep my cattle healthy since the only thing I buy for the cows is salt. I put up one cutting of hay off this patch so far, but I didn't even get the hay hauled off before it rained. My dad always said not to complain about rain, so I just let it go. I'll bring home the cattle and they'll spend the winter here. I bring them home the end of October, they graze here through January, February and if we have heavy snows, I'll supplement with hay. I try to put the hay and bale feeders out in the field, to keep the shit out of the corral. The cows will stay out grazing until the alfalfa greens up, so they are only in the (confined) lot a month or so, then they have their babies. It's a nice system; it is starting to flow together.

Here is the pond I had built. Originally, there was no pond on the place, only spring creek (at the other end of the farm.) It runs year-round, is spring fed, its nice. This pond is a pit pond, it looks small but it's very deep. Isaac and I planted this switch grass around the sides (sounds of footsteps on the grass on the recording). This is perfect, it's nice for the cows to come and graze, and have this water source. It doesn't look grand, but its "good enough." My sons go fishing, they get fish stock and put it in here.

We pause to notice a large number of dragonflies.

I think this prairie patch is the coolest place! It's a continuous bloom, just a constant bouquet; I can give myself flowers every day, just going for a little stroll.

We make our way to the highest point of the property, the northeast corner, in the middle of an alfalfa field being mown by her neighbor, Dave. We walk past a badger hole . . . no occupants visible today. Nancy continues to discuss her cropping plan and some of the challenges in meeting the crop rotation requirements of OCIA[2] organic certification and realities of dealing with the weather. Organic certification requires that one have a "soil improvement plan," and use soil improving legumes as much as possible in the rotation. Nancy's heavy use of alfalfa in the rotation exceeds these requirements.

There are 160 acres here, about 76 that are crop-able, but I keep them all seeded. When you think about it, this is about as no-till as you can get. The ground is never barren, even when I take a green manure cutting, on my oldest field. As soon as I plow it under to feed the soil, I come back in and drill a cover crop of oats. I usually plant red clover for two years, and then back to alfalfa. I was going to do rye and turnips this fall. This field needs to be taken under but we got the rain. What I don't like about requirements being too regimented is when you are qualifying to be certified organic, and they have these folks come out from OCIA. It all has to be so black and white with them. My dad said that nothing is cut in stone except for your gravestone. When you farm with nature, when it rains, or when there is drought, you have to go with the flow. You can't go "oh shit, look at my crop rotation, and I have to plow this under," or when folks at USDA[3] get the paper work look at the rotation, and say I'm doing it wrong. If you use chemicals you can control things from on top, you can do that, it doesn't matter, or didn't matter for a long time. It does matter, because you are killing people in Iraq for oil to supplement it. Sometimes I wish I could have farmed with my Grandpa and his horses!

Chapter 4—Nancy Vogelsberg-Busch

I wish I lived closer to Jackie (a good friend of Nancy's, also an organic farmer), have you seen her equipment? I'd love to have that equipment.... Especially her drill; my drill is old. I need a better drill and I need a seeder attachment to sow oats, alfalfa, rye, turnips in the fall. I really wanted to do it this fall. This spring my drill broke, and my youngest son drove the tractor while I rode the drill, hand seeding the alfalfa. I guess that sometimes you just have to do whatever it takes to get the job done. I have a seven-year rotation. My fields are all in alfalfa right now, but are all of different ages. I know which ones go under, which ones have oats, then red clover, and then I inter-seed alfalfa right into the red clover hopefully.

Her neighbor Dave on the swather drives by. She has hired him to cut, rake and bale her hay this year

Isn't that way cool?

We watch many butterflies coming up from the hay, and flocks of birds taking advantage of the easy meal of insects as the mower passes by.

With chemicals, I wouldn't have these insects, or the birds. Look at the life here! I'm getting more than enough hay to feed my cattle, so what does it matter if I have these insects, they feed the birds? I just love them; they are very cool.

At the far end of the farm, the western-most edge, we reach a ridge overlooking Spring Creek, bordered by mature oaks, with healthy perennial grasses between the oaks. Nancy notes the edge of the alfalfa field we are standing in. I had seen it several years before when I was on her farm to do some soil sampling for a research project on green manures and soil quality. This spot was bare from some unknown reason, and we were unable to solve the mystery with soil samples and a simple nutrient analysis.

This here is the barren spot…it looks fine now, has a stand of alfalfa. I just kept working at it. It healed itself. Impressive.

We change topics, and start to talk about farming and work in general. Nancy has been working at an envelope factory in nearby Marysville for over nine years, for the health insurance benefits and help to pay the mortgage on the farm.

Most of the men that I work with in the factory have lost their farms. They come in to town to work a job to get health insurance, and the money. Then they find they don't have time to farm, to do their chores. The first to go are the milk cows, because you can't run a dairy part time, then the hogs. Soon the only things they have left are the stock cows, which is about it for chores. It's sad, they give me a lot of shit but they respect me. I sense that they don't want to see me fail, but I don't want them to fail either. Most people want to see others fail, because they feel like shit themselves. I sense they hope to some day 'just farm.' I wish we all could leave the factory and just farm!

I have a neighbor who keeps his bees on my farm. I am grateful to have the bees to help pollinate my crops, and I get the honey! He likes that I farm organically for his bees. He got into bee keeping when his kids were in 4-H. Even though he is younger than me, he no longer farms. His kids are grown and gone. I am glad he still has his bees and his cow dog.

(Looking across the fence to the neighboring farm to the East of Nancy's little prairie.) The neighbor, across the road, is a big farmer. He was an only child, had two bachelor uncles, he has fared better than most, a little longer, but he didn't last just farming. Recently he went into town to work as a mechanic. His wife told him about 15 years ago, 'you know, this isn't working,' she was keeping the books.

Farming in the Dark

Now she has just joined me in the factory, it's the saddest thing ever; she is a sweet, sweet woman. She had to get a town job for the health insurance. Getting bigger is not working. Even though they have more than most, it's not big enough.

We continue to talk about the struggle of working at a job in town, and farming at the same time. Nancy recently switched from working nights to the daytime shift. This ends up requiring more nighttime farming.

I have to leave at 5:30 in the morning, and have to get up at 4:30 to do these chores. So one day I was doing this whiney shit, wah wah me, and thought, what the hell am I afraid of? Am I afraid of dark? And what can I do about that? I have the chore buckets, scooping out the bunks, it's dark as hell, the weather conditions are god-awful. I was trying to hold a flashlight and carry buckets of grain. Isaac came up with the idea, he said, "Mom why don't you wear my headlight" (for coon hunting). I'm thinking why do I need a big grant for this expensive yard light so I can see, when all I need is a coon light? It is all about attitude . . . mind over matter . . . if I don't mind it doesn't matter!

At that point, a neighbor stops by with lunch for his son Dave, mowing her hay.

That's just a real good family. Dave is married, has two little kids, and he's trying to keep his own place going. I'll shuffle envelopes (at the factory) to pay him. It used to be that not everybody had to own everything. We've grown away from that. I like the sense of depending on them, because they are very good at what they do, I just have to delegate responsibility.

I commented that my grandfather and his brothers used to put up hay for neighbors as a way to supplement their farm income in the 1920s and '30s.

We continue to talk about the financial struggles, including the price of land. There is resentment that the only people able to afford land are the hunters, but also opportunities, as some of them are Nancy's beef customers.

One thing that really angers me is that the only people who can afford this land used to be doctors and lawyers. Now it is the great white hunters from the city. They paved their land, built their houses, now they want to come out here for their recreation area. Mike Hayden (head of Wildlife and Parks in KS) could do something here. I know he has reconsidered, he believes in riparian corridors, wildlife habitat. I'm a part of nature too, I'm a caretaker, take care of me, put me up on the scale with the wildlife! Wes Jackson[4] used to say, humans are 'somewhere between the apes and the angels.' We are right in there, no better, but no less. I am trying to fit in by taking care of this place.

My most important crop is now my youngest son. There is a wonderful article in the magazine 'Orion' about getting kids out of the classroom and connecting with nature and giving them honorable work.[11] My eldest son teaches 7th grade life science He was chosen as "distinguished classroom teacher." He was highlighted for being raised without television on an organic farm! I joke with folks that I raised my kids in the creek. Noah is more valuable as a teacher than as a farmer. If he can teach his students to respect life and those that raise their food, then there is hope for small farmers like me. Being a farmer needs to be seen as very honorable work. Becoming a farmer needs to be an honorable profession.

I'm thinking of building a guesthouse, keeping it very simple. I would offer it to my beef customers when they travel a distance, a little guesthouse, stay here the night before. We had a professor of cardiology at KU medical center, he's 75 years old, loves to hunt, wanted a little red beef, not a lot, because he cooks, came here to get some meat, wanted to know if I had any turkeys. My youngest son is a hunter. The Doc and his wife stayed at the Super 8 in town, but there was a party going on, they didn't sleep. If I'd had the guest house…"

They got up at 4:30 in the morning, came out here and I fed them breakfast and they gave Isaac $40 to take him turkey hunting out in the timber. Isaac was just going 'cha ching!' They didn't get a one (a turkey), but they got more because they developed a relationship. They are coming back in the fall, and I want him to meet Noah (Nancy's older son). To have that kind of really connecting, in a human way, there is more here than just to put food in your face. He is going to be a wonderful mentor for Isaac, He is semi-retired, does a little teaching, gives a month of his time in Mexico, like Habitat for Humanity. I would love Isaac to go to Mexico with him at some point in the future. I want him to see how much he has at home.

Marketing "Soul Sticks"

It started when I was selling beef, my first steers. As time went on, I started to look at that, especially with my older girls (referring to her cows), the worst-case scenario is what if I didn't sell my steaks? What's the worst thing that could happen? I would have to bone them out into hamburger. Then I decided that I'd make it convenient, cook it for them, do everything but eat it. My homemade hot dogs, I call them my "soul sticks." It's a way to give thanks to those cows that gave so much of their lives in reproducing the offspring that I have. The older the better, because it means they've had lived a good life, for a long time.

The hot dogs are made at the Frankfort locker, with a custom blend of organic spices, cooked, and hand cut to fit Nancy's packages. They sell for about $10-$11 for a one- pound package of five hot dogs. She gets excited thinking about her new marketing strategies.

You know, working in an envelope factory at night, there's all this paper lying around. That's when I do my farm planning. Although, once I got wrote up for being "inattentive to my envelopes." It is hard to look at envelopes for 12 hours when I know I should be home working in the fields and taking care of my kids and cows.

I don't want the business to run me. I don't want it to be very big. When the article came out in Gourmet[5] (magazine) people asked how I could keep up. Fortunately, I came to terms with that about 5 years ago. Working in the factory, I knew I didn't want to treat people the way they treat me, so I never thought I could have a company. I talked to people who said you don't have to be big to have a company. I need a balance . . . to be home with Isaac, pay the mortgage and have that be "good enough." I got my own website[6]. You would be amazed how many people read that magazine, in Colorado and Texas. You would think they would have their own damn meat, but obviously (if they call me), it isn't very good! They like the story (about my farm), they like the notion (the hot dogs/soul sticks), but they also want my beef.

What did you tell them

Don't you have any farmers in Texas? Find your own local farmer! I have this spiel: If you ever come to Kansas, follow the yellow brick road, there is no place like Home. Folks are always welcome on my farm." (Home, Kansas, her hometown.)

Farming in the Dark

On local marketing, dealing with consumers

I'm getting involved with the Kansas City Food Circle. They are good folk. They do this local thing (defining 'local" as from within a 50 mile radius). Fifty miles, where did that number come from? My dad, to get a better market sold all his cattle to St. Jo, (St. Joseph, Missouri), but that's 100 miles from here. I'd love it if my whole market was in Home City, but you have to go where there is a market. They opened up a new café in Lawrence. That woman contacted me when she first opened, called "The Local Burger." [8] She was a chef at the Merc[9] for a while. She opened her own little place. It's kind of cool; she has elk burger, bison, some local hamburger, none of that is organic, but the local fruits and vegetables are organic for the most part. I don't want to compete with anyone else; I don't want to hone in, but maybe she would be interested in my hot dogs, which would open up another market. I'm starting to buy cattle from other Kansas organic farmers, as long as it's certified organic and I know them. One farmer was saying we should buy old organic dairy cows from Wisconsin, but he's not getting it (the idea that we should stay as local as possible). I say no . . . it's like shipping our corn to Texas to the big dairies. I like this farmer, he's like a grandpa, I want to help keep him on his farm and I would like to keep the cattle here in Kansas. I do not want my company dependent on shipping across the country.

The demand for organic is coming from those that are most displaced from the farm, the ones that have moved away, paved their land, lived in cities. They are starting to look at all of the money they have, and the necessities: food, clothing, shelter. They still put a lot of money into their houses, their cars, the clothes that they wear; they put very little into food, and what has declined is their health, they are really starting to look at that. They are starting to look at food first, what goes into their mouths, and they are willing to pay a little more now, but maybe not enough yet, because they still want to maintain all that other stuff.

I live in a real poverty-stricken area; more farmers are leaving this place, seeking jobs in town, some for their health insurance. That's why I have to leave my small farm, and go to where the customers are, to educate my customers, and to bring my product to them. I want to turn that around. If they truly want me as their farmer, if they honor that, if they still want to eat red meat, they need to come to the farm. They are so accustomed to going into a store, on their paved land; they expect someone to cater to them. For me that is a hardship, the time to get it there, the money, the fuel, the expenses. I have to pencil that in, and unless they are willing to pay for that.... they have to acknowledge that this is where I live! They have to come to where I am. I can't help that they left their place generations ago, so now they need to come to my place.

I like the idea of farmers' markets, but my market is here on this farm. All they have to do is come here once a year, to get my product, and they need to get a freezer. I like the idea of helping them to create their own home pantry, the comradery of supporting other farmers.

I believe in home freezers, I believe in crock-pots. I know as a single mom what it is like to have a hectic lifestyle, but I also love the notion of the slow food movement.[10] It is a good idea, but who has time to stay home and cook all day? You can put something in the crock-pot, come home and the meat just falls apart. I give them as wedding presents, gave one to my son. It's a nice way to cook. My mom, she used to cook. It used to be you should eat breakfast like a king, lunch like a prince, and supper like a pauper. You were supposed to eat light early in the evening, and now we got it back-asswards. The obesity is part of this. I know the meal is sacred, and if you can't get together in the evening, get up early, have a big breakfast, maybe start on a Sunday.

I have hope in the notion that as folks start to honor me as their farmer, they can cuss me like the doctors and lawyers; just pay me the same! Well, I do not want to be over paid, but paid enough to stay home and farm. I don't know many doctors and lawyers that have to work nights in a factory! I feel hopeful. My kids helped me with my chores, to help keep the farm early on. When they leave, it is going to be the customers that support me, and pay me enough, so that I can pay someone well enough, or an apprentice, to come here, to help. The bottom line is they can't do it for free - we can't always have slave labor. You can lower your lifestyle, but there is a certain point of land cost and farm expenses. For me right now, I'm still hopeful, because this really has worked. I've been on my own for 9 years, and I've never missed a mortgage payment, even though I've never received maintenance or child support payments. That's amazing. I'm only dependent on my kids, my cows, my customers and myself.

Another concern I have with the organic movement is size. I begged Wayne Martin (manager of the People's Grocery Co-op in Manhattan) to not keep expanding. It's a wonderful store, and its "big enough." When you reach the scale like me, I have this little farm, I'm never going to get any bigger, this is enough. Instead of getting bigger, why not open a sister store, but don't compete, don't undersell each other.

Farmers are their own worst enemy. I'm trying to maintain these prices, and they undersell every time. When I talked at KU (University of Kansas), I challenged them (the other farmers), and said if you don't need that much money, just charge the same and give me the extra!"

I recently bought cows from a local organic farmer, and I asked him what he needed. I gave him more than he would have gotten at the sale barn. It is a good thing; it is the right thing to be doing, to keep the other farms certified. A few years ago, many of the local organic beef producers sold out, when Dakota beef had this hot little market. They offered them a better price, and they sold all their feeder calves to them. The supply here is going to suck right now; we'll have a skipped generation. They could have held out, but no one is willing to work in a factory at night like me—what is it with these weenie guys? You were one of my first beef customers. I had those 10 steers, and I was the only one that kept mine back (to feed for the direct market). I feel like the little red hen, somebody has to do it. Sometimes you have to just go ahead and get started, and then see. I don't want to make promises I can't keep either. I went on blind faith. My business has continued to grow. I just felt that more Kansas folks would continue to want better beef. Now I need more Kansas farmers to join me!"

When I sell my beef, most of my customers are not real big red meat eaters. I'm more of a vegetarian than most people realize, but when I do eat beef, I want to know that it's mine or that I know how it was raised. I usually never sell more than a quarter of beef, so I have four different customers for each one, so if you lose a sale, it's not that much. I called this one person's cell phone, and he said pick up might be a little difficult; he had moved to Colorado Springs. I used to freak out about that, but it's just one quarter, it's not like having one buyer for a market for 20 head of cattle and they say they aren't going to buy them, or pay me less. So what am I going to do, sell my cows and then come home and cry because I lost my farm? I have more control over my market, I really do.

My customers are sweet people. I like when they are of gentle nature, and they say, 'gosh how can you eat your cows?' For me, it's just a transformation of one life form to another; it's this wonderful part of a cycle. My cows utilize this (grass and alfalfa), the cows become beef, I eat and become part of that and I'm going to die some day too. Life goes on. When I look out my window and see my cattle circled around a bale of prairie hay I envision my beef customers circled around a table sharing this beef with family and friends. I can see the circle of healthy life is complete and I am happy.

The Transformation of Life with a Knife

She holds her butcher, Ron Hards, in high regard. "He transforms that life into an entirely different life with a knife. I just really want to support him." *Ron is the owner/operator of Welch Brothers Meat Company; a small state certified packing plant in Frankfort, Kansas. They process about 20 head per week of beef, pork, lamb and occasionally in-season deer, for local farmers and hunters. Ron employs about five full time staff. His smoked and aged meats are a specialty, and Nancy is his only certified organic customer/patron. Nancy has done the paperwork to get him certified, so that her beef can be sold as certified organic. The changes for Ron mean that he has to process Nancy's beef first, before the other clients, when it comes in. He also uses a special organic spice mix with no additives such as MSG for her hot dogs. He uses cleaning products on the approved standards lists rather than the conventional ones. His record keeping must also be exceptional, like the farmers who are certified, and each cut of beef must be traceable to the individual animal. This is not difficult for his operation, since many of his customers also require this, since they bring animals in by twos and threes, not two to three hundred, or thousands, like the corporate giants. These two or three animals might be split up among four or more customers, and each client expects to receive the meat from the animal they brought in or purchased directly from the farmer.*

Nancy continues to talk about her high level of respect for Ron and the work he does.

I was joking with Isaac, my youngest son (age 11), who's very smart, that he could be anywhere and run our family business. All he needs is a laptop and a cell phone. He has an appreciation for this place, he likes money, and he's intuitive about what it takes, but I also told Isaac, always put farmers first. It's not just the farmers, but also the people who do this transformation from cattle to beef, the one that holds the wand, is the man with the knife, like ahhhh! That's Ron Hards. I'm going to nominate him for that sustainable agriculture award. He's been involved in this for 5 years, he's such a good family man, his devotion to his wife, its right up there, very important, He lives in Holton and bought the locker (in Frankfort) in 1999. I've been working with him from the beginning. He drives an hour each way from Holton to Frankfort, and drives home every night. I asked him, have you ever had to spend a night in the locker? Any time you haven't made it? (We can have some horrific snowstorms in KS in the winter, combined with wind and icy roads). He looked at me and said, 'I always made it home, after I got home, I sometimes thought I should have stayed.' He's going to go home no matter what. To me, that value is more important, when it comes to success, and running a successful business. I only want to work with people like that, who put their family first, and want to sustain that.

We talk about some of the competing brands in the local health food stores that advertise their product as "fresh," and "natural," but not necessarily organic.

The concept of "fresh" beef really irritates me. The only time meat is fresh is the day you kill it out in the field. From that point on, life becomes death, and that death takes on a different life form, with microorganisms starting to decay, beginning this wonderful cycle. What Ron does is unique, and it's a time-honored art. Ron does the dry aging, gives it that robust beef flavor. Many lockers do the wet aging, and then it is vacuum-sealed. It doesn't taste as beefy. The longer you can dry age it, the better the flavor. We do 2 weeks, because my cattle have the size, they can shrink down. Its even better after 3 weeks, it gets a real beefy taste, nutty almost, but has to be so temperature controlled. Ron's damn good at knowing how to age the beef. I'd like to educate consumers more, that when stores proclaim it fresh, it's probably just thawed; it isn't fresh at all. If you watch those steaks, they get old, then they cube it, then when it turns brown, they turn it into hamburger, and no one wants to tell them it is really rotting. Now they even use carbon monoxide to keep the beef looking fresh. How stupid and sad to ruin perfectly good beef.

We also discuss problems with the fact that the Frankfort locker is only state certified, which means Nancy can't sell her beef across state lines, and also a recent rule that livestock older than 30 months can't be packaged with the bone in, but only as boneless cuts. The rule is an attempt to prevent mad cow disease, (or Crutzfeld-Jacob in humans). The disease only occurs in cattle fed dead animal parts, particularly the remains of other cattle, which is currently illegal in the U.S. Organic standards specify that only plants can be fed to animals, while chicken manure and non-cattle protein sources are still used in conventional non-organic cattle feeds. Thus, organic meat has a virtually zero chance of having the mad cow/CJ disease organism in it.

It's so irritating to me; that rule is for people who have fed dead animal parts to their livestock, to prevent mad cow disease. Right now the steers you saw, they are going to push 30 months. I told Ron, we need to get them butchered, but they aren't quite ready. There is no way they'll have mad cow disease. I've never fed them animal parts, but USDA has a rule. Some of your most tender beef, its bigger, is dry aged longer. I don't push them with probiotics or antibiotics and a hot ration of corn. In China, they want animals to be 3 to 5 years old; there is no way in hell you can do that here. It's ludicrous because I already meet the standards; I do NOT feed dead animal parts. USDA organic should be an exemption, the way Bison are for Native Americans. A Native American can raise bison, can butcher it in a state locker and ship it across state lines, because of a federal exception. Maybe if I mix bison with my beef, I can ship it across the border. They put the rules in a square box, and it doesn't fit everything. I'd like to take those bureaucrats and their red tape and hang them up to age some common sense into them.

Farming in the Dark

I sit here on the border of Nebraska. I'd love to ship no further than 50 miles, but I'm only 10 miles from the border. Lincoln (Nebraska) is up there 50 miles away. I'm as close to Lincoln as Manhattan (Kansas), and the Kansas City markets are in Missouri. I sit here in this wonderful little niche. Am I in a good place...hell yes! I went to talk to a group of churchwomen, wonderful women in Nebraska, gave a little talk about my hot dogs. They were so wonderful, they said, "We'll just come down to the state line and get your hot dogs. Are they going to arrest a little old lady at the state line?" I do not have the time to change stupid rules. I will just keep selling in Kansas.

Definition of Sustainable

Sustainable to me means to *nurture*, particularly with organic as I know it. I think this is closest to the land ethic, the organic standards that USDA now acknowledges. They were started by a grass roots movement, written by farmers for farmers, and that in itself is good. I think they saw some successes; they keep their soil fertile, keep the soil in place, keep the water clean, the cattle healthy, and for me, to raise my kids on the farm. That's important to me. Of all my crops I raise, my kids are the most important. When I asked my oldest children, including my 20 year old daughter how to raise my youngest, they said to do it they way you raised us. My son who teaches school was nominated for a young teacher award, he won it and they highlighted that he was raised on an organic farm without TV. They thought that was impressive, to get your kids outside. I joke about looking at my chores as my farm aerobics, as something more positive.

In *Orion*[11] magazine last month, there was an excellent article, particularly about computer technology, how detrimental it is to the young people, locked in the classroom. Here we are thinking we are widening their horizons, but they are so programmed. It's just like the damn TV.

The article reminded me of my dad, who only had an 8th grade education, and they never had a science book. So the teacher at that time would take them outside, they would head down to the creek, identify all the plants and animals, going all the way down, and they would have to get down IN the creek, down and dirty, that was the best education ever. He only had an 8th grade education, and was one of the smartest men I know, because he continued to READ. Now my son is a science teacher down in Wichita and teaches seventh grade life sciences. Now teachers have a hard time getting out of the classroom. What are they afraid of?

In town, where my kids went to grade school, they just paved the whole playground, with asphalt; they only left a little patch of grass. Then they put up this big huge wrought iron fence, its eerie, and looks like a cemetery, above on the fence are names of the people who contributed, and this is supposed to be playground for children? I'm glad Isaac is out of there now and in middle school. That's sad. Why didn't they leave a little bit for a garden? I like what Alice Waters[12] is doing, is really moving (towards more schoolyard gardens), and doing this on the west coast. Donna Prizgintas[13] the chef this year at the Land Institute Prairie Festival this year, is also a supporter of that.

I've worked with the poor, in Appalachia, in the inner city in Rhode Island, where you try to do something for them, like housing projects. If you just do it for people, if they don't use their own blood sweat and tears, they don't develop their

54

pride. Habitat for Humanity is a good thing; they have sweat equity. On the Navaho reservation, you crowd people into a place they don't want to be, they don't put down roots, have a home. (It needs to be a place) where you can plant a flower, and not just annuals, plant long term, you can feel good, where you can feel secure, feel safe; that is homeland security, building a safe place. What I would love more than anything is to have a neighbor that would call me if my cows were out, they would be here to help me, if my child misbehaved. They would let me know."

Nancy has some good neighbors, but there are fewer people living on the land in her area every year.

What do we need to do in the next 25 years?

Educate people about where their food comes from, and how important it is to be keepers of that essence; to protect those that are raising their food, not just keeping their soil in place and water clean....but providing a healthy place, mentally, physically, spiritually, and financially, a place to make a living and to raise children on a farm. I feel strongly, its food, its putting food first. Somehow if you can make that chic, not just the damn designer jeans, and the cars . . .

I don't believe we can just maintain the status quo. I'm concerned about it, whether we've maintained a lifestyle that's worth saving. I question that. We need to live a lot simpler. Recognize that oil is a finite resource, we can't keep using that, we can't continue killing people for that. Wind is a good energy, is a good thing, alternative energy, but not if you are going to put it in a certain place. You can't rip up the Flinthills.[14] Some people who bought that land originally didn't buy it just for farming and ranching, but they bought it as an investment. It was for the oil under the hills. One rancher told me that his ancestors bought the land, and his grandfather said, "The oil will run out some day." To rip up that land to put a huge wind farm is a sin. The ranchers show you land that has been ripped up, where the grass is woven into the flint, and there is only a little bit of soil, and it is never as good, it never goes back, even though they call it go-back land.

I was here with a rancher; he looked at my little prairie, where my soil is deep. I asked him how long it would take my row cropland to go back, and he said, "I don't know how long, maybe 100 years?" You know, CRP[15] (conservation reserve program land), it looks good from the road, but if you walk out there and look at it, there are bare spots. On his ranch right now, they are ripping the grass up where there is very little to no soil to put in a huge wind farm. He uses the Montezuma (a location in Western KS with a wind farm) wind farm as an example (of the land healing), but there they have soil, and maybe you can get grass to grow up to the turbines. I question whether grass will grow back in the Flint Hills. Besides, they are putting that wind farm energy back into a grid where there is no separation; it will go right into a city to light up the Super-Wal-Mart stores. (It would be better) if someone could put up an appropriate wind farm, light up the other little ranches, or other little towns, or cottage industries.

At my son's school, they are locked in to a school lunch program, and that creates obesity. One thing I've noticed about fast food, when they whip through that, there are no leftovers, and it tastes like shit if you warm it up. There is no respect. I don't know if anyone says grace, I don't see people praying at McDonalds, especially if you go through a drive-in.

You can talk about cheap food all you want…it sucks…it has sucked the life out of people. Because it is so cheap, they have no respect for it or for the farmers that grow it; people have no respect. Food is a necessity. You can live without big houses, you can live without a lot of clothes, you can't live without food but they keep that very cheap. If they really wanted to do something, they would flip that around, they would get rid out of their big-ass houses, shut doors to rooms they don't use, use less clothing. How many pants do you need, how many can you wear? One, and have one that you wash. Park your cars and take a bus. People think I'm taking things away from them. They are all materialistic things that don't matter. (Yet) they (do) matter because they have kept people hungry, hungry. If folks did not have so many things, they would have more money for food.

A Critique of the Role of the Universities and Non-Profit Organizations

(The universities) need to go out to real farms. Their replicated test plots are not real. The thing I like about my soil tests, you know when you are on a tractor, tilling, or looking at plant growth, the soil changes. They need to get out on small and medium sized farms, do their research out there, get out of the damn classroom, get out on real farms. Do the whole analysis; how do you make a living at it? How can you encourage anyone to go into farming if you can't make a living? People ask me, are your kids going to farm? Well, hell no. Noah (her oldest) is a lot more valuable in the classroom right now teaching those kids where their food comes from, vs. coming back here and farming with me. The next generation of people farming won't have been raised on farms, if they do (farm) at all. Someone like Noah has to teach them the value of his farm life.

Do you think the land grant system has a future?

It needs to change drastically, because it is too compartmentalized, it needs to be interacting with the other departments, the sciences need to come together with the social studies of it, how it really affects people. They need to bring in economics, it all has to be blended together, how people raise their food and how they take care of each other.

At the university level, they could make a difference with their food systems. Now they only glamorize it, like a little niche. For example, when K-State did the organic meeting last winter, I found it a little pathetic. First, they brought up a big farmer from Texas to say that organic farming can be done. They served my beef at lunch, but where is the follow-up? I got excited, I thought "we are in," but we aren't. It's not "good enough." They know we are raising organic fruits and vegetables here. We could probably raise more and different varieties. They used to send peaches out of here in boxcars. Nobody seems to know about the Kaw valley, the river valley (a former vegetable growing region).

I don't mean to give Wes Jackson a bigger ego, but why isn't he the dean of agriculture at K-state? His research shouldn't be on little plots at the Land Institute, let's bring it up to speed, let's bring in a perennial agriculture. Why aren't we doing more research on perennial agriculture? The best way to do agriculture is not to disturb it at all. So far, he says it's another 25 years. I don't think we can wait that long. The reason I push heavier on K-State is that it is our land grant, and it is our tax dollars."

Chapter 4—Nancy Vogelsberg-Busch

I gave a talk at K-State a while ago. They made fun of me. They think I want to go back to "the cow, the sow and the walking plow" but you cannot go back; you can't roll back time. You *can* fast forward to what you left behind. K-state needs to coordinate their programs with research in other parts of the country. The only mistakes you make are the ones you make twice, so why not learn from other's mistakes. K-State should definitely be offering local organic food in the cafeteria, as well as promoting it to elementary schools. We should do this challenge. K-State and KU should have a "friendly food fight." People like to have a rivalry, like in sports. We could continue that, and have my hot dogs at the stadiums! Maybe it can happen at Washburn, they might be a little more open to something like this. Washburn is becoming progressive, really expanding there.

On training

Like poverty. I think there are people who are truly poor, and you should just give to them, but like the saying, 'don't give your brother a fish, teach your brother to fish,' set up an apprentice program; get young people out on farms. (The recent Growing Growers) intern program, they wanted me to pay them a minimum wage. That was a slap in the face. If I could afford that, I would have done it a long time ago. I really want them to address this. This (farm) should be treated like a school, if they want food security; they should set up a stipend, tuition, get a scholarship through the university, get college credit.

I tried to get community service, as a high-school graduate requirement, give kids something to do besides just be in a soup kitchen for a weekend. I did that for a summer in Appalachia. Most kids are given way too much. I guess I ramble because it all comes back to appreciate what you have and learn to live on less. I would like for kids to really know what it takes to raise food—WORK!

We had this couple visit from Denmark; they had to go to a four-year college, then do a 2-year apprentice program, all to get a certificate that said, "I am a farmer." You have to be a farmer, before you could get entitlement to land there, to live on the land, raise their kids, not just be an investor. I do not know if birthright should be entitlement to farm. Maybe you should have to prove you are worthy to become a farmer. My children's entitlement to our land is through sweat equity!

About the farm bill and Washington

I think there are good folks who like to live in Washington, D.C. and lobby and pay attention to what is going on, but I've never relied on it. The new program, CSP[16] (Conservation Security Program),[1] I went in and submitted my farm plan, since I have a certified organic farm. I did qualify, but to tell you the truth, they have this program, its new, the people administrating it didn't know a whole hell of a lot. They had to learn about it. When I went in it was slam bam thank you ma'am, and the person in the office was talking 100 miles an hour!

One thing irritates me . . . my farm was accepted at tier 1, and I know another organic farmer who's farm was accepted into the program at tier 2, and he got a lot more money. Someone should have been educating farmers about how to fill out all those forms. I can upgrade next year to tier two, if I have time to go in and do all that paper work. He told me just to keep doing it. That is one good thing (the CSP program), that I have been acknowledged by USDA for doing something right." Generally, the USDA farm programs only reward farmers for growing specific commodity crops, setting aside acres or installing conservation structures. CSP is new in the sense that it rewards farmers who are *already* doing 'the right thing' in terms of environmental stewardship.

So anyway, that's a good program. Consumers need to realize that food is not cheap. There are wonderful statistics that prove these programs that subsidize the cheap grain prices right now, those grain programs could go into paying smaller and medium sized farmers, they could set size limits. I could go on and on. I know they can do that. I'm not going to waste my energy doing a lot of that. In the mean time, I have a mortgage to pay, and kids to feed. By direct marketing, I can go around that. There is a story that illustrates a point I'm trying to make. A person asked God, "Why do you let poverty and famine go on?" A second person said, 'and what did God say?' The first person said, 'God just looked back at me and asked the same question!' We have a responsibility, those are the beliefs we need to take care of.

I asked Nancy if she considers herself part of a movement, since I was starting to wonder about this after one of my previous interviews with someone that questioned whether we are in fact a movement. Nancy's enthusiastic response was, "Hell yes, we are the shakers and the bakers . . . we LOVE to stir up s*** and make it RIGHT!" *(Nancy uses a lot of expletives in normal conversation.)*

Family Background

Nancy has the honor and privilege of coming from one of the better-known, historically organic farms in Kansas. Her father, John, was an outspoken and respected advocate of organic farming, not because he was a recent convert to organic, but because he never went to conventional. Nancy recalls her early years.

I was born and raised on a farm. Since I was the third born, there were no expectations. My older sister helped mom, and helped with the house. The second was a boy, to help my dad with the farm. Being the third freed me up to do what I wanted, help my mom, and then go outside. I had a great childhood….ran barefoot outside. I didn't have any expectations that I would be a farmer, because girls weren't farmers, they could become a farmer's wife. I went to school to get a degree to go into social work. I worked in the Appalachian Mountains for a while, then in "the projects" in urban Rhode Island, and then ended up on the Navajo reservation. The grandmother still spoke Navajo. She was a weaver, did a double weave, gathered herbs for the dye, walnuts for the dark brown, red onionskins for the red. When they weave, they put a little thread of red, as a "way out." No one was paying attention to her. It reminded me of my Dad, no one was paying attention to the way he and Grandpa were farming.

When I went back to school to finish my degree (in Salina), I was riding my bicycle back and forth between two Universities, Mary Mount and Wesleyan, and I stopped at the water fountain. This person was there and I asked him what he was doing, and he said "I'm going to have an organic picnic," and I said, "That's interesting, I grew up on an organic farm." The sparring was between Wes Jackson and me. I ended up being one of his first students in 1976. At that time he was into alternative energy, wasn't into perennials yet. The first thing we built was the out-house, and we paid him for the privilege of doing it. Then we built the Indian house. Dana Jackson instilled in me the stepping-stone (philosophy), what we need to do now while we wait on Wes. It's a good thing we didn't wait on him. We need to do everything we can today to take care of our soils and water, and also support Wes's work (over the long term.)"

Wes would often talk to the interns about his 'long term vision' of agriculture, somewhere at least 50 years down the road, while Dana would take a more prag-matic approach.

After being at the Land Institute, I went back to "study" with my Dad. I farmed with my dad, found a farm to rent on my own, started to put it into clovers and al-falfa. Then I got married, had my children and I wanted to be at home with my kids. First, we rented a farm, not this one, accumulated machinery and then this place came up for sale. We offered a price, and sold the soybeans from the year before to get the down payment. About nine years ago, I became a single woman, and what I got was the kids, the cows and the mortgage....and wanted to keep all 3, so I started to work in the factory at night. I worked out a business plan, what could I raise on my farm, and how I could get a good price for it? That's when I started to direct market my beef.

I have 160 acres here. When my parent's land had to come up for sale I bought 160 acres of the pasture, and I also rent some pasture in Pottawatomie County. My great-grandparents, homesteaded near here in 1875. All they did was build a shack, and got 160 acres, then they deeded it to my grandpa. He built a barn and a new house. Then along comes my dad, he gets the land, the house, the barn...all he had to do was mend the fences. He had total "freedom to farm." He sure never had to go into town and work in a factory.

Then I come along, and my generation didn't get the land deeded over. My dad died of prostate cancer, my mother went to a care home and the lifetime savings went to pay for her care. Even though the land was paid for, they couldn't deed it over. Consequently, we got first right of refusal. The land was appraised at market prices, which was over-inflated. Even at $100 per bushel, the land couldn't pay for itself." *[Note: this is an exaggeration, but one that makes a point].*

It used to bother me, as a child, living in the sticks, like that is something bad. People made fun of farm kids, as if we were all red necks. Now it's starting to feel like a good thing. People come up here once a year to get their beef, they love this front porch, have some sun tea, pick some peas. *I heartily agree, as we've spent the rest of the interview on her porch, watching bluebirds weaving in and out of the walnuts and oaks in her front yard.*

A concern I have in raising Isaac is that I only have him six more years. I want him to seek higher education. He may or may not come here, who knows. Farming is one of the few family businesses I know that lay on a guilt trip on their kids, they are so dependent on their labor, don't know how survive without them. If I can de-sign the corrals so they are easier for me to mange on my own, and get a yard light." *Nancy is currently working with the Kansas Rural Center to get a grant through the Clean Water Farms Project to upgrade her corrals, put in a new feed bunk, watering system and yard light.*

Farming in the Dark

What is the most important thing you learned from your dad?

I probably actually learned it from my mother. My dad was the farmer. My mom in that era was just an extension of him. What I learned from her was that she was vivacious, she came from Kansas City, and had 100 pairs of dancing shoes. They used to have live bands, came up here on highway 36. There were live bands on Friday and Saturday nights. Back in those days women had escorts, the men would bring them gifts like flowers, chocolate, gifts. Well, my Dad, farmer John, would come in on Saturday night and eat the chocolates from the other guys. He must have had some charm though, since they were married in 1951.

When she got married, she was a different religion (Methodist). She had to change her religion (to Catholic). She changed her political party, she was a republican, she probably should have been a democrat anyway. She only went by his title, was Mrs. John R. Vogelsberg. I didn't even know her first name until I was - I don't know how old I was. Her name was Loretta. She was only "Mom" or Mrs. John R. Vogelsberg. She had six kids, and only wanted two. There was a club, a social outlet for the women, but with the kids and the house, she gave it up. Of all the things she gave up, her religion, her politics, her name, the thing she regretted the most was that she gave up her women friends, she became isolated. She said that hurt her the most. She had to give in so much that she finally just gave up. She lost her spirit. What I learned is that I'm not going to give in. That includes the standards I have set on how to raise my kids and my cows, I'm going to hold on to that because I think it is the right thing to do. I've been doing it for 30 years and it worked, it is working. My family has been doing it for 125 years and it worked, so I'm not ready to give in, because I'm not going to give up yet. I am going to keep my Mother's spirit alive along with my Dad's knowledge. If enough folks want my beef, I will keep carrying the chore buckets! I will dance for my Mom and honor her spirit by referring to my chores as my farm aerobics.

Blue River Customers: Reviving the Big Blue Valley—Saving Soil and Souls Upstream

On a follow-up phone call, Nancy tells me a story that she said she'd like to include in this narrative. The Blue River, northeast of Manhattan, used to flow through a beautiful valley with several towns, farms and communities along its banks. The Corp of Engineers (and others) in the 1950's and 1960's decided it was a good idea to dam up many of these rivers in Kansas (and the rest of the US?) to provide more flood protection downstream, and recreational use reservoirs. A few are also used as public drinking water supply, and/or farmland irrigation, but most are not. After much protest, Tuttle Creek Dam was built. The towns, farms and even cemeteries were moved to make way for the water. Now, about 50 years later, the lake is silting in as predicted. Eventually, it will become a large, atrazine-laden mud flat/silt basin.

Nancy's sister, Barb, recently moved to one of the nice, upscale houses that have been built with a view of the lake and the surrounding upland prairie. At a party the previous weekend, Nancy had a chance to meet Barb's neighbors, with some trepidation, remembering her Dad's opposition to this project and all it represents, and all that was lost to make way for the lake. When the neighbors found out that SHE was the one producing the Bossies Best Beef that they had been

Chapter 4—Nancy Vogelsberg-Busch

buying down at People's Grocery, they were very excited to meet "the farmer." These are Nancy's customers. Nancy's comment after the experience is that in some odd way, justice is being done, if those that live with a view of the lake are supporting her farm, and other small independent producers like her.

I am grateful for these folks. Most of my sister's neighbors do not know how many farm lives were destroyed with the construction of Tuttle Creek Dam. They do not know how much precious top soil lies beneath the trapped water because of unhealthy farming practices upstream. If these folks are beginning to buy local organic food, they are supporting a farming system that can revive the valley.

It isn't that *shit* happens; it is that *life* happens. This is the way life is. We lost paradise when Adam grabbed that damn apple from us and chomped down on it. Maybe Eve needs to inspire women to entice more men to eat local organic food!

With this mortgage, and the struggles I've had, I've been practicing 'fear based farming' for over 30 years. I want to change that. Your book, *Farming in the Dark*, needs a sequel. The sequel will be called *Dancing at Dawn*, because that is what I want to do. I want to dance in the light of the dawn. I want to wake up early feeling good because farming pays physically, mentally, spiritually, and financially[17].

Notes

[1]The Land Institute. Founded by Wes and Dana Jackson in 1976 to work on sustainable agriculture, in general, with a specific focus on breeding and designing edible perennial polyculture grain systems. Also, have an important project looking at alternative energy, called "the sunshine farm." Nancy was in their first class of interns. For more information, see www.landinstitute.org. The Prairie Festival is their annual conference/celebration/outreach and education event.

[2]OCIA—Organic Crop Improvement Association, one of 70+ organizations accredited through the new USDA organic certification program to certify organic farms.

[3]USDA—United States Department of Agriculture, oversees the organic certification program, food stamps, several conservation and commodity payment programs.

[4]Wes Jackson, co-founder and president of the Land Institute (see footnote 1 above).

[5]Gourmet—*No Place like Home*, by Kerri Conan, is an article featuring Nancy's beef and farming methods. July 2004, Pp 90-91. www.Epicurious.com.

[6]Bossies Best Beef—Name and label developed by Nancy for her product. Website: www.BossiesBest.com

[7]Kansas City Food Circle—founded to promote local food, community gardens and access to local food by people at any income level. See website at: www. Foodcircles.missouri.edu.

[8]Local Burger—located at 714 Vermont, Lawrence, Kansas. Features burgers from ranging from vegetarian to buffalo to elk. "Healthy fast food." www.localburger.com.

[9]The Community Mercantile—an organic/healthy food cooperative located at Ninth and Iowa St., Lawrence, Kansas. www.communitymercantile.com.

[10]Slow food USA—*Taste, traditions, and the honest pleasures of food.* Slow Food is about taking the time to slow down and to enjoy life with family and friends. Everyday can be enriched by doing something slow." http://www.slowfoodusa.org/change/index.html

[11]"Charlotte's Webpage," by Lowell Monke. *Orion* Magazine (Subtitle: Computers are dramatically altering the way your children learn and experience the world—and not for the better.) Sept. Oct. 2005. www.oriononline.org

If children do not dip their toes in the waters of unsupervised social activity, they likely will never be able to swim in the sea of civic responsibility. If they have no opportunities to dig in the soil, discover the spiders, bugs, birds and plants that populate even the smallest unpaved playgrounds, they will be less likely to explore, appreciate and protect nature as adults.

[12]Alice Waters—Executive Chef and Owner, Chez Panisse, since 1971, based on a philosophy of serving only the highest quality products, only when they are in season. In 1996 she created the Chez Panisse Foundation to help underwrite cultural and educational programs such as the one at the Edible Schoolyard that demonstrate the transformative power of growing, cooking, and sharing food. Chez Panisse restaurant was named Best Restaurant in America by Gourmet magazine in 2001. http://www.chezpanisse.com/alice.html

[13]Donna Prizgintas—organic advocate, chef and culinary consultant, and inaugural chef for Paul Newman's newly opened Painted Turtle camp where nearly all food served to campers was certified organic. Chef for the Natural Products Expo in CA 2005, Organic Farming Research Foundation fund raising dinner 2002, Land Institute Prairie Festival 2004-2005, also see article in "Delicious Living, Sept. 2005, about seasonal food.

[14]Flint Hills—a region in east-central Kansas. It is one of the largest continuous areas of native tall-grass prairie left in the U.S.

[15]Conservation Reserve Program (CRP)—A program authorized in the 1985 farm bill to pay farmers to take highly erodible land out of production, and plant it into either perennial grasses or trees. The land must be "retired" for a 10-year contract, which may be renewed. No harvestable crops may be obtained from the land, including hay, unless specifically authorized during a drought year. The government pays the equivalent of a "cash rent" on each acre taken out of production. The rent price is determined through a bidding process, meant to reflect local land and rent costs. Much of the land in Kansas signed up for the CRP program looks from the

road as if the grass is well established, as it is tall. However, if one walks through a CRP field and a virgin prairie field, the difference is as stark as night and day, with the native prairie forming a complete sod, and the CRP ground showing more bare ground than cover.

[16]Conservation Security Program (CSP)—A new program authorized in the 2003(?) farm bill to pay farmers on a per acre basis for various conservation, soil improvement and water quality enhancing practices. It is novel in that it is independent from the commodity payment programs, and will reward farmers who are already "doing the right thing," not just to put in a new structure or planting, the way earlier conservation programs were set up (see CRP above). Similar in some ways to "green payment" programs popular in European agriculture.

[17]Dancing at dawn—Nancy has told me about times when she was working the night shift at the factory (from 6 pm to 6 am) that she and the others would go outside to the parking lot in the middle of the night on their break, turn on a radio, and just dance for the fun of it, to move around, and help keep themselves awake. Though her comment was referring to the title of this book, no doubt, Nancy has actually danced in the dawn.

Discussion Questions

1. Nancy and many of her farming neighbors have jobs in town, in part, to help pay for their health insurance. How would a national health policy/single payer system impact agriculture?
2. The price of land keeps going up, partly because non-farmers buy it as an investment and/or for hunting purposes. On the other hand, Nancy might be able to build a guest house and gain from hosting hunters. Is this trade-off worth the higher land values?
3. Instead of trying to expand her farm size, Nancy's marketing strategy has been to have a unique product, and connect with her customers. Is this a model that could be replicated by other farmers?
4. Nancy talks about wanting to have an apprentice on her farm, but has realized that even if she paid minimum wage, the apprentice might make more than she does! Could programs be created as training opportunities, or subsidized for farms like Nancy's?
5. In Denmark, people who want to farm apparently must complete a degree and apprenticeship program to earn the right to buy land. Would similar requirements be a good idea in the United States? Why or why not?

5

Jim Bender
Weeping Water, Nebraska

"How did you first get into sustainable agriculture?
I had the opportunity to have a very good liberal arts education."

We decided to meet at Mahoney State Park, east of Lincoln, Nebraska, rather than at the farm. I first met Jim a number of years ago, at a sustainable agriculture training being conducted at his farm, before his book[1] came out. I was impressed then by his systems, and the thinking that went into his farm planning, and was delighted when I was asked later to review his book. On the day of the interview, I found myself enjoying a three-egg omelet with a gorgeous view of the Platte River from our restaurant window. Jim impresses one as a thoughtful person. He was wearing a grey sweater and jeans, has a tidy haircut, and his eyes match his grey sweater. His manner of speaking is somewhat unusual, as he pauses after each question to gather his thoughts and present a considered answer. Later in the conversation, I learned he majored in philosophy, and his speaking pattern is much more like a philosophy professor than the usual stereotype of a farmer. He also has this in common with others that I've interviewed for this book.

After chitchatting about the weather, about Nebraska, and a more serious discussion about the state of the world, we began the interview. I started with the positive: what do you consider are the major successes of sustainable agriculture throughout the past 25 years?

Sustainable agriculture, for me, began in 1975. It seems there was a critical mass for change in the agricultural community; also, it is when my project started. In many stations in agriculture, there was dissatisfaction with conventional agriculture at that time. Much was launched—individual farming projects that went against the tide, organizations that were dedicated, the Center for Rural Affairs for example, and Warren Sahs[2] at the University of Nebraska. Looking in a most general way, from then to now, I see what I consider to be a paradox that has unfolded in the history of American agriculture. On the one hand, individual farmers and those in research institutions who are sympathetic to alternatives have gone about their job of creating and demonstrating the viability of a different kind of agriculture. I think that particular mission has been a spectacular success. The viability, with all its components, the agronomic soundness of it, the practical workability of individual farming systems, the economic viability, even in the context of difficult obstacles, some artificially created, gives us a level of success that surpasses what would have been necessary for us to be confident about sustainable agriculture.

I sometimes tell groups for example, that almost everything has worked out much better than it would have had to for me to remain committed and motivated. So, in that context, as I see it, we've had an extraordinary success. Another way of characterizing part of that success is that I'm not aware of any conceptual, practical or agronomic obstacle to the continued development and problem-solving nature of this vision of agriculture. Alas, there is a paradox. On the other hand, when you look back at the things that the trend, the practices, that we were acting against—and there is a long list—they have only intensified. Not only have they intensified, but also the list has grown longer, adding genetically modified crops. But looking at the original list, the general reliance on chemicals, and synthetic fertilizers, pesticides, soil erosion, issues about soil quality, the infrastructure, the separation of livestock and crops, the increasing size of farms, the trend towards separation of ownership from those who actually do the farming . . . all of those trends, have intensified, have gotten worse. And, I would say that broadly, the situation is worse than it was in 1975. That for me provides a very dramatic paradox.

Now, an analysis of why we are in this situation is a huge subject, but as a start, I see different components at war with each other. The first is what I referred to a moment ago, the demonstrated viability of alternative agricultural systems, which seems rather undisputable. Another component is, for a variety of reasons, the temptation for conventional farmers to embrace a kind of agriculture, which is easier. The third component which from my perspective overwhelms the other two, is a long list of policies that support, subsidize, induce and encourage, directly and indirectly, the other kind of agriculture, the conventional agriculture, the easier one, the one that is destructive and causes the environmental problems. I see those three components as interacting, but the latter two, the temptation towards the easier farming and policy as prevailing.

Do you think the corporations play a significant role in this?

It would be hard to overstate the role of corporations. When I look at my neighbor's farming practices, and try to understand the etiology of it, I go back ultimately to campaign finance laws, and the configuration of contributions, and how that translates in myriad ways into policy in Washington. For a variety of reasons, partially because farmers rather trap themselves, they fall in step with these policy options. So yes, in so many ways corporations affect this equation, directly and indirectly. As an example of how they influence directly, a few years ago, the farm bill program used to have a "use it or lose it" requirement for the feed grain base. Such that if a farmer decided to introduce another crop on a temporary basis, which we all know is at the very heart of thinking about alternative agriculture, a formula kicked in that permanently reduced the base acres. That was a direct and profound policy. As for indirect ones, an example would be something that the Center for Rural Affairs[3] focuses on, the real implications of the bulk of farm payments going to large farms and the variety of things that it promotes. It's indirect, but the consequence for the immediate future of sustainable agriculture is just as profound.

When you think about what we should do, what is on your list? Probably more than just tweaking the farm bill? How do we practice sustainable agriculture in the context you were just talking about? How do we change it?

I'm going to begin by speaking in generalities in answering that. The temptation towards an easier, less demanding chemical farming has been promoted by a variety of policies, and the very least that we could hope for in the name of fairness is a level playing field. Beyond that, we need to find it within ourselves actively and explicitly to promote sustainable agriculture, and there are many ways to do that.

Chapter 5 – Jim Bender

In your introductory explanation, you mentioned the role of social science and wondered whether it is as prominent as it should be. That is a large issue. I have noticed that my neighbors are not adept at appreciating all of the things, all of the advantages that a farming system such as mine, at this moment, is offering. I'm not just referring to environmental considerations which might require an ethical sensibility that is not as common. I'm referring to things that are more practical, like agronomic and economic advantages. An example is, the very elegant way the system manages risk. I'm finding, to a surprising degree they don't get it. And if I'm right about that, that would seem to be a call—well, a call for a social science approach to try to unravel what is going on here and what can be done about it. I understand some of the components, policies' influence in a variety of ways, and there is a sense that many of them are trapped. They have made a lifetime of infrastructure decisions—from the perspective of my kind of agriculture—that have been systematically disastrous. Now they don't have many options, and I think that tends to affect their perspective. Again, this is in the domain of social science.

We discussed the relatively new tool of framework analysis, and considered whether it would have anything to contribute. Jim then related another experience to illustrate the perplexing dynamic between him and one of his neighbors.

An example of this from a couple of days ago, which is related to what you just mentioned, I was visiting with a neighboring farmer, who knows, or is quite aware that his corn is apparently selling for about $1.35 per bushel. He also knows that this fall I've sold several semi loads of corn at $6.25 per bushel.

What he thought made sense to bring up between us was to poke fun at me for having to go out into my cornfield this summer with a corn knife to cut sunflowers manually. Now I'm seeing him really in economic meltdown, and what he wants to focus on is that in his system he was able to side step this bit of manual labor. There are a variety of things here that he's not getting. I mention this economic component with circumspection, in that I don't think that should drive major agricultural decisions, but it is there, and we all understand that whatever else we do as practitioners of agricultural systems, they have to be economically viable.

Would you be willing to comment more on the economic aspect? Sustainable agriculture has promoted a variety of ways to create economic viability: by reducing costs, bargaining for better prices, specialty markets, organic certification, direct marketing, etc. Have we actually helped people stay economically viable? I see many farmers taking off-farm jobs, farming on weekends, including those practicing sustainable agriculture. Is there more that we could do?

I wasn't aware that it is somewhat characteristic of sustainable farmers that they are part time people . . . But as far as that tendency is there, I think we can hearken back to our discussion of a moment ago, of the perceptions of sustainable agriculture that are in dissonance with our interpretation of what is actually happening. The bearers of these perceptions include landowners, a tremendous source of bias is that they're in another world. They generally don't have a clue as to what's going on, and not a clue about our economic performance, which is something that would be foremost in their mind.

I'll try to say something about what I see as the major components of the battle. But there is also a joker in there which relates to your question, and that is the increased public consumer respect and demand for organic products. That is something our opponents can't quite control; it's been a bit unpredictable as to how far it's going to take things. Clearly, that is a great assist to those individuals you refer to, who are in a marginal situation with regard to their agricultural projects. It's fun to watch this unfold. The general statistic is that the demand for organic food increases by 20 percent per year.

It has been for 10 years now. Are the sustainable agriculture farmers that you know able to succeed as full time farmers?

I'm afraid I'm pretty isolated. I don't know any. Another situation in my county raises some larger issues. This is a situation driven by the convictions of the landowner. She is fervent. She's had a renter whose heart wasn't in it, and now he is going to give it up. She is in a very difficult situation. It is not clear what is going to happen. The operation is in need of some extensive infrastructure. It's very difficult to undo a whole series of infrastructure mistakes. The practice, in my judgment, that stands more at the headwaters of everything else that has followed in conventional agriculture, in terms of its direct and indirect effects, has been separation of livestock and cropping. It would be hard to over estimate the consequences of that. I understand why it happens. It's seductive. But, when you make that, what I consider to be a catastrophic mistake, and systematically undo the infrastructure of that, and in addition lose your intuitiveness about that kind of agriculture, and have offspring whom you make utterly disconnected from it, and suddenly you're a middle aged man or woman, it's a daunting situation. I don't really know how to construct a viable system without livestock. I have a neighbor who is close to fifty, farms several thousand acres and is aware that he is on a treadmill; many things are going wrong. He is beginning to see some of the profound benefits of livestock in my system, and I think he was wishing that he could go back to that, but he is at that age where it is not clear how to handle that problem.

My livestock infrastructure, which is intricate, has been a lifetime project. It was the last thing I was doing last night, before dark. In addition to fences around four non-contiguous pieces of land, fences around pasture, fenced fields, there are hay management facilities on each of these four places, elaborate livestock working facilities on each of the four places, and watering facilities, feeders for grain, it goes on and on. It's been a lifetime project. It's not exactly complete, it's on going and it requires maintenance.

The other side of the equation is, it would take much too adequately explain the benefits of livestock upon my system. The economic side is important to society in general, because it serves society to have stable farming systems. From my view, my farming system has positively addressed every single environmental problem that afflicts contemporary agriculture—and it's a long list. In every subject in that domain, at some point, the livestock are crucial to me having figured out an effective way to address that environmental problem. If I'm right about that, if livestock, organized in an interesting way in the system have that much capacity, that much efficacy, in dealing with our environmental problems, then perhaps we need to commit ourselves to restoring that kind of agriculture. If we can commit ourselves to rebuilding New Orleans from a swamp, I think we can also come up with incentives toward that. That can take place in a variety of ways; of course, it also involves some cultural changes. But, we try to influence opinion all the time in other domains; it is do-able. The new employees of the Natural Resource Conservation Service (NRCS) have an orientation at a facility near here several times a year. One of their stops is this farm. We spend at least half of their time in a plenary session, where I provide an introduction. After I present an overview of our goals, what we've done, where we've been, I pick a special subject to cover. But it comes back over and over again to an attempt to convey to them they way in which livestock are the lynch pin in that system to making all sorts of other desirable things happen.

Another common subject with them—one I tend to pick when we are climatically stressed—is the tremendous resiliency related to risk management in a system like this. You can't go very far in telling that story without invoking livestock, directly or indirectly. Then I point out that my neighbors' lack of resilience about risk management, and the way that society steps in to cover that, illustrates one of the subtle but non-trivial ways that conventional agriculture has managed, as I put it, to externalize risk. In my system, the risk management is integral to the system, and I'm not rewarded for that, it just has to be a personal virtue. I'm not rewarded by policy. Many good things come to me by that, but I'm not rewarded by policy to the extent that my conventional neighbors are.

Is there any thing else you want to say about successes in sustainable agriculture?

I've taken the farm through transition, and now I'm in what I characterize as fine-tuning a mature alternative system. That has proven to be just as rewarding and just as educational to me as the initial conversion. I am therefore continuing to discover possibilities and benefits. What I'm seeing is an extraordinarily viable system, but with challenges that are not going away. That is one reason we have an ongoing public relations problem. I think the challenges are blown out of proportion, and the farming system has potential, if widely adopted to solve a long list of problems. I might remind you at this point that almost all of my land is highly erodible, so I have at least my share of agronomic challenges. I tell the groups that come to my farm that I had two goals. One was organic methods; the other goal was soil and water conservation. Therefore, the overall objective is to blend those two separate goals into one operation.

The highly erodible land, at the western edge of the dryland Corn Belt, has given me at least my share of challenges. I'm glad; I wanted to have to work through those things. Not only am I responsible for land that is highly erodible, it has been highly eroded. On every knoll where there could have been erosion, most of the topsoil is gone. One part would have been considered a disaster maybe 60 years ago. Those challenges have been meaningful to me, and have given me confidence in talking about viability.

Do you still have bindweed?

Indeed!

I still use your chapter with the bindweed example for people who ask me about how to deal with bindweed.

I feel that is my one unsolved problem; I used to beat up on myself more about that than I do now. The reason is that this is an unsolved problem for ALL of contemporary agriculture. I'm realizing that one of the important contrasts between my conventional neighbors and me is that I'm facing up to it. I'm not solving the problem, and it hasn't been for lack of effort, but at least I'm facing up to it. My neighbors are using prophylactic herbicides that suppress the problem in one particular year, so they can get a crop, but it is a classic form of addiction. If for some reason the inputs they use are interrupted, they'll only find that it has gotten worse under their watch. We're all in a mess. It's a very, very serious problem. Are weed science departments engaged about this?

Let me get back to the question you asked about of challenges that don't go away. I have witnessed weed science departments have whole conferences on problems that I consider to be simply a byproduct of incorrect farming, An example would be a conference on what to do about shatter cane in corn. Shatter cane is a non-existent problem in my system, and I know exactly why. It has to do with crop rotation in general, and alfalfa and wheat in particular. It devastates shatter cane. They're contending with shatter cane problems because they are trying to have 50 percent plus of their acres in corn. Yet problems like bindweed go begging.

I had an opportunity to talk to a group at the University of Nebraska about the evolution of my weed control situation from 1975 to now. I chose three different categories. One is what I'll flat-out call unsolved problems. We've just been discussing that. The category at the other extreme, the other one we've been discussing, are problems that I've solved. I am troubled very little by such things as hemp dogbane, shatter cane, grass in my row crops, milkweed, and other weeds . . . and I think I can say why in each case. There is a middle category, where the challenge, the pressure, the susceptibility to mistakes is never going to go away. Prime examples would be buttonweed (velvetleaf) and pigweed. If in any given year I screw up, I don't pay attention to what I'm doing, or am victimized by unfavorable weather for my procedures I will pay., That challenge is not going away. The year that I retire, I'll be battling it out with those two weeds, and with morning glory and with cocklebur, unless I'm able to have new insights.

In the other areas, well, let's take fertility management for example, while I don't consider my fertility situation optimal, it gives me very little concern.

What is your fertility management system right now? Do you use local sources?

There was a period of about 10 years where I applied about 300 tons of rock phosphate, and that was a direct purchased input. Now my inputs off the farm tend to be a variety of sources of biomass. I recruit a variety of things . . . woodchips, sawdust, leaves. I get all the leaves from the cemetery, it's covered with oak trees, manure from neighboring farms. I import over 100 loads per year, and I have a large manure spreader.

Why don't your neighbors use it themselves? It seems they wouldn't want to give up these resources.

I don't know . . . I can't imagine. Well there are places where there is no land. For example, this afternoon I might go to a horse arena and pick up bedding with sawdust, and he has no place to spread the bedding. I do have neighbors with hundreds of acres and they don't want to mess with it. Others, I don't know . . . and with the price of anhydrous.

Other components of my fertility are the huge numbers of acres in legumes.

What percent of your land is in legumes?

At any particular time, it is about 25 percent alfalfa and grasses. In a typical year, I'll often have more acres than in soybeans. One of the things, if you take a long view of the strength and viability and the future of this kind of a system, is that you don't chase markets. You proceed with conservative disciplined rotation.

You mean, don't put it all in soybeans just because there was a good price last year?

Exactly. To bring up another subject here, its amazing, the extent to which corn is a problem. The basic problem is the separation of livestock and cropping, but there are subsidiary problems. And corn—corn comes at this subject from every angle. Now I'm not anti-corn. Corn in a very small amount has a place in my vision in a disciplined rotation. For example, if you build up soil, you want to use that for something, it is not an end in itself. To plant 50 percent plus acres in corn—the economic stress, the environmental stresses, the agronomic stresses—every thing about corn takes and makes things worse. Almost every other crop that I can think of that I plant does something good, even grain sorghum. It has a wonderful allelo-pathic effect. Some of my cleanest soybeans have been in the years following grain sorghum. We know the advantages of soybeans, e.g. its nitrogen fixing capacity. Everything about corn takes. Even in the Brady[4] soils book, important to many students of soils, he has quantified how fast and the extent to which corn de-granulates soil, in essence, un-does soil structure. It promotes weeds, it takes nutrients, it's . . . many things are tied together here, when you think about what policies to promote. It's corporate agriculture's dream!

Let me say parenthetically and briefly, because this gets away from the larger issue—there has been a revolution of sorts in corn breeding. The latest hybrids—they are making it a different crop. Its more nitrogen efficient, more drought resistant, the dry-down is wonderful compared to 20 years ago. I've typically combined 13 percent (referring to moisture content) corn. So things have been happening there, it's not quite as bad as it used to be. And this nitrogen efficiency issue is huge. This year in a drought, I've raised 1,700 bushels of corn on 12 acres (142 bushels per acre). This is in a drought where I've had to cultivate twice. Corn breeders deserve lots of credit there.

Don't underestimate the value of improved soil quality either, along with improved water holding capacity!

Are you organic?

Yes, I've been certified organic for 15 years.

Do you worry about pollen drift from GMO corn?

Of course. I think the trespass problem there is even more difficult than with the other substances we worry about. We're fighting a loosing battle, and that alone is a self-contained critique of what they are doing. My reasons for not wanting to be involved in that may or may not be sound, but be that as it may, I respect their prerogative to conduct the kind of farming operation that they want, and they do not reciprocate. No questions asked; I'll tolerate the kind of farming system that they want, and understand that in our society, that's how it works, but they do not confer that upon me.

To me this seems egregious! At some point, someone is going to have to take that to the Supreme Court?

Yes, so I'm reduced to creating barriers. In the case of corn, I think it is about 160 feet out for certification. In a huge way, I'm forced to subsidize their form of farming, and can't grow my organic corn adjacent to their conventional.

Can you use timing to avoid cross-pollination?

Yes, to the extent that I can. I have my hands full figuring out my own crop rotation, without having to figure out what my neighbor's plans are.

For a long time I have wanted to welcome a book that tells the history of the USDA war against sustainable agriculture. I've even outlined how it would go.

Farming in the Dark

Here is an example of the kind of policy story that needs to be told about USDA. Let's go back to the 1980s, the legislation that mandated that farmers had to comply with certain soil conservation procedures to continue to get government payments. Compliance, ultimately, could be obtained by practicing pesticide intensive no-till. So a program with a very different intent ended up subsidizing what, as far as I'm concerned in key respects, is a very different program. I know one of the people who helped author that legislation, Maureen Hinkle. She was a lobbyist for Audubon, and an environmental agricultural specialist. She was involved as anybody in the origins of that legislation. It was not the intent there to subsidize chemical no-till. She and her kindred souls there had in mind conservation practices and structures of the old school; terraces, waterways, crop rotations, cover, so on, NOT chemical no-till. It's a huge, sad story. This is a prime example of what deserves to be a chapter in this book.

What else would you like to say as a critique of sustainable agriculture? What should we have done differently?

This takes us back to the beginning. I think we've done what we could, in our respective stations, but we need to appreciate the magnitude of the forces we are facing. It's as if it's been an opposition to a cultural movement. It would be a little bit like critics of television saying, "Well, what could we have done?" What could we have done? Television is something that just over-whelms the culture. The seduction of an easier, simpler agriculture is something that has overwhelmed rural America. I suppose that it's going to have to take its course, and we'll have to learn the hard way. To re-iterate, the farmers in the movement have demonstrated viability. You in the universities have assisted with that, and have done your own battle against a bizarre educational culture, which is another whole, well, important story. People in organizations such as the Center for Rural Affairs have done what they could to pinpoint bad policies, and identify, articulate and push for change. We've all done what we could in our own stations, but we are up against a larger cultural context.

An example: industry developed herbicide resistant plants, and the companion herbicide, and the Faustian bargain with these farmers that all they have to do is work a few weeks a year. That's too big of a temptation for us to stop. I think there are huge problems that will become apparent as time passes. It has simply been too much for us. Back in 1975 one of the reasons I was exasperated with what I was seeing my neighbors doing, was that there were more problems inherent in the system. The way they had to manage herbicides, for example, and carry-over, the way it would jeopardize their plans for the following crops, and the amount of work they had to use for incorporation. I was thinking at the time, just on agronomic terms, this doesn't add up. I was watching my neighbors till a field once, maybe twice, only for incorporating herbicide. Just think of the implications here. We've developed a sound critique of excessive tillage, what it does to deplete organic matter, we have the cost, the wear on the equipment and the time—I didn't like it.

This other stuff, Round-up ready crops and post-plant herbicide application, on the surface, this is sweet. You can see the seduction. This thing has really changed, but it's just a matter of time, and it will have its own internal problems too, setting aside for a moment the broader environment. There are the ethical issues such as the subject of chemical and GMO trespass mentioned previously. There are going to be inherent, self-contained problems such as rapid weed resistance to Roundup. The system will become a lot more entrenched until that awareness surfaces.

What do you think about the future of agriculture? Of sustainable agriculture?

Idealism about agriculture springs eternal. It just amazes and energizes me to see generations coming up in waves. They care so much about agriculture, but there aren't enough to man and woman all the stations. There are enough out there perhaps to maintain a few representative projects such as mine, if we can get them positioned. So at least there is that, and they aren't going away. At one time, I had the opportunity to have all of the graduate students from the Iowa sustainable agriculture program at my farm, wonderful group of kids. A few more them, positioned, could transform things. I see the same kind of dedicated kids at the Center for Rural Affairs. That's not going to go away. At the moment at least, I don't have a broad projection. Although much of what we oppose will collapse under its own weight.

How did you first get into sustainable agriculture?

I had the opportunity to have a very good liberal arts education. When I had an opportunity to rejoin the family farming operation, I had already been very committed to environmentalism and its broad manifestations, and I saw this as an opportunity to become a practicing environmentalist. It was a chance to express some prior convictions. Your question raises some larger issues. We were talking a moment ago about idealism in agriculture. Many times we see that our most energetic and most idealistic practitioners have been educated and around topics or ideas very different from agriculture in and of itself. Many times, they come from liberal arts background; you see this with Wes Jackson's interns. Not to say an agronomist can't, but many come from political science, sociology, religious studies . . .

I had a student in my sustainable agriculture class last year in philosophy.

Surely, you probably know that's what I studied at three institutions, and that's much more than a coincidence. What we are on to here is something that has implications for how we are to understand the best form of agricultural education. I remember one time being able to participate in the development of curricula in agroecology at the University of Nebraska. My input stressed the importance of the college of arts and sciences.

Jim then went on to tell about a scholarship he set up for students at the Weeping Water high school with the primary criteria being a desire to study liberal arts. When asked why, he responded:

For a variety of reasons, I think this is the most effective route, if there is going to be one, to this kind of agriculture.

What would be your advice to young people, thinking about sustainable agriculture as a career, in farming, working for NGOs etc., any specific advice to them?

Well, my educational advice follows from things I've been saying. I realize especially, that if you want to be involved in some way in production agriculture, you have to know certain techniques and areas of science. It won't do here to say, "Study Melville." [laugh] That won't quite work. It may be as important as everything else is, but what we are going to have to promote is a blend. Surely, they are doing that at ISU, where they have the most prominent and conspicuous major in sustainable agriculture. They must be receptive to large numbers of credit hours in the liberal arts, so it is going to have to be a blend.

Regarding your last question, that's a very good one, maybe I should start by saying what I need to do, in addition to continuing to fine tune the system and putting myself in a position to report on that. I think that I have to strive to continue to get better at explaining what that system has to offer. I don't know if this is going to be helpful, but I don't know how much, further we could go without a more broadly educated public. If the public is going to continue to be willfully and profoundly ignorant about agriculture, eventually that is going to put parameters on what we can accomplish. To put this a little differently, maybe it has come to the point that THAT is the limiting factor.

So what do you think about the future of land grant universities?

A future in serving our purposes?

In serving Nebraska, for example.

Your largest question is too big for me at this moment. The smaller question, as it applies to what you and I can visit about, what we care about, while trying to be respectful as I can to individuals such as you who are crusaders within these organizations. In the main, these institutions are very, very much part of my problem. One example of the issues is research priorities. I've seen nothing to change that while wanting to be respectful of some wonderful, courageous individuals working within that system.

Notes

[1] Future Harvest: Pesticide-Free Farming, by Jim Bender. 1994. University of Nebraska Press.

[2] Warren Sahs, University of Nebraska. Started the first organic/conventional replicated farming systems trial in the US, pre-dated the Rodale trial by about 4 years.

[3] Center for Rural Affairs, located in Lyons, NE. For more information, see http://www.cfra.org/

[4] Brady soils book, used historically for most introductory soils courses. Current edition *The Nature and Properties of Soils* by Nyle C. Brady and Ray R Weil, 14th Ed, Prentice Hall, 2007.

Chapter 5 – Jim Bender

Discussion Questions

1. Jim has obviously fine-tuned his farming system with years of experience. Should other farmers try to replicate this system, develop their own or adapt aspects of the system to their own conditions?

2. The question of chemical and (genetically modified) pollen trespass came up in this interview. What are current protections for farmers along these lines? Is the protection sufficient to protect against significant loss? If not, what sort of policy or changes in law should be made to protect farmers, especially organic farmers against such trespass?

3. Jim believes that a liberal arts education is important to understanding and promoting this type of agriculture. Do you agree or disagree? Why?

4. Some farmers apparently haven't adopted sustainable agriculture practices because conventional agriculture seems to be easier, at least in the short run. Do you think this is true? If so, what could be done to limit the amount of harm done by conventional agriculture and/or promote sustainable agriculture in such a way that it is appealing, or at least adopted by a larger percentage of farmers?

5. Jim talks about the problem of corn, for example weed pests that are induced due to lack of crop rotation, the destruction of soil quality etc. Do you think the United States should produce less corn? If so, how could this be accomplished?

6. Jim repeatedly returned to his sense of the critical importance of whether livestock and crops are produced together in one operation. What are some of the reasons? Given that a long standing trend in American agriculture is separation of crop and livestock production, what does that portend?

7. Jim has problems with the historical relationship between the development of organic agriculture on one side, and the USDA and agricultural colleges on the other. Why would there be points of contention between organic agriculture and these institutions?

6

Ed Reznicek and Mary Fund
Amerugi - Goff, Kansas

"You can't treat farmland as a commodity and achieve sustainability."

I first met Ed and Mary in the mid to late 1980s through their work at the Kansas Rural Center. I was living on the east coast, working in the area of sustainable agriculture first as a student at Cornell, then at Rodale in Pennsylvania, and on one of my annual trips home to Kansas I stopped in to meet them at their office in Whiting, Kansas, population 200. The Kansas Rural Center was founded in 1979 and early projects included land and resource ownership, control and access and on farm energy issues. Mary was writing about water quantity and quality issues; and Ed was involved in farmer advocacy work especially during the farm crisis of the '80s. We stayed in touch, and our paths continued to cross at meetings. When I moved to Kansas I was able to collaborate with both of them on the Rural Center's "Clean Water Farms" program, and help develop a whole farm planning tool still used by the project.

I have another, almost equally important relationship to them and their farm; they are my chicken feed suppliers. My grandmother raised chickens, and I raised them in high school. Although it was interrupted by the transience of college and graduate school, I began raising them again as soon as I had a place to put them. When I moved from Pennsylvania to Kansas, part of the flock came with me, in a cage in the back of the Subaru, and I needed to find a source of organic feed! Ed helped found the Kansas Organic Producers (KOP) Marketing cooperative, and is one of the few area members with a grinder-mixer required to blend feed, so every few months I drive the 60-plus miles to their farm to get a truck load of feed. This gives us a chance to visit, in addition to seeing each other at various work-related meetings around the state. I always find value in these interactions and learn about their farm, though the visits are somewhat infrequent.

Their farm is 400 acres with about 160 acres in pasture; it has a stream and several farm ponds. Ed has the cropland in a seven- year rotation, has fine-tuned it through the years and helped many new organic grain and hay producers around the state develop and fine-tune theirs. About 10 years ago, Ed added a watering system and cross fencing to their pasture, and has their beef herd rotate through 26 paddocks. They do some on-farm feeding of the beef, and sell some meat products locally, though not necessarily labeled as organic, even though they've been certified organic for 16 years. They market all the grain as organic through KOP, often receiving premium prices.

On one of these visits, Mary mentioned that she was trying to discover the meaning of the word "Amerugi." She said the neighbors refer to the area around their farm by that name, but no one knew where the word came from. At first it was something that she had heard only her dad mention, but later she also heard others talking about it, so she began asking questions. They seemed to use it to define an area of about six square miles near their farm, but as far as anyone knew, there had never been a town or settlement by that name. When she asked what it meant, they say things like, "over there, beyond those hills." At a writing workshop in Emporia that Mary attended a few years ago, the mystery deepened, when someone at the workshop doing historical research in Missouri mentioned a place south of Kansas City, now a state wildlife area, also called Amerugi. It too had this history of being a place that was "over there beyond the hills, somewhat mysterious." This particular place in Missouri was also known as "a place where everything from horse thieves and witches and bank robbers and people hid out, in this rather rough area." Mary hasn't found a connection to the Missouri location and the one where their farm is located, though it also includes some "rough country," in terms of terrain, by Kansas standards. Though there are several possibilities, she also hasn't figured out the linguistic roots of the word.

What does this have to do with sustainable Ag? Some of us that have been working in this area for more than 20 years may feel that our goals are also illusive, or "just over the hill." We also sometimes define sustainable agriculture in not-so-precise terms, but more as a destination, or goal. Perhaps it is Amerugi?

I interviewed Ed on a warm day in July 2005, and Mary a few months later on a surprisingly warm day in November. Though they work together at the Rural Center and farm together, they each bring their own thoughts to the questions I posed about the past, present, and future of sustainable agriculture.

Ed Reznicek

Ed stopped in at my place in Wamego for this interview on his way home from a KOP meeting in Manhattan. I started by asking him how he first got involved in sustainable agriculture.

Well, I never intended to go into farming when I was in high school and in college. I grew up on a diversified grain and livestock farm in south central Kansas. We raised wheat, milo, alfalfa, had cattle, and for a while when I was small we milked cows. We also had beef cattle, finished steers, we did farrow-to-finish hogs and we had chickens.

Sounds like you had everything?

It was a well-diversified farm. We farmed largely without chemicals, without pesticides, although my dad did buy fertilizer. I grew up running a tractor with a cultivator, so I knew what that was like. While in college, my plan at the time was to get a Ph.D. and teach at a university. (Ed's degree is in philosophy). But I always liked working outside, and had part time jobs that put me outside. My wife grew up on a farm in Nemaha county. At one point, I was starting in graduate school, her dad was sick with cancer, and eventually died from it. At that point, we decided to move into the area, and were interested in moving back to a rural area to see what kind of connection developed with that farm.

I came out of college interested in doing advocacy work. I had a job as a paralegal in legal services; I was interested in poverty issues in rural areas with regard to farming. At that time, the farm debt crisis was developing. I began working with farmers doing credit counseling and debt restructuring.

This was in the 1980s?

Chapter 6 – Ed Reznicek and Mary Fund

It started about in 1983. During that period I helped my mother-in-law cut and put up her hay, help her take care of her cattle, feed cattle, doing the hay operation,. She had the cropland rented out to another farmer in the area.

The credit counseling work gave me a good idea what farm finances looked like, where they were spending money. At the time, I could see estimates of 80 percent of the expenses going for chemicals, fertilizers and interest.

Wow, 80 percent ! And that didn't scare you off from farming?

Well, it scared me away from just the chemicals, fertilizers and interest! I became interested in sustainable and organic farming the late '70s. I think the first time I went to a KOP (Kansas Organic Producers[1]) meeting was probably in '77 when we were living in Lawrence and I was going to grad school there.

That interest developed out of the energy crisis of the '70s when interest in farming systems that were less dependent on fossil fuels was coming about. I remember visiting some organic farms that were farming very much the way we were farming when I was growing up, except they more systematically rotated legumes through the system, which was the way they reduced or eliminated fertilizer needs. The farm I grew up on, the one thing we weren't doing was systematically rotating alfalfa throughout all of the cropland.

But you were rotating?

Alfalfa would tend to stay in five or six years, might grow wheat a couple of years, milo a year, cut silage, as one way to make a transition from a summer annual crop to a wheat crop. We applied manure after wheat harvest, when it was usually dry and you had time to haul manure. We weren't doing soybeans at that point. It wasn't a matter of working soybeans into that rotation. The rotation needed more diversity and more systematic use of alfalfa.

We had a chance to buy a part of my mother-in-law's farm in 1981. We decided we were going to stick around the area. We bought our first 40 acres so we could start making some improvements, invest some money that we were saving into making those improvements. In '86 or '87 the tenant that was farming the cropland gave it up, so at that point I had a chance to farm the crop land. Prior to that we started getting a few head of cattle of our own, got a few items of equipment. In high school, I was doing some mechanical work, so I knew how to fix up old machinery. Basically, I started farming with no debt. If I used any debt for equipment, it was short term. I never did borrow money for crop inputs. I continued to buy land on contract, direct from the owner, not through a bank, which ended up saving a little money on interest.

So you started out farming without using that 80 percent that most people use on interest and inputs?

I used some fertilizer until I could get some alfalfa established. I didn't grow up farming with chemicals, and I decided I didn't really want to bother learning that stuff. We certified[2] (organic) our first ground in 1990.

When did you first hear the word organic?

We first heard about it in '74 or '75, out of *Organic Gardening* magazine. At that point, we were going to college, but also gardening, and grew up gardening. We knew and learned quickly it was gardening or farming without chemical-based pesticides and fertilizers. With my farm background, I knew what that was about, so it was no mystery.

It didn't seem far-fetched?

No. First, I didn't think too much about it, because we applied some of those techniques in our gardening. As I became interested in farming possibilities that way, I was interested in seeing it applied on grain and livestock farms. By '77 and '78, I saw some of those examples, they were good examples, and could see and understand how it worked.

Examples such as?

I saw John Vogelsbgerg's, Jack Dwerlkotte, Al Ketter, Bud Bauman; there were quite a few in the area at that time--older generation organic farmers in North Central Kansas.

Given current pressures, what do you see as the future of agriculture?

Actually, I think that the importance of farming becomes even more vital as there is more pressure on natural resources, and in the context of climate change. It looks like there will be increasing pressure on resources. To have the ability to produce food, in a resource conserving and efficient manner, will be important, and may reward well. A part of the rewards has to do with market concentration by large corporations. It looks to me like the days of big food surpluses are not going to go on indefinitely. There are possibilities for good demand, good prices for farm products. The challenge will be to produce good food in a way that is not heavily dependent on high-energy use. I think biodiversity becomes more important. Diversified farms are going to be better able to cope with climate change, and with escalating costs on fertilizers and other oil based inputs. I think from a farmers' point of view that looks somewhat bright.

What about new open borders policy, where U.S. farmers compete with places where land and labor is cheap? Will farmers in the United States be paid what their food is worth?

I think free trade will hurt farmers, but not just farmers; it hurts the whole economy, in terms of the people who build the equipment, manufacture the inputs, haul the grain, what they are paid. To compete against low-wage economies that don't pay for the social and environmental cost of what they are doing, will be harmful to farmers and others here. That has to do with the social sustainability of agriculture, and currently it's not sustainable. I think in the future food security is going to mean making sure that you have the production capability in terms of resources in your own country and probably in your own region. You already see how on our KOP marketing there is a surcharge on shipping. We've seen that come about in the past year, and it ranges from 19 to 24 percent to cover the cost of diesel fuel . . .

What is a surcharge?

We're quoted a price, and then there is the fuel surcharge, increasing shipping costs by 19-24 percent. When they bid shipping charges, they are factoring in the increasing cost of fuel. It's particularly high with trucks, not nearly as bad with rail. At some point, it starts to change where you do the processing, where you do the feeding. Currently, with dairy, it's cheaper to produce the milk closer to the consumer population, but as the price of shipping feed to those areas becomes higher, some of those things will start to shift around

As demands on resources increase, and we factor in climate change it creates instability, it will affect farmers as much as anybody else. I think the social need for access to renewable resources will be there, and farmers who've learned to operate

Chapter 6 – Ed Reznicek and Mary Fund

and manage more in terms of renewable resources instead of oil will have an easier time adapting and be in a better position to succeed in a more unstable environment.

Would you encourage young people to go into farming?

Yes. To me, farming is primarily resource management. You are managing a diverse array of resources, not just natural resources, but financial resources, social resources, markets, your own ideas and intelligence, and others around you, and associated with you. If you are stimulated by that, you should consider it. You still need a practical plan for acquiring access to the land, the equipment and other components; so you need to do the planning to have a decent shot at it.

So it's not a lost career?

I don't think so.

In past 25 years, what do you think have been the biggest successes for the sustainable agriculture movement?

From my standpoint, there are a number of achievements. One of them is the establishment of an organic market. Twenty-five years ago that really wasn't in place the way it is now. In most major commodities there is a national market, and to some extent a strong international market. For the last three to four years, and looking down the road for the next few years, the challenge is to come up with adequate production. There is a strong market for organic grain, and to some degree, livestock products, with the challenge of producing more. There are more opportunities for farmers out there. That's one area.

Another area is, 25 years ago would have been 1980, the land grant universities were paying virtually no attention to sustainable agriculture. That was right at the beginning of the farm credit crisis, which took most farmers by surprise, although there were signs before that. In 1977 and '78, the American Agriculture Movement[3] was taking tractor-cades to Washington because they couldn't generate enough income to pay their debts to support their families. They were rewarded mostly with lower interest loan programs, which helped for another three or four years, and then were no longer a benefit.

I think those two areas, and for adaptation of sustainable farming practices, there has been some success there, although one would have thought there would be more.

You mean just learning about practices on the ground?

When I talk to other farmers, I like to point out the application of sustainable agriculture practices. They wonder where I'm getting my nitrogen. Those naïve about sustainable agriculture like to criticize it by saying there is not enough livestock manure produced to provide the fertility for crop production. What they fail to recognize is the importance of legumes. What is it, 78 percent of our atmosphere is nitrogen? We tend to think of it in terms of oxygen, but it is a lot more nitrogen than oxygen. We walk through this pool of nitrogen, and with legume plants, we have the means by which you can take nitrogen from the air, fix it in the soil where it can be utilized by plants. And it doesn't take genetic engineering to do this. In Kansas, there are perennial, biennial, winter annual and summer annual legumes that fit many niches of crop production. Farmers generally still fail to take advantage of these. I would think that by now we would see more of that taking place.

In terms of farm policy, the recent approval and now the beginning of the implementation of the Conservation Security Program[4] is the first big step we've seen in federal farm programs that start to reward sustainability. That, I think, is a recent achievement. People that have been involved with sustainable agriculture for the past 25 years have been talking about it for a long time, and working on it, and it has finally bore some fruit.

So you think it is the fruit of their labor, not some other force in Washington?

Yes, I think the sustainable agriculture organizations are largely responsible for it. The World Trade Organization agreements on farm programs and so forth probably have helped some because they are moving in the direction of eliminating subsidies to crop production. I don't think their purpose was to move farm programs toward sustainability, but to lower grain prices. That movement helped build support for programs like the conservation security program.

How do you view the national organic standards? Do you consider it a success, or a mixed blessing?

On a practical level, in terms of certified organic crops, the national program has made certification more cumbersome and expensive than what it was, but it has helped eliminate fraudulent labeling of organic products. I think it is also helping build broader support for organic both among consumers and in federal and state agencies. It does build recognition that this is a legitimate farming system, and deserves some public support. So yes, that has been a positive for sustainable agriculture.

What do you see as sustainable agricultures failures, or things we should have done better?

Nothing that stands out as some big error that we omitted to do, or something we should have tried and didn't.

Then what are the forces that are holding it back, or keep us from doing more?

Farm size is a limitation. As farms get larger and larger, they are less able to adapt to and adopt sustainable practices. I don't know how we are going to address that issue. In this region, you need a certain amount of those premier forage legumes—you know, alfalfa, red clover—to make an organic or a sustainable cropping system that relies on minimal fertilizer inputs to work well. As farms become really large, dealing with that large of a component of forage legumes becomes harder to integrate into the overall farm operation.

I think concentrating livestock production off the farms is a limiting factor. In dairy, beef cattle and hogs, the dairy component on an organic farm works particularly well, because they have the need and are able to utilize well the legume forage production. Forages in the cropping system helps build the fertility and is a good weed control practice. As agriculture has to move more in the direction of sustainability, to be able to produce without such a heavy dependence on oil inputs, we are going to have to move livestock production back on the farms, and back to more integrated crop and livestock production systems. As you move livestock back onto farms, the total farm acreage required to make a living will begin to come down, because they can generate more dollars with fewer acres. Those farms that still integrate livestock well, beef finishing operations, family farm dairy operations, hog production, those livestock may actually be generating the majority of the receipts. I think we'll have to move in that direction to achieve greater sustainability. One of our failures might have been not keeping livestock as part of the cropping operation, particularly in the past 15 years or so.

Are you talking about organic farmers, about farmers in general?

I'm talking about farmers in general. I think my sense would be that a higher percentage of organic farms tend to have, particularly beef cattle, as a part of the operation than do their conventional counterparts. I think organic farmers tend to be a little bit younger on average as well.

One thing people seem to be moving towards is value-added. Do you see this helping many farmers?

I think value-added does help people in sustainable ag. In organic, they are looking at the organic certification as a value-added component, because you can get more money for your crops. I think that it fits a value-added paradigm. Probably there are farm related services that can go along with that. From the standpoint of most conventional farmers, they don't see value-added as doing much for them. For example, with the opportunity to invest in ethanol plants, or other food processing, they re skeptical what kind of return they will actually get on that investment. The jury is still out on how well some of that will work.

People who have fed livestock, who see the important need to have livestock dispersed across farms, and maybe most farming operations say that livestock are the value-added enterprise we ought to be looking at. I think there is a lot of truth in that. We didn't help agriculture when, as a society, we passed laws to permit corporate ownership of livestock, and facilitated concentrating livestock production on areas that really aren't farms; they are hog factories, or they are very large feedlots, or dry lot dairies. If we want to benefit agriculture broadly through value-added initiatives, it would be to decentralize livestock production and reform livestock marketing. Along with that can come cooperative development in terms of providing some of the specific characteristics, processors; things consumers may want. The reasons argued for concentrating animal production maybe can be done as well as in a decentralized system, organized somewhat around cooperatives.

What would be some recommendations that you would like others to hear?

I think one is to continue the effort with the Conservation Security Program, to accelerate the implementation of that. There may need to be reforms in that program. There is a tendency to look at things that farmers are doing now, practices that do not lead us to greater sustainability. There should be more incentives and enhancements. We need systematically to increase the use of legumes on crop production farms.

Have we done a good enough job of educating consumers?

Have we done enough? Probably not. In terms of sustainability, a more regionally based food system is an essential part of sustainability. More consumer education on what that is and how you support and encourage that is needed. Some of that debate with my peers is about whether concentrating on consumers alone is enough, whether that that will force the change. I don't think it is enough. We need to continue to work on farmers who are not necessarily predisposed to thinking about sustainable farming and making changes in that direction. I think the opportunity is there with younger farmers.

Another person I interviewed suggested that farmers that are already organic don't necessarily want a lot of other farmers joining them, competing in their markets.

Farming in the Dark

Well, we don't need to worry about too many farmers jumping on it right away. In organic there is a three-year transition period, so that gives some time for markets to adjust. If farmers on a wide scale adopted sustainable practices, it would mean a lot of land that is currently in program crop production—wheat, corn, soybeans—a significant part of that would go into legumes, kind of like the old farm set aside programs. Organic farms typically have 15 to 25 percent of their land in soil building crops like alfalfa or a cereal grain-clover mix. If that practice was adopted on a wide scale, we'd see 10-20 percent of land shifted in to those kind of crops, which would do quite a bit to bolster prices in conventional markets. If adopted on a wide scale, organic might not mean as much, but I think prices for conventional crops would strengthen. I don't think it is any thing to worry about really.

One reason more farmers don't farm that way, some would say, is it's more difficult, primarily to control weeds in row crops. As the average age of farmers' increases, the average age now is 58 or 59, as farmers near retirement, they aren't as interested in making fundamental changes in the way they farm. That is one factor inhibiting wider adoption of sustainable practices. It will come with younger farmers. Some reforms will need more incentives for younger farmers; reform of livestock markets. The primary reason farmers take off-farm jobs is health insurance, which becomes a big component of the family budget. I think fundamental health care health insurance reform would benefit young farmers and benefit sustainable agriculture. Eventually we'll have to go to a single payer plan. A year ago, I saw that Canadians are paying one-third less than we do, and their life expectancy exceeds that in the United States. France's single payer plan is paying half, nationally, of what we do for health care, and they've got everybody covered. Our big manufacturers have a huge disadvantage competing with the Japanese and other industrial countries, because of the component that health insurance represents in the cost of production. We're hurting our manufacturing base, our economy with our failed health care system.

What are the major barriers to going into organic farming?

For older larger farmers, farm size, at a point in their life where they don't want to make fundamental changes in the way they farm. You need to move through the crop rotation system at least once before things start working well, which takes at least four years or more, so it takes time. Bankers generally don't understand that, so if a farmer needs to work closely with a banker; they probably aren't going to get any encouragement there. There's an educational component, although not necessarily a difficult component, just understanding what an organic system is and how to put it in place. In KOP, when we work with farmers interested in making the transition, the first thing they have to do is figure out the cropping system that will work for them, what kind of production they will get, how that will affect their financials.

The marketing is different; they are used to just hauling it to the elevator, and selling at harvest or sometime afterwards. The organic market is not that way. They'll need some on-farm grain storage, and that's usually a barrier. The farms with beef cattle usually start the transition with the crop production side and use the livestock sales to stabilize cash flow as they figure out when the grain sales are going to occur. Getting markets is not a difficulty, at least with corn and soybeans;

you can pull down contracts at planting time. That's not so much the case with wheat.

I think the first barrier is the desire to make the change, and then becoming comfortable, learning enough, that it can work. Depending on the type and size of farm, how to phase it in. Particularly on a large farm, don't try to do the whole thing the first year.

What do you tell people to do about bindweed when starting a transition to organic?

Bindweed is the toughest weed to control, and there's not a good organic way to do it. You can use chemicals to control it before you start your transition. Bindweed is not a problem when you are in forage legumes, or even cereal grains. In row crops, it can be a serious difficulty to deal with. Biological control for bindweed would be of great benefit.

In general, do you think sustainable agriculture is itself, sustainable?

Generally, I think so. It's moving ahead in the area of organic, it's made some progress on policy. In terms of application, there is more application than what some of us think, particularly on the organic side of it. There are more legumes used, somewhat. It will be interesting in the next year or two to see if energy prices stay where they are. I've heard many farmers complain about fertilizer cost. We'll see if they integrate more legumes into their cropping systems.

Potentially the integration could happen in no-till, there is the rhetoric on the importance of crop rotations, the ability to reduce fertilizer and chemical application costs by cover crops. I've done some crop rotation planning work with a few no-till farmers, but where we put the cover crops into the crop rotation plan, they don't quite seem to get planted in the field. What I've heard from the extension specialists, is that the costs of chemicals and fertilizers are higher on no-till fields than on conservation tillage or conventional tillage systems.

So the no-tillers will feel the price pinch first?

It could be. I could see how no-till and organic systems could be combined in ways that could rely on minimal amounts of fertilizer and chemicals. Probably one of the weak areas of organic systems is the tillage. I could see how incorporating some degree of no-till would greatly reduce tillage. Whether it is organic or whether it is no-till; to go 100 percent, that last 10 percent is the toughest, and imposes the greatest limits.

How do you integrate systematic applications of livestock manure in no-till?

No-till is adapted to an agricultural system that removes livestock from the farm. If you just apply manure on the top, you leave it open to water pollution. I don't know that if with 100 percent no-till your gains are worth what you give up in terms of other practices. Maybe that's where, at some point, we'll see a lot of change in sustainability; when you start to merge no-till systems with organic systems. What the no-till has going for it is the mechanical technology. They have the equipment for establishing cover crops. I think some of those mechanical benefits could benefit other farmers.

Twenty years from now, what do you think the next generation of people going in to sustainable agriculture will be sitting around the table talking about? What are they going to be worrying about?

Well, I think we will be dealing with the high, escalating cost of energy. I wonder what will happen to the price of farmland, as it becomes more costly to farm in terms of energy inputs. That should help stabilize land prices, or bring them down. Without farm programs, land prices would be significantly lowered, which could lead to another financial crisis in agriculture, because that's an important part of collateral.

Do you think the production of energy crops will be a big part of the future?

Perhaps soy diesel. That fits well with agriculture, because agriculture uses diesel so heavily. Ethanol is more of a market to non-farmers than soy diesel would be. I'd like to have better access to soy diesel.

I think the future will include looking at ways to create and capture biodiversity on farms, in terms of stabilizing agricultural environments. I think the issues of climate change will be more prominent. It depends on what kind of progress we make on social issues like health care. It could be a time of revitalization for rural communities, just as people are interested in getting away from the hectic life of the cities, with internet and telecommunications are making it easier to have home-based businesses and jobs. It depends on how we resolve the issues as a society, what we are willing to invest in education. In the current debate about school financing, we should be asking, "What would be the benefit of developing a first rate education system?" What would that mean 20 years down the road?

How would we benefit from putting in a first rate education system, so anybody who has a desire to attend college can afford to do so? Those would be critical investments. It seems to me like we've let the rail system decline; at some point, it looks like it would make sense to reinvest in that. Too many agricultural products are transported by truck than there should be.

Twenty years really isn't very long. We probably will be doing mostly the same things, wondering where the next generation of farmers will come from.

Where do you think they will come from?

I don't know. Some will be people with other professional capabilities and combine other career possibilities along with agriculture. It may be that the next generation that actually does the work of farming will be mostly Latino, at least if current trends continue.

What do you think about that model? You've combined farming with your work at the Kansas Rural Center[5], I've combined a little bit of farming with working at K-State. Is that a bad thing or a good thing?

For us, it was a way to finance the beginning of the farming operation. Plus, it's an opportunity to be involved in a wider social circle, more exposure to ideas, different views, more diversified income source. Unless farming generates more net income, both for the farm and the family it is supporting, combining careers is necessary. Fundamentally changing our health care system would be a substantial benefit, particularly to young farmers, and other small business folks.

People who want to be self-employed?

The entrepreneurial types who want to start businesses, unless they are wealthy, are inhibited, because if you don't have health insurance you can be ruined financially. I can think of several farmers that I've worked with that don't carry health insurance. They are just one major chronic health issue away from financial ruin.

Chapter 6 – Ed Reznicek and Mary Fund

The Cubans say that land can't be treated as a commodity in a sustainable farming system, that buying and paying for land every generation is too big a financial burden. You can't treat farmland as a commodity and achieve sustainability.

How do they handle land?

Land is publicly owned, with some restrictions on how you treat it. They pass on the right to farm it. Some of it is cooperatively owned. There are some lessons for us to learn from Cuba. They've had their collective farms, they've had their cooperative farms and they've had individual plots. When the Soviet Union collapsed, and the importation of chemicals and fertilizers ended, they were faced with a sudden transition. The large collective farms that were heavy users of fertilizer, chemicals and machinery had the biggest problem changing. The individual plots could adapt most quickly, but they were so small and had a limited amount of production, they needed to produce more than that.

The research agenda changed rather quickly, to do more research on biodiverse systems, the use of legumes, and the integration of crop and livestock production. It seems like from what I've read, they also did quite a bit of research on the biological fertility products, enzymes and some forms of bacteria. Not on rhizobia bacteria that were associated with legume plants, but on those that could pull some nitrogen from the atmosphere and fix it into the soil. Of course in that climate, managing for pests was a challenge, so for them diverse systems are a critical component of managing pests. While it was still an underdeveloped system of agriculture, some of the challenges they faced in increasing food production in a period of severe resource limitations, mainly chemicals and fertilizers, would be applicable to us as well. Of course, this is just completely contrary to our economic system, and the way we think about buying and selling. What if it is true, that treating land as a commodity that is purchased every generation inhibits sustainability? It's a big chunk of money that most farmers, young farmers in particular have trouble paying. You are looking at 20 to 30 year mortgages in some cases.

I've seen that in my own family history. My father attributes some of his success in farming to having had low prices and interest on the land he purchased from his father. That gives an advantage to multi-generational farms where that can happen.

If we keep looking at underlying issues, and possible radical transformations, should we look at the original legislation that allowed for the existence of corporations, rather than farms as entities owned by individuals or partnership? Moreover, are large corporations controlling agriculture?

Well, I wouldn't put that much emphasis on corporate control. First, I don't think they do it that well. It's a central system of control, it is difficult to achieve, slow to change and if they don't take care of people's needs, they are subject to radical change and attack. I think that people finding common interest isn't necessarily inhibited by a strong presence of corporate interest. But that's difficult to achieve. I guess the answer is no, I don't see that as that big of a threat. I think it inhibits markets, and they've done pretty well in allowing farmers to become dependent; we've seen it in fertilizer, we've seen it in chemicals; we've seen it in seed. There are still options out there by which farmers can break that dependency.

What do you think about the large organic companies, some of which are buying up smaller companies?

I don't see that as a big threat. The organic market has to exist in the current marketing system. What some of those companies can do for the organic market is to get much broader distribution of organic products. I think they can perform a positive service in building that market. What that does is help create more farming opportunities for farmers who are interested in that. It's a contradiction, or paradox, in some ways in what you think of as an organic food system, but it still has to exist within this larger market, and for now, I think it is a positive benefit—not to say that it will always do that.

What is your opinion about the role of the Land Grant Universities? Has the most important research been done? Is there more research to do? Is there continued public support?

In the Cuban example, the change they faced in terms of sustainability seemed to create a huge need for their university agricultural researchers and farmers. It was *urgent*. I think part of the reason land grant universities should be involved in sustainable agriculture is because sustainability is a *public* need, and the role of the land grant universities is to serve those public needs. The trends we see in genetically engineered seed, limits to the availability of seed, the (lack of) diversity of varieties—there is an important need for public seed varieties. We are rapidly loosing a public resource. The land grant university is the first obvious place to look to rebuild that resource. There is probably a lot more to learn about the role biodiversity can play in agriculture, and how do we integrate that? I think the university has a role in identifying where to find the next generation of farmers and the generation after that.

Probably there is an important role for the university in terms of how we maintain a thriving and adaptable rural culture. Regarding some of the fundamental farming practice issues, we know enough to move ahead, but there are probably ways those can be improved. Some work with mechanical technology that would help smaller and mid-sized farmers adapt practices; some plant breeding outside of genetic engineering that could be beneficial. The issue of bio-control of bindweed, and if you talk to chemical farmers about it, they say that chemicals don't work either. They just mask it over. In the area of policy reform, there is an important role for the university there. The problem that universities need to address is "who sets their agenda." It's up to the public to weigh-in heavier on that. We probably should be making a more organized effort at the state legislature in support of the land grant universities. That may take some political reform, in terms of who can be elected. Public financing of political campaigns is probably an important component of sustainability.

One of my goals with these interviews is to get to the underlying issues like health, education, international policy, are there any others?

I've found Herman Daly's[6] work on no-growth economies very interesting. His argument is that it is impossible for economies to grow indefinitely, because they are limited by the ecosystem within which they function and there are resource limits within those ecosystems. I think his arguments are compelling. For economics to function well without growth means you have to address social justice issues. We tend to deal with poverty by growing the economy, it's always, "grow the economy, grow the economy." His argument is that at some point the resources aren't avail-

Chapter 6 – Ed Reznicek and Mary Fund

able to continue growth, and on the other side, the ecosystem becomes unable to continue to function properly as a waste sink, so you compromise your water quality, water resources, have problems with land, and so forth. It sounds radical, but I think those are solid arguments that at some point we're going to have to deal with as well. What happens if China and India adopt a lifestyle like the Americans? I don't think anybody can see if that can actually occur. What does it mean for the rest of the globe?

I'm trying to picture what it would look like if in the United States we had a different way to pass land on to qualified farmers. In Europe, there are places where you have to pass a test in order to have the right to farm.

It seemed like when I was in Bolivia there were no or very few legal records on land ownership on the sides of those mountains. It depended on an oral history to determine who farms which plots. It showed me how treating land as a commodity played a big role in the development of our agriculture system. They don't buy and sell land in the Bolivian mountains. Of course it's a different quality of land too, but being able to purchase and pay for it, it's a difficult hurdle to overcome, and when it has to happen with every generation, basically you just put a big hurdle in front of people. I mean, it would change everything in agriculture.

Do you have any siblings that have gotten involved in farming or agriculture?

I have a brother that raises forages, a little grain and cattle in British Columbia. I've got another brother that farms the land that I grew up on, and two sisters farm, so there's five out of 10 directly involved in farming. In Mary's family, there are four kids, and she's the only one farming.

Anything important that I haven't touched on?

There were some books, *The Family Farm*, published in the late '70s and early '80s, which was an overview of agriculture and sustainable agriculture. Would any of those have anything to offer? In some ways, it might be worth revisiting those 25 years later. During the Carter years, they did those reports, "A time to change." Some interesting stuff came out of that.

We continued to talk about some of the older writings, what we knew of land reform movements in various parts of the world and sustainable agriculture efforts in Europe, such as the Slow Food movement that started in Italy. I asked Ed if he thought any of those had an impact on trend in the United States?

I don't know that it's had a big impact on sustainable agriculture here. To me it seems that in the '70s it was in response to the escalating energy crisis, and in the '80s it was in response to financial problems. In the '90s, it was market opportunities in two things: market opportunities provided by organic and converting cropland to management intensive grazing. Grass-based dairying started coming out of that. In the '80s there was also the CRP (Conservation Reserve Program[7]). I don't know if CRP had a big effect on sustainability except from a resource conservation standpoint, since conservation compliance[8] was required. The 1985 farm bill was a milestone in conservation legislation, with CRP and the cross compliance requirement for conservation. I think it finally finished the job, in terms of some kind of terraces or practices on highly erodible land. It forced the issue, and I think it succeeded pretty well. Even with the River Friendly Farm[9] assessment, generally with the crop production acres, there aren't big problems in those areas.

Farming in the Dark

You think the '85 farm bill helped with that?
I definitely think it did.
Do we need a similar bill to deal with livestock waste issues, especially in terms of improving water quality?
Well, farm programs typically haven't done much for livestock producers. Some would say that when you do away set-asides and let the prices fall it helps the livestock feeder. It does help the feedlot feeder, but it doesn't mean a lot to the on-farm feeder because they are going to put a certain amount of their grain into livestock anyway. The reform in livestock production needs to occur because of sustainability and to help diversification. If that were ever to occur, if with public policy we tried to distribute more livestock production, decentralize across farms, we'll have to make sure that it's done in such a way that it doesn't threaten or impair water quality. One thing the concentrated livestock production has for environmental requirements, some of those do an adequate job of protecting water quality, it appears.
You mean like the big lagoons?
It is yet to be seen if they impair water tables due to leaching, but the necessary engineering has gone into most of those.

Where do you think whole farm planning fits in the big scheme of things? Has sustainable agriculture contributed this to agriculture in general?

I think more farmers need to do planning, just to go through the questions about quality of life, community, and landscape goals that are outlined in Holistic Resource Management[10]. Many times farmers don't think about those things. To have to consider those and have goals would be beneficial. It seems like the sustainable agriculture community readily saw the benefit of that and adopted a planning approach that looks at not just money, but the community, and future of the landscape. Working with crop rotation planning, it's a design tool. In agriculture planning, we don't usually think about what design can bring to agriculture, particularly landscape design. We need to be talking more in agriculture about agricultural systems design, in terms of what do we want out of this agricultural system, and how can looking at design and applying that help us reach those goals. In organic, in field crop production, our cropping system has to be stable, it has to produce the various commodities we need and be suitable to the land we have; it has to provide the fertility. I think we miss opportunities by not thinking about what we want out of our agricultural systems. Conventionally, what we want out of our systems is that it has to be profitable for those efficient farmers. It has to provide a lot of relatively cheap food. Conservation is a secondary goal that has come in from outside pressure. Do we really care whether it supports the same number of families we have now? In the past, the purpose of it was to take people out of farming and put them into other occupations. The farm programs were designed to facilitate that. We still aren't asking what do we want from our agricultural systems; is churning out a lot of grain all that we want?

Do you believe that statement that we hear repeated a lot of the time that agriculture in the United States is the most efficient, safest food production system in the world?

No, I don't believe that. I think the Sunshine Farm Project[11] that the Land Institute did, and some stuff that Marty Bender handed out showed that the 1940s style Amish farms did a lot better job of producing calories based on calorie input than what our modern farms do. There is more energy efficiency in, not the highest yields, but somewhere below that, with the modest use of inputs, and modest levels of technology. During the farm crisis, it was some of those diversified farms, older equipment, fiscally conservative, but able to consistently get, not highest production, but decent production you could count on even in a bad year; they would have some production due to diversification. They were the ones that didn't have a lot of debt to begin with, and when weather was bad and prices were low, they were in a position to weather those problems. Some argue that agriculture is over-capitalized, there is too much money in agriculture and that's why there is such a low return on capital. Farm programs are part of that. They reward the biggest farmers, and some of that money goes into machinery and farm expansion. If they are looking for a tax break, they put it into machinery. If they are looking into expanding their production base, they put it into land.

So our tax laws push farmers into inefficiency and over capitalization? If we are going to reform health care, we might as well reform tax laws too?

That's right. They should be reformed. The Center for Rural Affairs[12] has always argued that tax laws are a driving force of what happens with agriculture.

I can hear our sheep outside, letting us know it is past their feeding time. I thank Ed for his time. His closing comment:

It will be interesting what Mary tells you.

Mary Fund

On a warm fall day in early November, Mary and I were sitting on the back porch of my house in Wamego. We had a nice view of the herb garden just off the back steps and the sheep in the pasture further out. After exchanging small-talk about how our respective mothers were doing, about their health issues and how her kids were doing, we began our discussion of sustainable agriculture. I've known Mary for a couple of decades counting the time I lived in Pennsylvania, and we've been friends since I've moved back to Kansas, so I thought I knew quite a bit of background about sustainable agriculture in Kansas. In this interview I learned some interesting pieces of history that I hadn't heard before.

Throughout the past 25 years, what do you see as the main accomplishments of sustainable agriculture?

The main one that I see is that environmental issues in agriculture generally have become a lot more mainstream. I think 25 years ago, and certainly before that, people did not understand what was happening to water quality, what was happening to soil, didn't realize there was a negative impact. The science wasn't there. I have a liberal arts background, not science, so I was doing interviews with people at K-State, asking my "dumb" questions. Really, they weren't dumb questions, but they couldn't answer them.

One of them, at the time was the head of the Kansas Water Research Institute, a joint KU-K-State entity that got state money for doing research. Not a lot of money, enough for six to eight research projects annually. I asked him what happens to nitrates in the water. What happens to chemicals that are applied, where do they go? He sat there, and he leaned across the desk and said, "Mary, I'm just as shocked as you are that we don't know." He was being honest. He said, "We need more research." This was in the early '80s when I was doing research for 'Water in Kansas, a Primer."

There has been a lot that happened during the mid-'80s farm crisis, the infamous Rural Center meeting in the basement of the Catholic Church in Frankfort. One of the questions that was asked there by K-State faculty and administration, was, "Just what is it that you want us to do about water quality?" We said, "We want you to look at some of the farming practices and what they are doing, and look at the impacts of water quality. You guys know how to do that, we don't."

There has been a huge shift if you think about water quality issues especially. It wasn't environmental groups, even though they certainly had their impact. In Kansas, sustainable agriculture advocates and organic farmers were asking questions, and trying to get those things elevated in the researchers' and regulators eyes. We wouldn't have seen rules on atrazine, and the creation of the Delaware River Basin Atrazine Management Area. We wouldn't have seen any attention to those things without the sustainable agriculture perspective out there. I think that is a big plus. I'm not sure some of the answers from the research were necessarily heeded the way they should have been. I'm still not convinced that the BMPs (Best Management Practices[13]) on the Grasshopper watershed are still being used, but more people are aware that there is a repercussion, a reaction to their actions on the farm.

So, what else? What other milestones—things we are proud of and can hold up and say that we did this?

Well, probably, and this is a double-edged sword, the federal organic standards. There is good and bad in that. There are big battles right now, how organic is organic, what is pure enough, what is allowable, that will go on forever. Still, we achieved a milestone when the first version of the standards came out and there were these 400,000 letters and responses, this huge outpouring. There has been acceptance of organic in the mainstream-level politically, at USDA, and with consumers, even though there is still a lot of misperception out there. Probably the rise of farmers markets and the interest in local and regional food production, those are outcomes. People are realizing, for many reasons, that it is a true food security system. If you look at energy cost, transportation and raising it in the conventional system, it starts to come together, you understand that you have to come up with a different system.

You can point to the Center for Sustainable Ag[14] at K-State as a success. I think we pushed some buttons in the right places. You are here in Kansas that was one of the successes. That was so funny, after that one spring sustainable agriculture meeting at K-State, we had great speakers, and great turnout of farmers, the dean, the chairman, a couple of senators, standing in the back. I overheard them saying, "Now what the hell are we going to do? We have to respond to these people, they sound legitimate!" There has been a lot of credibility gained in that 25 years. It hasn't quite filtered from department to department in the land grant here.

Chapter 6 – Ed Reznicek and Mary Fund

At the federal policy level, you could say that CSP was a success, but right now, it is so screwed up, it is on the chopping block, and it is just a shadow of what it was meant to be, because of congress and budget cuts and what they are doing to it in the implementation. I don't know if you can count it yet exactly, but I wish we could. I think the rise of Holistic Management in general, whether it is Alan Savory or some other offshoot, those all came about because of the sustainable agriculture influence.

I think we've had some impact on the kind of research that takes place on the "back 40[15]." There are more people out there, more farmers, trying things a little bit differently because they've seen it or read about it. They are still very conventional, but they're making a gradual adjustment. There are the early adopters, the laggards, and then there are the people who start chipping away at thing. The incrementalists just making a few changes like maybe they'll put an electric fence across 40 acres of an 80-acre pasture, rotate them from side to side, just to see how it works. Things like that. I've seen that in the neighborhood. There are no lights going off, no bells are going to ring.

It isn't going to show up on anyone's statistics. I can't document anything. I struggle with how can I document this statistically, especially changes to the land.

We were talking about that today. KDHE has the same problem we do. Any 319 project[16], they have to do some bean counting, quantify the results. What is the impact of the BMPs you helped get on the ground? It is also a dilemma with the River Friendly Farm notebook and process. Initially it was perceived as a whole farm plan and process and it wasn't specifically water quality. Even the questions were geared for understanding the farm, for educating, it was a systems approach. Somewhere along the line, since KDHE was funding it, everything has to be geared toward the impact on water quality so they can measure it, so they can justify the federal funds. Accountability is fine, except that you are probably getting benefits beyond how many pounds of this or that are entering or not entering streams and reservoirs.

At times people have suggested taking out the family goal planning part. I said no, that is what makes it what it is, or what it is supposed to be. Still, the real thinkers out there, the ones that are actually doing the whole farm plan and are thinking in the broad sense, they aren't thinking piecemeal, just thinking about putting in that cattle waterer or how many tons of something has been stopped. I understand we have to get a hold of that information and data, but that is not all there is.

We could use that to segue into the gaps, or things we should have done better?

We missed some opportunities organizing wise, at least the Rural Center did, around corporate hogs, factory hogs. Things didn't go well and we just lost some opportunities to take it another step further. I know we didn't have the time or the staff to pursue what we needed to pursue. Some of what happened was that the market kept some of the expansion in the state out, but we also lost the small producers anyway. There was a question about whether there was a critical mass left of small producers to do anything. There should have been some follow-up building on sustainable ag. Many people were worn out from the battle, I think that happened to a lot of hog producers.

Farming in the Dark

I think too that something we could have done better was help people understand that sustainable agriculture is not something that is achievable in a couple of years. This is a long haul. Understanding that going in makes a difference.

Are you saying that we did or didn't understand this?

We didn't understand this. Well, some of us did. When some people first get involved in a political issue, when it doesn't go well, they abandon all political activity. I think that was something that happened concerning corporate hog stuff, people wore out on it, "we don't win anyway," what is the point. So lost our organized base.

I always thought it was too bad that we were able to organize Heartland Network[17] clusters around practices and marketing, around "selfish," which is a negative word, "motives," but not around political action. We found out we had some diverse political views and values. I think that is a problem today. There are a lot of conservative folks involved in sustainable ag, and alternative ag, and when you start talking about bigger issues, whether it be the dissolution of the family, lack of values in our society, gay marriage, premarital sex, all kinds of things, people aren't going to agree, very different perspectives.

So people might agree around the table about a farming practice, or even farm bill discussions, but not anything broader?

Even the war in Iraq and what is right or wrong there, but the farming things are all part of bigger decisions being made.

I don't know how we would have done it, but we haven't done a very good job to bring sustainable agriculture discussion into a discussion of sustainable society in general. It seems clear to some of us sustainability is the only way the planet can survive. That is a pretty tall order, but some of those things stand in the way of progress, those other issues. Education in this state, for example, my child's science teacher was at parent-teacher conference, joking that "they'll have a minister up here with me pretty soon, telling me what I can and can't say." At least he is laughing about it; he hasn't quit teaching yet. He is a good science teacher. [18]

Sometimes I struggle with the question, "How do I accomplish sustainable agriculture in an unsustainable society?" Do we have to make society sustainable first, or just keep plugging away at sustainable ag, hoping that the other will come along too?

Some of it almost has to for sustainable agriculture to work. If you talk about energy, local and regional production, then that requires some kind of infrastructure, relationships among people in a community to make things work. I still think those kinds of things can happen. All the other stuff just makes it incredibly complicated, if you have the "thought police" coming down on you, what religion you are or aren't. I don't know.

So what do we do for the next 25 years? Where do we put our energy?

Well, I don't know about sustainable ag, but I think we are going to have to see more energy in general go into re-defining democracy. That sounds somewhat abstract, but I think for certain things to happen, we are going to have to safeguard such things as our rights to free speech, etc. some of those things. This to me means we get more involved in the political realm. We don't just accept the view of the world that says my values are the only right ones, and the right wing is going to prevail.

Chapter 6 – Ed Reznicek and Mary Fund

I think we have to deal with some big issues—the sustainability of the planet and how we feed ourselves in a very changing world. You can look at a year in Kansas, and say I don't know what is normal anymore, in terms of drought, heat, and how that affects raising a garden.

When do you plant radishes anyway? [laughter]

Like today, it is 90 degrees in November—when *do* you plant radishes?

This was Mary and Ed's son Evan's first year of really being a gardener, being in charge of some things, and Mary said that he sounded like an old farmer complaining about the weather.

Even some of my conventional wisdom about gardening doesn't apply, when the weather is so strange. It is hot and dry, and you get this kind of day in November, it affects all the bugs, affects how big the weeds get. We are going to have to be very adaptable.

To do that, we have to stop some of the people who are trying to squelch education and growth and information. It is ironic that in an information age, we are going backwards at a pretty fast clip. I've had a recurring dream in my life, I'm in a car going down the road backwards and the brakes don't work, so I'm trying to stop, and somebody is in the car with me, looks at me and asks, "Is something wrong?" And I say "No." [Much laughter.] As if going backwards is normal! I think society is like that.

In the short term, I think we are going to have to try to get more people involved in the policy-end of things. I know that people don't like it. I've been sending out these legislative Action Alerts. It's frustrating; politicians will tell you one thing and do something else, or they will lecture you. I really hate those. When you try to give them your opinion, and your view of things, and then they try to tell you why that is wrong, and you should . . . But, didn't I elect you? I think we need to get away from that sense that it's a bad thing, politics and policy. All those people who don't vote at all, that could screw us all.

Do you see the new WTO[19] rules as a bad or good thing for shaking things up?

I think it is a good thing, ultimately. It is interesting again that in the information age we aren't more global in understanding the impact of what we do. On the one hand, I work on projects that try to teach that what we do at the local level has an impact downstream, that we need to change, one farm at a time. We also need to be aware that it is bigger than that one farm. That farmers just like you in other countries are struggling as well. Do world courts have any credibility, or do we just thumb our noses at everybody else, and do whatever we want, or what?

Then it seems like the Bush administration has argued for doing away with subsidies. That has really upset the mainstream farm organizations. They say, "Whoa, wait a minute, what are you talking about." Yet there are people out there that don't understand food production. We are looking at increased oil dependence when we have decreased supply; the cost is sky-high. I forgot his name, the person from California who said we are going to import all our food. He argues that we can do away with farmers in this country because we can just import cheap food. We aren't going to be able to do that. Not if you look at the energy issues, all the other things aside. It is just not going to work. I think that as individuals we have to work on the local issue, try to understand policies.

Farming in the Dark

Do you think sustainable agriculture will benefit from the commodity payment programs either being reduced or going away?

I think that if it happens in tandem with many energy issues it could help. Because, back in the farm crisis of the '80s it was still alternative agriculture and it was called low-input agriculture. Many people made changes, not because they wanted to, or because of the environmental impact, but because they had to. They looked at those old BMPs, legumes, cover crops, finding ways to renew the soil and provide their own resources from the farm. I think we are going to see more of that. I hear farmers saying they are going to more no-till with the high-energy prices, because you are reducing the number of passes through the field. But, they are using more pesticides, fertilizer, and relying on the inputs to pay for huge equipment. At some point, it doesn't balance out. If we reduced commodity payments, or we shifted payments to more of a green payment or conservation payment, pay more attention not to how much they are growing or what they are growing, but how they are doing it, what the ultimate impact it, the multiple benefit is, then we could see sustainable practices benefit. That is what you are going to have to do to grow anything in an agricultural system less dependent on fossil fuels.

How did you personally get involved in sustainable agriculture?

I come by it somewhat genetically. [Laughter]
You are the first one to say that!
My Dad was a kind of a Luddite. He did start using chemical fertilizers, but it wasn't anything he wanted to do. He did not use herbicides or insecticides. He hated them. He was of a sensitive nature, physically and emotionally I think, and didn't want to use them. So he hauled manure, and he was not known as a good farmer, because he was considered old fashioned. I remember him talking about he never used to grow milo because he thought it was hard on the soil. He didn't want to grow it, but that is what the commodity crop programs supported, so he ended up farming for those. But he always had a lot of alfalfa, clover, a small dairy, and pigs. It was a very traditional small farm.

He told a story of being out in the field one day when a spray plane went over. There probably weren't any regulations about anything at that time; it was near the border of the neighbor's field. Anyway, they sprayed him. He got down under his tractor to try to get away from it, but he got drenched. It wasn't any protection but it was the only thing he could get under. He didn't like the smell, didn't want them around.
Did he get ill from the spray?
Not immediately, not that time. He died of cancer years and years later, but not related to that specifically. That knowledge, he was the kind of farmer that let the field borders get bigger and bigger. He had some health problems, so he wasn't farming every inch. His turn rows and field edges, the grass and brush started creeping in. Other farmers would have bulldozed it all out. That is the genetic part. What I saw there, it seemed like it made sense.

When I was in college in Emporia, I worked on the science and technology floor of the library. We had everything divided different from they way they do it now. We had all the periodicals on that floor, all the topics. That covered everything from *Advertising Age* and *Harpers Bizarre* to *Organic Farming* magazine to *Prevention* to *Science* to *Nature* to the *New England Journal of Medicine*, all these things. I checked them all in and read things that interested me. That was probably where I started to read about world hunger, food production, and some of the Malthus predictions, and those kinds of things. I was intrigued by gardening; I've always been a gardener. Rodale's[20] stuff came through the library, and I started looking at that. I started raising a garden in college.

Did you have a garden as a child?

I worked in ours. We did a lot of gardening, canning and preserving. We always did a lot of that. I learned to drive a tractor, drove the hay wagon in the fields for my brothers to throw hay on. Graduated to throwing on bales myself, but could only could throw two tiers, and then I had to drive. I couldn't go above that. That's how I got started.

When did you start working for the Rural Center?

How did that happen? We ended up in Lawrence. I had a degree in English. Ed was testing graduate school, which didn't work for him, and we wanted to move back to the farm. My father was very ill with cancer. So we were up there every weekend, helping with work and stuff, and I got a job with the community action program, which was the anti-poverty program that grew out of former President Johnson's war on poverty. I had no clue that this thing had been around in northeast Kansas for years. Eventually I worked as a planner. I could write, so they figured out, "Put her to work writing grants, get her some training." I wrote grants and several things happened. I learned more about where I had grown up. The demographics, the income data showed that poverty was a problem. I started putting two and two together. Here are these vast natural resources, and yet there's 12 to 19 percent poverty in the county. What's wrong with this picture? What's wrong with this?

There were other people around; Fred Bently, Vaughn Flora, Jim Lukens, putting together the Rural Center based on the Center for Rural Affairs. We did some visits up there. We met Francis Moore Lappe on one of those visits[21].

Was she involved in the work of the Center for Rural Affairs?

No, she was just a fan of their work, and made connections between what she was doing early on with food issues and where it is grown. That is the short version of how it came together. I can remember sitting here thinking that I always thought it was just my family was poor, but there was a whole segment of the population that was. The benefits from the resources weren't going to the people who lived where the resources were, or even to the people who owned them. A lot was being denied to people who lived there.

Even then, I came up with the "total family concept." It was holistic for thirty years ago. We had the community action program that had a piece-meal approach. So we tried to come up with a program that would deal with all the issues a family had, not just a health issue, or a problem, or they needed Head Start for the kids. We tried to offer the clientele something that would address all their problems—whether it was training for work, being an advocate for the family. I didn't work directly with people, but I wrote proposals and the plan for all that.

Even at that time, we realized that people were not very self-sufficient. People did not know how to cook from real food. Give them a box of dried potatoes and they can make mashed potatoes.

It seems like there was an era about when I was learning how to cook where all the recipes began with "start with a package of Jell-o gelatin . . . ."

Farming in the Dark

[Laughter]. Yes, I always blamed Cooperative Extension for that but it probably wasn't fair. My mother belonged for years to the extension unit, and she has a whole drawer of those recipes.

To link the poverty issues back to the agriculture; one of my frustrations with sustainable agriculture is that we haven't addressed the income issues with farming. We are accepting the fact that we have so many part-time farmers. Have we really solved the rural poverty problem?

The rural poverty problem is no longer an agricultural issue. You go to these rural communities, and you have people, they have no ties to the community, they have no ties to the land. They end up there because it is a cheap place to live. Housing is cheap, so those communities have a hard time with cohesion, getting people to work for the community's sake. They may be there a year or two. A lot of them are on welfare, lot of young families. They think it is a better place to raise their children; it's cheaper and so forth. That is why you'll see more meth labs out there, what you think would be urban social problems in a tiny community like Goff or Whiting. We had a bust down the street from our office a few years ago

We haven't addressed rural poverty at all in sustainable agriculture. That is one of my concerns for a long time. We are encouraging people to adopt this kind of farming, but we are also saying you'll have to accept a poverty lifestyle! That just doesn't cut it. I've also been a little averse to the idea that we try to funnel our production to the top restaurants, the fancy chefs. I've never been comfortable with that, that doesn't seem right to me.

I like what is happening, not just in Kansas, but also with these food policy councils, linking low income and nutrition issues back to local food and education on how to prepare nutritious meals, linking it with children's health, the health of the elderly. The problem with how it had to be set up in Kansas is that production ag, field crops like wheat and corn, were left out of the equation. They tried to bring it in, but it wasn't made a part of it. We're not making policy for big production ag, but that ultimately is what has to happen. You take what you can get, and try to work in to it.

At that point, I had covered all my questions, the sun was starting to go down, and Mary said she needed to save the rest of her voice to "yell at the kids" tonight. I'm sure she was joking, as I've never heard her raise her voice at anything, including her children, but with over an hour to drive, dinner to prepare, her evening would be full enough. The first time my husband Raad went with me to Mary and Ed's house to get a truckload of chicken feed, he asked me why anyone would live so "far away." Though they are in the eastern part of Kansas, which tends to be more densely populated than the western half, they are about seven miles from the nearest very small town, and 18 miles from a grocery store. Farming the "home place" is a huge attraction and one that draws many farmers and offspring of farmers in Kansas. I can see exactly why Ed and Mary live where they do, in Amerugi, and they are fortunate to have the skills and opportunity to combine their farming with their important work at the Rural Center.

Chapter 6 – Ed Reznicek and Mary Fund

Notes

[1]Kansas Organic Producers (KOP)—is a marketing/bargaining cooperative for about 60 organic grain and livestock farmers located primarily in Kansas, with some members also in bordering states. KOP's purpose is to help build markets for organic grain and livestock and to represent its members in negotiating sales and coordinating deliveries of organic products. KOP was first organized in 1974 as an education association for organic farmers to promote organic agriculture, develop organic certification standards, and establish organic markets. http://www.kansasruralcenter.org/kop.htm

[2]Organic agriculture—certification has been available since the late 1980s through several private organizations. National Organic Standards were adopted by USDA in 2002; see http://www.ams.usda.gov/nosb/index.htm for details. Ed and Mary, and several other crop and livestock farmers in Kansas certify through the Organic Crop Improvement Association, which provides certification services in the United States and internationally. See http://www.ocia.org/

[3]American Agriculture Movement—founded in the fall of 1977 after congress enacted a farm bill they believed continued to pay farmers below their cost of production. Their most famous event has probably been the "tractorcade" (and subsequent "strike") in Washington, D.C. in December 1977. Additional tractorcades and protests followed in 1978 and 1979. AAM continues today to develop farmers as leaders and spokespersons on behalf of grassroots agriculturalists and to advocate for parity pricing. http://www.aaminc.org/history.htm

[4]Conservation Security Program (CSP) - is a voluntary program that provides financial and technical assistance to promote the conservation and improvement of soil, water, air, energy, plant and animal life. The program is available (on a limited basis) in all 50 States, the Caribbean Area and the Pacific Basin area. The Farm Security and Rural Investment Act of 2002 (2002 Farm Bill) (Pub. L. 107-171) amended the Food Security Act of 1985 to authorize the program. CSP is administered by USDA's Natural Resources Conservation Service (NRCS). http://www.nrcs.usda.gov/programs/csp/

[5]The Kansas Rural Center is a non-profit organization that promotes the long-term health of the land and its people through research, education and advocacy. http://www.kansasruralcenter.org

[6]Herman Daly, currently professor at the University of Maryland. Has written numerous papers and influential books ranging from "Toward a Steady-State Economy" (1973) to "Ecological Economics: Principles and Applications" (with Joshua Fareley), Island Press, 2003.

[7]Conservation Reserve Program (CRP)—Allows farmers to enroll highly erodible acres in the program. In exchange for seeding acres to perennial grasses and/or trees, and not producing a saleable crop or graze livestock for 10 years, land owners receive annual "rent" which is determined by a bidding process. http://www.nrcs.usda.gov/programs/crp/

[8]Conservation compliance in the 1985 farm bill—required all recipients of farm program payments to have a soil conservation plan in place, and to begin implementation of that plan by a certain date. Prior to 1985, program payments and conservation practices were not linked.

[9]River Friendly Farm Program—A whole farm planning and assessment tool developed jointly between KSU and the Kansas Rural Center. It is still being used in the Clean Water Farms pro-gram by the KRC. See http://www.kansasruralcenter.org/CWFP.htm and www.oznet.ksu.edu/rff for more information.

[10]Holistic Management®, founded in 1984 by Alan Savory, offers a whole farm and ranch management-training framework to practitioners and educators. See http://www.holisticmanagement.org/

[11]Land Institute Sunshine Farm project—From 1991-2001 this project collected comprehensive data on the energy, materials, and labor going into 50 acres of conventional crops and into 100 acres of cattle-grazed prairie pasture. The Sunshine Farm's goal is to calculate the amount of productive capacity a sustainable farm must devote to its own fuel and fertility. Several publications written by Dr. Marty Bender, the project leader can be found online at http://www.landinstitute.org/vnews/display.v/ART/2000/08/01/377bbca63

[12]Center for Rural Affairs, based in Lyons, Nebraska, does sustainable agriculture and advocacy work for family farmers, see http://www.cfra.org/ or more information.

[13]BMPs (Best Management Practices) refers to recommendations from University and NRCS sources that help farmers conserve water, soil, and reduce run-off and pollution from nutrients and pesticides. These include things like reduced tillage, cover crops, rotations, and other more specific practices for specific situations.

[14]The Kansas Center for Sustainable Agriculture at K-State, (KCSAAC) was established by Senate Bill 534, passed by the 2000 State Legislature, out of concern for the survival of small farms in Kansas. The Center works in partnership with state and federal agencies, nonprofit organizations, environmental groups and producer organizations to assist family farmers and ranchers to boost farm profitability, protect natural resources and enhance rural communities. http://www.kansassustainableag.org.

[15]Research on the "back 40" refers to informal on-farm research, often done in a field not easily visible to neighbors or relatives who might be critical.

[16]EPA 319 funding includes federal water quality block grants going to state agencies such as KDHE (Kansas Department of Health and Environment) in Kansas. The funds are then distributed to projects and communities with the purpose of improving water quality through education, research, and voluntary efforts (Mary, fact check this statement?)

Chapter 6 – Ed Reznicek and Mary Fund

[17]Heartland Network—a multi-year project funded by Kellogg to the Kansas Rural Center to create farmer networks of clusters. "The Heartland Network empowers rural communities to develop production and food systems that effectively balance profitability, quality of life, and land stewardship. The Network seeks to build leadership and working partnerships that integrate farms, food systems and institutions into a sustainable future." http://www.kansasruralcenter.org/heartland.html

[18] This is a reference to the Kansas Science Teaching Standards questioning the validity of teaching evolution in science classes. By 2006, the state school board changed composition, and reversed the ruling that took evolution out of the state-wide assessment tests.

[19]WTO rules are under discussion that would limit the amount of direct support for agriculture that any country could provide, which would include our current commodity support payments. Instead, countries could opt for more "green payments" in the form of subsidies for conservation practices, which would include things like the (Conservation Reserve Program), and the Conservation Security Program (CSP).

[20]Rodale's publications include the magazine *Organic Gardening*, and several books on the topic of organic gardening and self-sufficiency. From the mid 1970s until the mid 1990s they also published *New Farm* magazine, about organic and sustainable farming practices, and this publication continues today in an online version at www.newfarm.org.

[21]Francis Moore Lappe, wrote some of the early influential books such as *Diet for a Small Planet* and *Food First*.

Discussion Questions

1. Ed makes the comment that in Cuba, land is not a commodity. It can't be bought or sold. Would that framework help farmers in the US? Are land trusts a viable model for that to take place on a limited scale here? Are there any other models?
2. The statement was made that "land grant universities should be doing research for the public good." Do you agree or disagree? If you agree, how should that research program or agenda be set, and who should decide what is in the public good? How?
3. The comment is made that working in sustainable agriculture, one needs to be in it for the long haul. How long have other major shifts in society taken? How much longer should we wait for sustainable agriculture to make a difference?
4. What did Mary mean when she said we need to "re-define democracy." Do you agree or disagree? How would we do that?
5. Can something be "good" and "bad" at the same time? The discussion about the impact of the WTO on US agriculture might be a good example. Explain your views on this.

7
Fred Kirschenmann
North Dakota Farmer and Leopold Center
Distinguished Fellow

"If they don't change, the Land Grant (Agricultural) Universities will be complicit in their own demise"

Ames, Iowa—It was November 14, 2005, the first really cold, cloudy, blustery day this fall in Iowa, or most of the Midwest/corn belt. We had all been enjoying the warm fall days (65+), but I was also feeling apprehensive, deep down, about having "too much" nice weather. In fact, only 48 hours before, nine tornados had been sighted over Iowa. Three of them touched down in the Ames area and caused considerable damage to two small rural towns. Ironically, the Iowa State University football game kickoff, for "the Cyclones," had to be postponed thirty minutes Saturday night due to area tornado warnings. The unseasonably warm weather, combined with a cold front had set off dozens of tornados across the United States that weekend from Maine to Texas.

I had come to Iowa that weekend to attend the fifth annual Organic Agriculture Conference sponsored by Iowa State University, and to visit with Fred Kirschenmann and others. Iowa has been a magnet for good people in sustainable agriculture these past few years. First the Leopold Center for Sustainable Agriculture was created in 1987, then in 2001 the first coordinated multi-disciplinary master's and doctoral programs in sustainable agriculture were begun, not to mention attracting many individual faculty noted for their expertise in sustainable agriculture and scientists at the USDA/ARS soil Tilth Lab. I was fortunate to have some time to sit down with Fred during a coffee break after lunch.[1]

What are the high points you've seen in sustainable ag throughout your career? The successes?

I think that at least one of the high points has been the accumulation of a body of literature, not all of which is uniformly great, but has within it virtually everything that we need to know to proceed in the right direction. They certainly include Wendell Berry and Wes Jackson[2], those luminaries and others. To me, it really goes back to 1840 and Justus Von Liebig's[3] essay, which formed the beginnings of the industrial way of thinking about agriculture. It took us 100 years to put that in place, and there were people almost immediately after its publication who reacted against it. They didn't have the benefit of that body of literature of some of the things that we know now. The principles were there, we have a rich heritage that provides what we need to be doing. That's one of the great success stories. Then we have some individual farmers that have been enormously innovative and creative. We have those models are out there, farmers like Joel Salatin, and Takao Furuno[3]. So that is the other great strength.

I think the weaknesses are that our culture of agriculture has been so indoctrinated with the industrial model, that it is still difficult for both farmers and non-farmers to see how they can do this in a different way. Even though the sustainable literature informs us, making that transition has been slow. A second disappointment is that our university systems have been indoctrinated with that same culture. They have not seen the need to move in a different direction.

I think we are on the tipping point of some major transitions. I think they are going to be driven, first and primarily, by our new energy future. This industrial model built on Justus Von Liebig's theory and implemented essentially 50 years ago, is enormously fossil fuel dependent. Fertilizers, pesticides, farm equipment, irrigation, in addition to (actual) diesel fuel, are all fossil fuel. As most industry experts now admit, we have reached, or will reach in the next decade—give or take 10 years—peak production. At the same time, the demand for fossil fuel is increasing dramatically. Maybe you have seen that ad Chevron oil put out, saying that it took 125 years to use up the first trillion barrels of oil, and will take only 30 years to use the second trillion. That's never going to happen, and I'm sure they know that once you reach peak production, you can't produce it that fast anymore.

This is obviously going to change agriculture. On the short term, we are going to see more of what we see now, assuming we can solve that problem with alternative sources of energy,

When you look at it from the point of view of energy efficiency, it's not going to solve the problem. Marty Bender's[5] work points out that in the 1940s we were getting about 100 units of energy for every 1 unit we had to invest to make energy from oil and natural gas available to us. All the alternative energy supplies including nuclear, wind, solar, bio-fuels, only yield a 1 to 15 or a 1 to 20--conversion rate at best, most yield less than that. This energy equation is going to force agriculture to move in a direction that has to be more energy conserving rather than energy consuming. That is going to create a fundamental shift. It will give a strong comparative advantage to the kind of sustainable agriculture that Takao Furuno and Joel Salatin are doing, and things that Wes Jackson has been exploring. Whether or not we make that transition quickly enough so that we can do it in a rational deliberate way, a planned kind of way, or in a crash, I'm not prepared to predict. There is likely to be something of a crash along the way. Nevertheless, it creates an opportunity to move into a different kind of future.

When you add to that climate change, a more unstable climate, it also gives the comparative advantage to—leads us to—more diversified systems. Then we have our environmental situation—I was shocked by the UN Million Ecosystem Assessment synthesis report[6] last March. It was written by over 1,400 of our leading scientists throughout the world. Over the past 50 years, we have polluted, used or exploited our ecosystems to the point of collapse—60 per cent of our ecosystems—*and* we have reached this point over the past 50 years primarily due to the way we have procured our food, fuel, fiber and water. That means we have a short window. Then they go on to say that given the way we have reduced our biodiversity and our genetic diversity, our natural ecosystems no longer have the resilience to bounce back from this situation. Therefore, there is a real possibility of imminent ecological collapse, which could lead to what they call a different kind of structure and functioning of the planet. We've been through such major extinction periods at least five times before, and they always fundamentally changed the way the planet functioned. The only way that there will be an incentive for change is if our social structure brings this home to more people, that they understand it, and people see the need to make changes. Here again, our universities should be playing a lead role. They are very slow in doing that. Maybe they will come around to it.

The other thing is the farm economy. The industrial agricultural system has not worked for farmers. Net income now is less than in 1929 as an aggregate. Farmers are constantly on the edge of going bankrupt. It's not working for them. That has both a positive and a negative effect for change. The negative is that due to difficult economic circumstances, they aren't going to take economic risks. They'll just hang in there, and hope they can survive. The positive side is that if they step back and realize there is no future in what they are doing now, *and* if we provide real practical alternatives that are better for them economically, at least some of them will change. That is a long answer to your question. That is the framework within which we are operating, both the positives and negatives.

I had said very little up to this point, since Fred was answering my first two questions, about the strengths and weaknesses of sustainable agriculture without any prompting. I was glad that he was also mentioning issues like energy and the global environment, since they were weighing heavily on my mind. I followed up with my third question; what should the movement do? Where should we put our energy? Do you have any advice to students?

That actually touches another strength that we have, which I didn't mention, which is the next generation. Here at ISU (Iowa State University) in my experience we have two distinct kinds of students. One is here essentially for job training. They want to get a specific set of skills so they can be gainfully employed. From my perspective, they aren't the most interesting students; they are fine, they are ok, they are wonderful people, but not where my passion is. Then we have this other group of students; the ones in our sustainable agriculture graduate program, for example, are universally of this second type. They are not here for job training. They are here because they know that some changes need to take place, and they want both the depth of information and the skill sets so they can be effective and be the change agents to bring us into a new future. A core of these students is going to be a part of that future. They are absolutely incredible and exciting to work with. I run into them all over, not just at ISU. Without this next generation, this core of students, we won't make the changes we need to make . . . so how can we help them most? Some younger faculty members want to move in this direction.

Farming in the Dark

I don't know about other institutions, but the dampening effect here (ISU) is that the administration, in my view, is not showing the kind of leadership that lets these students and faculty know that this is the future, to show that this is critical to the future. It's more like they are tolerated as a marginal, fringe group. In order to be fair, our university is always getting budget cuts, I understand that, but they are still dealing with a lot of money, so it's really a matter of priorities. The priorities are still with the industrial model, the technological solutions. The problems that we have now are not going to be solved by technological solutions, but by ecological and social solutions. We have to figure that out; that we have to overcome, and I don't have any quick answers for that.

I somewhat hesitatingly ask Fred if he'd like to comment on Universities in general. Do the land grants have a future? Knowing that there are politics everywhere, I told him that he didn't have to answer that if he doesn't want to. He laughed and answered.

Use whatever you want. I don't have to protect a career any more. I think that is an interesting and good question. I don't have a clear answer to it. If change does not take place in the next decade probably, certainly in the next two decades, most of our public institutions will have outlived their usefulness, because the old model isn't going to work anymore. If they continue to put their energies into that, at some point the public is going to ask, "Why should we still support this?" We are already seeing the early signs of that. In a sense, our public universities are complicit in their own demise. They aren't showing the public the ways in which they are serving the public. It's a self-perpetuating cycle. And it's not just the private money; it's what it symbolizes and indicates. At ISU, there isn't that much (private) money involved, about 12 percent or so, but it serves as a lever; it is how it is leveraged.

Have the land grants ever shown leadership? Especially if you go back to the beginning, when they were first created?

Yeah, sure. Liberty Hyde Baily[7] is someone that comes to mind, immediately. He had vision and passion. He wasn't, in my view, taken all that seriously in terms of his longer view. As another way to look at the prevailing attitudes, Aldo Leopold[8] has a wonderful statement in one of his essays published in 1945, "It was inevitable, and no doubt desirable, that the tremendous momentum of industrialization should have spread to farm life," but that some day it would "die of its own too much." This was from 1945. He went on to say it would die of these extremes "not because they are bad for wildlife, but because they are bad for the farmers." When he said it was "inevitable, and even desirable," we have to take that seriously. We were facing a set of problems in the 1940s and 50s, especially right after the Second World War, that seemed like they were so easy to solve using the industrial approach. Especially with technologies developed in the war that we now could bring to agriculture. Justus von Liebig provided the theory for that. It was so inevitable in a way, and our universities were able to help develop the technologies.

Another thing that happened in the United States was that a science advisor to the Roosevelt administration, a person by the name of Vannevar Bush[9], was asked by the president to come up with a policy on how technology should be used in a post-war situation. He came up with an incredible document in which he said technology that helped us win the war. That's how he saw it, that it was our superior technology, not intelligent leadership. So what we need to do now is apply that incredible resource to our domestic problems. And food production was one of the core pieces. Roosevelt never saw the report, because he died a few days before the report was delivered, so it was delivered to Truman's desk. When you read that report, you can see that what happened from then on exactly followed the recommendations of the report. For example, the report even said that if we are going to do this, are to be successful, to use technology in this way, industry has to be involved. Our universities can't do this alone, so we need to give industry tax incentives to get them involved in this. So you can see how it all played out. If you look at what happened in terms of public policy from that point on, it all exactly follows that script. Can we use that lesson to reverse this trend? Yeah, if we can get a president (laughter) . . . that would ask for a new report. That is exactly what we need. We need a new vision for how science can be used in agriculture in the twenty-first century.

How do we reverse this? Do we need a more deliberate policy to promote organic/sustainable principles?

We need to shift from external or exogenous technical inputs into agriculture, to internal inputs, ecologically driven biological synergies—that's the fundamental shift. We've got at least 20 years of ecological and evolutionary biology science to support that. We pretty much know the science to make that happen. We would solve a lot of problems if we really took that seriously and worked especially with farmers that are doing this. I've been saying if we could get 30 percent of our current public research dollars, in agriculture—about 30 percent of two billion per year—we could come up with all kinds of solutions, but it needs that drive. In the mean time, we are still totally focused on the next generation of inappropriate technologies.

Have you seen Joe Lewis's report in the National Academy of Sciences Proceedings in 1997[10]? I think it is one of the most significant descriptions of what we need to be doing in agriculture that has been written. It's only a 10-page piece. He and about four of his colleagues wrote it. He is a pest management specialist, so he writes this from the point of view of pest management, but he points out in the article that it applies not only to pest management and to agriculture, but also to medicine and social work and to other fields. The basic shift that we have to make is from the current paradigm under which we are operating, what he calls "therapeutic intervention." You have a pest problem, you bring in a pesticide from the outside, as a therapy to solve the problem. What happens with that is that you never deal with the system and the weaknesses of the system that caused the pest to emerge in the first place. So the pest will keep re-emerging and re-emerging. That's one of the reasons farmers go broke. So, it will never disappear. But we have been able to make that relatively successful, because all of the inputs/solutions are based on cheap energy, and that is increasingly not going to be the case. So he argues that we need to shift from therapeutic intervention to natural systems, which is similar to what Wes Jackson is saying. That is, to find the strengths from within the system—the soil, the plants, all of the biological mechanisms—that are available. Find the inherent strengths; release those, as a way to deal with the problem. Farming in nature's image, I think that is exactly right. So we should use at least 30 if not 50 percent of our research dollars in making that shift. Joe Lewis got a fair amount of attention and an award, but almost no one is really taking it seriously within USDA or the land grant universities.

It sounds like something that students should be reading.

Anybody who is involved in agriculture in any way should be reading Lewis' article.

Do we need a similar paradigm shift for social sciences?

I think we need to move toward biological synergies in our production systems. And for marketing systems, at least farmers in developed countries need to move in a direction like what I was talking about this morning, to be involved as partners instead of raw material suppliers, including partners with processing, distribution and even retail. The basic idea is to enable farmers to participate in markets, where they can produce more value and retain more of that value on the farm, so farmers can have more flexibility economically. If they retain more of the value, it puts them in more of a position to take the risks. Farming with biological synergies instead of toxic chemical inputs has a better story, consumers respond to such good food stories. In addition to that, we need to have some policy changes that enable farmers to do this.

The other thing we are working on at the Leopold Center is to give farmers the tools to calculate their production costs. I'm one of those farmers that if you asked me what my production costs are on my farm for various commodities I wouldn't be able to tell you. One reason is that I've never been able to go into the market place and say, "gee guys, it cost me this much to produce a bushel of wheat, why don't you pay me that and a return on my investment?" Instead, I say, "how much are you going to pay me?" In the past, calculating our cost of production has seemed like a waste of time, so we haven't developed that skill. But we are finding now that some companies take seriously this notion of farmers being fairly compensated—it is in their interest to assure themselves of a reliable supply of quality products. And farmers are interested. We've done a couple of small grants and have extension working with farmers, and also make that information available to other people in the values-based value chain (marketing system). This has been met with a reasonable amount of success.[11]

Given that farmers need to make a living wage, do you think there is a way to turn the WTO negotiations into a discussion of 'fair trade' instead of 'free trade'?

I don't know. That is a really difficult question. The only way I can see that we can begin to make those kinds of changes is we start to mobilize—this sounds like liberal commie socialist philosophy—but we need to mobilize the power of the people, which I think is what <u>democracy</u> is about. The vast majority of people don't understand that half the world's people live on $2 per day. This is not an acceptable situation for a global village, which is what we are now. So how do we begin to address that? And we have to address it, at least to some extent in terms of how we do commerce on the planet. So how do we get people, get the citizens of the globe, as a body politic, to address these issues? This has to begin, and to some extent, that global conversation has begun. The reaction to globalization, the battle of Seattle, the current fiasco in Latin America, these are all parts of the conversation already underway to a certain extent. How do we focus that conversation now and create a kind of social synergy, to move things in a different direction and force the commercial world to begin doing business in a different way?

Have you seen Peter Senge's most recent book, called *Presence*? [12] He has been involved in organizational behavior for many years; worked with major corporations. He and three of his colleagues have written this new book, which is a recent view on how change takes place and what we need to do to bring about the social and economic change that we need. He argues that we are now in a period of such rapid change, that the old models of change where we decide on a set of goals, objectives and strategies no longer work. Things emerge too quickly, so he has come up with a concept symbolized by the letter "U." Over here, on the left side is the male right brain side of goals and objectives. Let that go, come to rest at the bottom of the U, let those strategies go, be present, pay attention to our colleagues, whoever it is that we work with, and to the greater world out side, and attend to what emerges out of that process. Come up on the right hand side of the U, and that is an on-going process.

It makes a lot of sense. One of the instructive things is that at one point he talks about a two or three day meeting at Marblehead, Massachusetts, with some of the CEO's of major multi-national companies. They all admitted that the current planning system is not functioning, but they feel caught. The companies, the systems they engaged in, demand quarterly reports, short term returns, push the pedal to the metal . . . but they know it is not sustainable in the long term. He works with them, to try to help them move to the new model. Again, there is hopefulness that those kinds of things are going on in these major corporations. The big question is; as asked by Paul Roberts in his book *The End of Oil*[13], the issue facing us is not whether things will change, but whether they will take place in an orderly peaceful way, or whether they will take place in a type of major collapse because we haven't planned. If anybody knows the answer to that question . . .

I comment that sustainable agriculture seems dwarfed by these larger issues now, more than when I was in graduate school, for example. However, I keep working in sustainable agriculture, because I believe it has something to offer toward solving those problems.

It's the reason I'm in it too. There are a lot of things in sustainable agriculture that are not where I think they need to be, dragging along, pushing, or whatever. But at the same time, the environmental community is the same way. There is no obvious social structure now that says, "Oh, that's what we have to work on." We are all struggling and figuring out how to do this. And again, there are people, particularly in this upcoming generation that are very exciting. One of the reasons I'm here is because of that group of people. I constantly feel disappointed because they don't get the kind of support they need from the administration. But they are still hanging in there; we have to see how it plays out. One of the things that is deeply disappointing to me, is that we have all this rhetoric at ISU about being at the forefront of sustainable agriculture, but we don't even provide the basic support the grad program needs; the disconnect between the reality and the rhetoric is pretty stark at this point.

What do you think about the statement by some that we aren't a movement?

I'm sure you've read Wendell Berry's disenchantment about movements[14] (laugh). I'm always pulled one way and then another on this issue. I think Wendell makes some good points about that which we need to take seriously— that movements very often become so invested in their own capacity to be movements that they don't accomplish what the movement ought to be accomplishing. We need to be careful in deciding whether sustainable agriculture needs to be a movement. Though I'm open to seeing if we could create the kind of movement we would need to be to be successful; my own personal preference at this point is to infuse the sustainable agriculture agenda into existing movements that are compatible with our goals.

Here is one of the things that I'm seeing, at the very, very early stages. I think this whole division, this dichotomy that we've operated out of for a number of years, with preservation over here and extraction over there is breaking down. Conservationists are starting to recognize that preservation isn't working in isolation; patch ecology doesn't' work. So they aren't fulfilling their goals just by protecting stuff. Farmers are beginning to recognize that they need the richness of ecological restoration that the conservationists talk about for their farms to be productive. The rate at which we've lost soil quality, especially— we've been able to mask that with cheap energy, and in the future, we can't do that. So farmers to some extent already are—and even more so over the next decade or two—will recognize-that they need the richness of ecological restoration in order for their farms to be productive. We have the ability to have farmers and environmentalists at the same table looking at the same issues. Things like Gary Nabhans's forgotten pollinators[15] suddenly become real for farmers. It probably makes more sense at this point—instead of creating another movement—to infuse sustainable issues into existing movements. Bring the diverse groups that are now beginning to see a common agenda together at the same table. My sense, at the moment, is that will give us more traction than to invest in a separate movement.

How did you first get started into sustainable agriculture?

How I first got into it? That's an easy story to tell. I learned about organic agriculture from David Vetter, who was a student of mine at the time, in Dayton, Ohio, at a seminary there. I had been invited out there to help start a consortium in higher education religious studies. A colleague and I recognized in the late 1960s, that students were talking about their desire to become engaged in some kind of socially responsible activity. But they were hugely suspicious of established social institutions like the church and social work etc. We were trying to understand what it was they were saying and realized that what they were talking about was the worker-priest model. They wanted to be actively engaged, earning their own living, but making a contribution to some social dimension in addition to that. So we started this program called the Dual Career Training Program. It was small, we only admitted 10 students to start with. They would have a career path, to make their income, an earning path. They would also have a dual career path that would be their social work contribution, whether it would be the ministry, or what ever they wanted to do. The entry into the program required that students had to come up with a prospectus as to what their dual career would be. We would invite them in for interviews, and then select the students for the program. David was one of the students. The prospectus that he had written was that he wanted to be a minister to the soil.

I grew up under my father, who started farming in 1930. The dust bowl hit North Dakota in 1932 and his early experience made him a radical conservationist. He knew that what happened to his land was not just due to the weather, but also how he was treating the land, and he didn't want it to happen again. I grew up under a father that said that taking care of the land was at least as important as taking care of your neighbor. So David comes along and wants to do this ministry to the soil. So I started talking to David, and asked him what motivated you, how did you get to this point? It turns out he did his undergraduate degree at the University of Nebraska under Warren Sahs[16]. As I understand it, Warren was one of the first extension specialists to take sustainable agriculture seriously in his research. They started these long-term plots where they compared conventional and organic. David wanted to find out what happened to soil quality. Of course, what they discovered over a period of four or five years was that the soil quality in the organically managed plots dramatically improved compared to the conventional plots. That was a brand new insight for David, and for me. But when you think about it, now we know it is obvious. When you return organic matter to the soil, it improves soil health. If you simply insert synthetic nitrogen, it makes no contribution to soil health. When David shared that with me, what went through my head was that we have all this land entrusted to us in North Dakota and we aren't caring for it as well as we could.

I started talking to my father about that, and my father immediately gets it. My father was also progressive, and therefore was one of the first farmers in our township to start using pesticides and fertilizers on the land in the late 1940s when they became available. He used them on the condition, with the assurance of the county extension agent and other farmers that this won't hurt the land, it is good for the land, improves the nutrient quality, and he was convinced by that, so he went that route. By now, he begins to see some of the weaknesses of that system, so I say, "are you going to start farming organically?" He said "Oh no," he was 68 then, "what you are talking about is an entirely different system of farming, that's not for me, that's for someone else." I said, Ok Dad, we won't go there. Then in 1976, he had a mild heart attack and announced to the family that the doctor said he needed to get out from under the stress of managing, he could stay at the farm, but needed to get out from under the stress. He was looking for someone to manage the farm for him. At that time, I was getting bored with academic life in Boston. I'm a prairie boy, the urban life was not serving me well, so our family talked about it and decided that if my father would let me convert the farm to an organic farm we would go back and manage it. So I called him up, told him that, then there was dead silence for about a minute and a half. He said, "Whatever works." So off we went.

I was communicating with David about converting our farm, and he said he could help with the principles, but couldn't help with the application of them since he said he only knew Nebraska soils. In the 1985 farm bill, when they first introduced LISA (now called SARE[17]) they didn't appropriate any funding. Rodale and couple of other organizations decided they needed to lobby to get some funding for it. Quenton Burdick at that time was a Senator from North Dakota and was chair of the senate ag appropriations committee. The common thinking in conventional circles at that time was that the kind of farming the LISA program represented was for hobby farmers and real farmers didn't do that, so there was not real enthusiasm for providing any funding for this. Rodale and other folks felt that they needed someone to testify to Burdick's committee who was a "real" farmer, doing sustainable agriculture . . . so I get this phone call. I said, "all right, you guys have to help me." I had never testified. We worked on that together. I was supposed to have five minutes for my part. I wrote part, they wrote part, we put it together, we all agreed it would be a good statement. I get there, something happened with the schedule . . . I only have three minutes; I'm going to be on in five minutes. I decided to do it from my heart, say it as I can, speak directly to Burdick, which I did. There were a couple of newspaper reporters in the room at that time, they picked up my comments, which were in the Washington Post or wherever, I don't know anymore. I've been on the speaking circuit ever since.

I had admired Fred's work for years, and always learned so much from his public speaking, eloquence and scholarly approach. I was surprised by these beginnings. I asked, so it all started with a three-minute talk!

That's really how I was pulled in, I had no intention . . . my intention was to just do the work on my farm, make it work. That's how I was pulled in to the national arena.

Now we've all benefited from your speaking! I asked him if he had any final comments . . .

I'm glad you are doing this . . . .

Chapter 7 – Fred Kirschenmann

Notes

[1]Dr. Frederick L. Kirschenmann—Current position is Distinguished Fellow with the Leopold Center for Sustainable Agriculture at Iowa State University, and President of the Stone Barns Center for Food and Agriculture at Pocantico Hills, New York. Dr. Kirschenmann came to the Center from south central North Dakota, where he operated his family's 3,500 acre certified organic farm, which he still manages. Certified in 1980's, it was one of the early operations to make the transition. The farm is a natural prairie livestock grazing system that combines a nine-crop rotation of cereal grains, forages and green manure. Kirschenmann currently oversees management of the farm and has an appointment in the ISU Department of Religion and Philosophy. He holds a doctorate in philosophy from the University of Chicago, and he has written extensively about ethics and agriculture. He has held national and international appointments, including the USDA's National Organic Standards Board. He is a board member for the Food Alliance, Silos and Smokestacks, and Whiterock Conservancy. In 1978 Kirschenmann helped organize North Dakota Natural Farmers that later became the Northern Plains Sustainable Agriculture Society. He helped found and for 10 years was the president of Farm Verified Organic, Inc., an international private certification agency. Dr. Kirschenmann was Director of the Leopold Center from 2000 through 2005. For more information about the Leopold Center and their projects, see www.leopold.iastate.edu.

[2]Wendell Berry and Wes Jackson—both have published several books on the philosophical underpinnings of sustainable agriculture. Wes Jackson is co-founder and director of the Land Institute in Salina, KS.

[3]Justus von Liebig—described as one of the most important chemists in the nineteenth century, and one of the founders of organic chemistry. He proposed the widely cited "barrel" approach to soil fertilization, illustrating that all nutrients must be present in sufficient quantity for optimal plant growth. For more information, see for example "Justus von Liebig: The chemical Gatekeeper," Cambridge Science Biographies, 2002.

[4]Joel Salatin and Takao Furuno—Joel Salatin has published several books about his experiences in direct market farming, including "Pastured Poultry Profits" and "You Can Farm: The Entrepreneur's Guide to Start & Succeed in a Farming Enterprise" and Takao Furuno has written "The Power of Duck,"

[5]Marty Bender, The Land Institute, Salina, KS. The late Dr. Bender produced a manuscript to be published summarizing over 10 years of research on energy and agriculture. Interim reports can be found in the refereed journals and at the Land Institute website www.thelandins-titute.org.

[6]United Nations, 4 April 2005: (summary from website) The Millennium Ecosystem Assessment (MA) Synthesis Report, a landmark study co-sponsored

and launched by the United Nations Development Program (UNDP), the United Nations Environment Program (UNEP), the Global Environment Fund (GEF), the United Nations Foundation, the World Bank, the World Resources Institute and other international scientific and development partners, reveals that approximately 60 percent of the ecosystem services that support life on Earth—such as fresh water, capture fisheries, air and water regulation, and the regulation of regional climate, natural hazards and pests—are being degraded or used unsustainably. Scientists warn that the human activity is putting such pressure on the planet that its ability to sustain future generations cannot be guaranteed. The poor will be particularly affected. http://www.undp.org/dpa/pressrelease/releases/2005/april/pr4apr05. html. The Millennium Ecosystem Assessment Report is available at http://www.maweb.org/en/Article.aspx?id=58

[7]Liberty Hyde Bailey—Cornell. See for example "The Holy Earth," Lebanon, PA: Sowers Printing, 1915.

[8]Aldo Leopold This quote appears in "The Outlook for Farm Wildlife" written in 1945 and published in the book, *For the Health of the Land*, edited by J. Baaird Callicott and Eric T. Freyfogle.

[9]Vannnevar Bush, director of the Office of scientific research and development, Roosevelt/Truman administration. "Science, the endless frontier. A report to the President." July 1945.

[10]W.J. Lewis, J.C. van Lenteren, Sharad C. Phatak, and J.H. Tumlinson III. "A total system approach to sustainable pest management." Proc. Natl. Acad. Sci. USA, Vo. 94, pp. 12243-12248, November 1997

[11]Ag of the Middle—a program created to provide research-based information to support the business development and public policy change components to "value-chain" based food systems. See www.agofthemiddle.org.

[12]Peter M. Senge, C. Otto Scharmer, Joseph Jaworski, and Betty Sue Flowers. "Presence: An Exploration of Profound Change in People, Organizations and Society." 2005. Currency (Press).

[13]Paul Roberts, "The End of Oil: On the Edge of a Perilous New World." Mariner Books, 2005. [another book with the same name and the same author? "The End of Oil: The Decline of the Petroleum Economy and the Rise of a New Energy Order." Bloomsbury Pub. Ltd, 2005.]

[14]Wendel Berry's critique of movements appears in several of his essays.

[15]Stephen L. Buchmann and Gary Nabhan, The Forgotten Pollinators. Island Press, 1997.

Chapter 7 – Fred Kirschenmann

[16]Warren Sahs, retired, University of Nebraska. For a summary of the long-term systems research he initiated and also the Rodale long-term trials, see Andrews, R.W., S.E. Peters, R.R. Janke and W.W. Sahs. 1990. Converting to sustainable farming systems. Pp. 281-313 in Sustainable Agriculture in Temperate Zones, C.A. Francis, C.B. Flora and L.D. King, eds. John Wiley & Sons, Inc.

[17]SARE (Sustainable Agriculture Research and Extension) program, through USDA, first mandated in the 1985 farm bill, funded in 1987 as the LISA (Low Input Sustainable Ag) program. Funds research, extension, and training programs throughout the U.S. See www.usda.sare.org.

Discussion Questions:

1. What did Fred Kirschenmann mean when he said that "our public universities are complicit in their own demise?" Do you agree or disagree? Why or why not?
2. On page 106, Dr. Kirschenmann reminds us of the Aldo Leopold quote: "It was inevitable, and no doubt desirable, that the tremendous momentum of industrialization should have spread to farm life," but that some day it would "die of its own too much." What is meant by that? Do you agree or disagree? Do we see this happening now, or see examples in history?
3. After discussing the report to Roosevelt/Truman that began the industrial era of agriculture, Dr. Kirshenmann says that: "We need a new vision for how science can be used in agriculture in the 21st century." What would that vision look like?
4. Instead of less than 1%, what if 30 percent of public research dollars in agriculture were spent researching sustainability. What would that look like? How would you prioritize those areas of research?
5. What does it mean to switch from "therapeutic intervention" to "natural systems" agriculture? Give an example.

8
Donn Teske
Wheaton, Kansas and Kansas Farmers Union President

"The corporations—they have no soul."

I've known Donn Teske for about six or seven years. I first met him through the Kansas Rural Center's Clean Water Farms program, when he was hired as one of the field staff to work one-on-one with farmers filling out the environmental assessment notebooks, and also to take their farm financial data through a somewhat complicated financial analysis software program called FINPAK. He also worked for Kansas State University for a couple of years as a financial advisor to farmers, especially farmers headed for financial trouble. Donn farms near Onaga, Kansas, about 15 miles north of where I live now. It is an area in the Northern flint hills with rolling pastures and good farmland. Donn has recently become more involved in farm policy work, and although he makes frequent trips to Washington, D.C., and to Topeka (our state capital), everything about Donn's demeanor and husky physique tells you that this man is no stranger to physical labor. He is currently the Kansas president for the Farmers Union. His quick wit and smile provide a counter-balance to the seriousness of the topics he is willing to discuss. We met in my stone farmhouse in Wamego in mid-July 2005, but it was not oppressively hot, so the windows were open, we could hear the wind-chimes outside, and we took a break to eat watermelon in the middle of the interview. Donn and I share a background in the dairy business, as my parents had a dairy farm when I was growing up.

Farming Background

How long have you been in farming? When did you start to milk cows?

I started farming and milking cows when I was a senior in high school. I didn't go very far, I still farm the multi-generational family farm. I live a quarter mile from where I was raised. I went to high school half days my senior year and farmed the other half. I thought my goal was to go to college, but my dad nixed it. I could do that and not come back to the farm, or farm. I don't really regret my choice. I'm proud of my family and proud of my kids; everything is good. I ain't got no money, but life is good. That was a defining moment.

I told my high school ag teacher I'd never milk a damn cow, but I milked from when I was a senior in high school as part of our family operation that included my father, uncle, brother, and brother-in-law, from about 1972 through 1979. We quit for five or six years, then in 1986 my wife Kathy and I started and continued until 1995, so I've been in dairy for 20 years over a 25-year period. I tell people I'm missing the cows, I'm ready to start again and Kathy says, "No you're not! You're doing just fine." She's been a lot happier with the cows gone.

Is there a future in it, to get back into it now?

Perhaps if one of my kids came back and started farming. I have some good examples to look at, what not to do. The last thing I would do is buy out my neighbors to be the biggest farm in the county. Some of my very good friends are the biggest farm in the county, but I wouldn't do it for anything.. I don't want the headaches. I work many of the big farms' financial books. They just look pretty when you drive down the road. Why make yourself an indentured servant to John Deere? I would tell my kids, here are the acres we have, make a living at it, do value-added if you need to. We're working a deal right now and we may be raising dairy replacements for an organic place in Colorado. There could be some neat arrangements.

Farmers Union

How did you get involved in Farmers Union?[1]

After high school, I got involved in Young Farmer Educational Association, which was grown up FFA (Future Farmers of America) boys. I eventually became a national officer. Then the organization went by the wayside, it had its hey day. I was a republican precinct committee member, a Farm Bureau member and a KLA (Kansas Livestock Association) member. Then in the late 1980s, when industry was trying to take on the anti-corporate agriculture laws in Kansas legislature, I got involved as an individual lobbying against it, not as part of any group. I watched the representatives of the organizations that I belonged to do their lobbying down there. I was so frustrated by the system I came home and quit them all the same day.

My dad was big in Farm Bureau, and was actually a Farmers Union member but I didn't take it seriously. That was an old persons group, and I was a young farmer. After I got so frustrated that I quit all the others, I was sitting there on the dairy farm one day, farming, and one of my neighbors stopped by with this guy and talked to me about it. I signed up, went to my first county meeting, and they elected me county president! [Laughter.] They probably said, "Oh look a young guy!" They needed a delegate to the state convention, so I said I guess I can go. Then they elected me to the state board! I'd just been a member for six months. I said, "Ohhhh, they're really hungry here. It was a perfect fit for me though. I tell people all the time, I don't have to read the policy manual at all, if I speak from my conscience, or if I'm out ranting and raving for what I think is right, it is usually what the Farmers Union is doing. I take great pride in that. It was the right fit at the right time. Maybe I'll evolve out of this into another organization, I don't know.

What did you find frustrating about your Topeka experience? Weren't Farm Bureau and KLA representing you?

Chapter 8—Donn Teske

My interpretation was that they weren't representing the farmers view, and I was frustrated by their extreme domination of the political system. I've seen legislators stand up on the floor, of the house or senate, and flat out lie. Even on small issues, such as, "Did anyone testify in opposition to corporate dairy?," and the house ag Committee chair said, "No sir, no one did." I had personally testified against it, that is just a personal private example. The heads of both the House and Senate ag committees forced that bill through their committees. Right after that, the head of the Senate committee resigned and took a position as a lobbyist for Seaborg (a national pork corporation), and the head of the House committee sold land to Seaborg for extravagant prices compared to the current land values. In my interpretation, they were both bought off. I used to know that Senator from Franklin County, and after that, he wouldn't look me in the eye. They both lost their souls. Just the arrogant manipulation turned me off so bad, to think that anybody had that much right to do that to another human being. It turned me away from those organizations, and it hasn't changed any, it's just gotten worse.

Sometimes it seems like it would be difficult and discouraging to be involved in the political process at all.

That's the truth! KLA—they are so well indoctrinated into Kansas politics, it's like a god out there, it's an entity. I asked one of the Farm Bureau presidents, "How did they get so much power?" He said, "I'd like to know too!" The KLA membership numbers aren't as large as Farm Bureau, but they have big money behind them. They throw socials at least once a week for the legislators in Topeka. They don't legally get check off dollars, but the check off dollars go into the same building, and then you've have these firewalls to keep the money separate. Maybe it can't go for lobbying, but it goes for other things to free up the funds for lobbying. Those check off dollars really mess up agriculture, really mess things up. Now all of a sudden these commodity organizations have all this money. It isn't doing the best for farming overall or for farmers overall, it is just for this specific commodity.

On Sustainable Agriculture

What is your definition of sustainable agriculture, and what do you think it has accomplished in the last 25 years?

My concept of sustainable ag would be an industry of agriculture that allows the manager to make a living on his operation, hopefully to leave the land a little better than when he got it, and to allow continuing generations of farmers to farm if they want to. I realize that doesn't say organic, but I don't know if organic is the true concept of that. Right now, I think organic has some advantages, because they still have more control of their markets than commercial ag does.

Is farming still a full time occupation? I know it can be, especially if you have livestock in the winter months, but most farmers don't anymore. Should a person who plans to farm also plan to have another source of income?

There you are getting into the concentration issue. Why have we concentrated our operations into monoculture, instead of the diversified farms that we were, that operation where you were busy year-round? I don't think it is healthy for society. The concentration of livestock is causing all kinds of environmental problems. The more that happens, the more we dumb ourselves down. If we had a total collapse of our system right now, I can still butcher, I can get by but my children probably couldn't do what they would need to do. All at once, you are at the mercy of the elements. My children range from 23 down to 14. The last time I butchered one child helped, but he was green most of the time. I suppose that not only says something about me but also about the society we live in and about how we devote our time. My grandpa taught me the basic skills of butchering. There's going to be a rude awakening if it ever becomes a necessity. I think we're vain to think it won't. We are in a fragile system now; we have gotten soft.

So far, every generation has lived a little better lifestyle than the previous generation. But, I doubt that the next generation will be able to do so. They will have a different lifestyle than what we have. Think about a role in that society— think about that situation. We will need a whole generation of people that are going to need to be taught certain things that have been lost.

Has the sustainable agriculture movement helped agriculture?

It has been marvelous! I came from a background of what's considered "modern ag." I farmed full time until 1995, and all at once I began working off the farm. What was so amazing to me about the Kansas Rural Center was they didn't try to change what others were doing, but make their own niche. They worked by setting an example. That's the reason behind the tremendous success the Rural Center has had; the fact that they are quietly going about developing their own system and setting an example, not raising hell like Farmers Union used to do. The Rural Center and the sustainable agriculture movement set powerful examples. For example, we have a state legislature that mandates a KSCSAAC[3] (Kansas Center for Sustainable Agriculture and Alternative Crops), and have a SARE[4] (USDA Sustainable Agriculture Research and Education) program at the national level—those aren't small items. For one thing, they are showing that there is an alternative. I think that people don't want to admit it, but they are starting to realize we are on a dead-end road now with big companies manipulating traditional agriculture.

I was shocked a few years ago when legislative listening sessions were being held around the country, and Marc Johnson, the former Dean of the College of Agriculture at Kansas State was testifying in Kansas City. He said there are two types of ag still remaining; large, growing, super modern ag, and niche markets, and the Kansas Rural Center is a perfect example of a group that nurtures the second system. I was shocked to hear him give the Rural Center credit at a legislative listening session, also shocked to hear him admit that. He said right then, that middle class ag is gone. I think he was honest about it.

This whole GMO (genetically modified organisms) thing; I don't like GMOs and what they are doing to the milk and dairy industry. I guess in the back of my mind, I'm not sure that GMO food is bad for us; but what is bad for us is the problem of market control by one company or certain companies, to control the whole ag industry. That is what I see as dangerous. Someday we may embrace GMO products, but right now the total market control is the danger, and whoever owns the patents has the control. That scares the daylights out of me. I can see regulations coming in that would prohibit someone from planting crops that aren't licensed through a specific company; they lose market access. I've been in meetings at the Department of Agriculture and told them I'm just waiting for their GMO to contaminate my crop, I'll file suit. Then they turn around and say they are waiting for my organic weeds to contaminate their crops. "If you bring a noxious weed into my GMO crops, we are going to sue you." I just say, "Right."

What is the latest on the pharmaceutical crops in Kansas?

We have more than 200 acres of pharmaceutical crops in Kansas on more than 60 farms, and no one knows where they are. Our Secretary of Agriculture does not know where they are! They aren't monitored because EPA is overseeing it, and if there is an outbreak, they will notify us. That they are allowed to self-monitor is just blatant arrogance. They say it is because the quality of their crop is so important, and security is so necessary. They are worried about "someone stealing the science," but off the record, they are also worried about vandalism.

What are some things that sustainable agriculture hasn't done so well? Mistakes that have been made?

I question in the long run, whether embracing a national organic program will be beneficial to sustainable agriculture. All at once, we've institutionalized ourselves.

The bad thing, which is also the good thing, is that the people that have embraced sustainable agriculture are the radicals and the eccentrics of society. I think that is neat, but that stigma is hard for conventional ag to embrace. I don't know that anyone in sustainable ag cares about that. It isn't something they've done wrong. The type of character who's willing to try something out of the ordinary isn't going to be your normal person. [Laughter] Since they are almost all my friends, I'd better watch what I say!

The Land Grant Universities

My brother in law told me "I don't need extension. Chemical companies can tell me everything I need to know." But what about needing a neutral third party, taking a step back and evaluating the whole thing, how do you know what is right? That relationship is being negated with the way Extension is going now; they are losing credibility.

Do you have any thoughts about the future of the Land Grant Universities?

About six or seven years ago, I sat down after one of those farm profitability conferences, to eat supper with Barry Flinchbaugh, an agricultural economist from K-State, and a few others. He was going on and on about the future picture of ag. I said, "This picture you show doesn't have any place in it for the Land Grant universities. What do farmers/society need you for?" He said, "That is a good point. There isn't a place if we don't evolve as an organization or as a structure." We're very close to that right now, I think.

The original concept of the Land Grant universities is worthwhile, to educate the uneducated. I'm just so proud of my children going through college; they have wonderful opportunities.

I think we could do a lot better in sustainable ag if we would work more closely with the land grant institutions. I realize the frustrations of that; it is such a rock wall. Of course, you know that, but until there is rock hard data that supports what we already know, we are going to have a hard time making it with the indoctrinated farmer of today.

At one of the farm profit conferences that K-State puts on, a K-State ag economist said that if you want to see someone making money in farming, it's the farmer in organic. Later I was at an organic field day, and told them what I'd heard at K-State, and they got very quiet. Later, one of them told me that they are having success by their quietness. "If all at once we become mainstream, organic won't have the markets, or be profitable." They don't want K-State talking about how profitable it is. We've allowed ourselves to be at the end of the food chain. Nobody recognizes how much, how such a small amount of the food dollar in the food item goes to the farmer. Nobody takes it seriously. You know as well as I do, something like $0.03 of wheat is in a loaf of bread.

The Downside of Sustainable Agriculture

The real down side of sustainable agriculture is that most of the people who are good at producing organic crops are not good at selling organic crops. So, you have a bunch of anti-social people out there, waiting for people to come to their door wanting their product. This independence and eccentricity is a limiting factor and it limits sales. We need a sales force. The KOP (Kansas Organic Producers Co-op) is trying to do that, but with bulk commodities, not the specialty commodities. You need to find people who are magicians at marketing. Pay them a king's fortune in salary, and turn them loose to perform.

At a meeting with other industry leaders in Topeka and Governor Sebelius, I brought up the question, "Why aren't we servicing the food system, selling to schools, prisons and hospitals?" The main reason is that we don't' have a supply system to supply the state. We don't have a structure for this yet, a marketing system.

Was she receptive to that idea?

I think we can make some headway there. All of industry was around the table, and this was a weird comment among many normal comments. Her relationship with Sarah Dean and Dan Nagenagst can go a long way toward that if you just want to talk the politics, but we need a supply structure to give it some credibility. I suppose, you know more about this in other parts of the country. I would guess they are doing better at it in California.

Chapter 8—Donn Teske

Our Farmers Union state president in Pennsylvania is a tomato grower, has a lot of respect for Campbell Soup. So why couldn't we hire an executive from Campbell Soup, find out how they do that, contract with growers, contract for the best, buy the best, work in the sustainable ag industry. Why do we have to be so individualized? This story, that has been rammed down our throat, "The independent farmer, nobody is going to tell you what to do by god," that is going to be our downfall. We don't do anything together any more. It used to be we helped our neighbors. If the barn burnt down, we helped rebuild a barn. Now we hope they grow broke so we can buy the barn. We're a little bit like that in sustainable ag. I think the more we can work together the better we'll be. Sometimes we work together; we saw that with the national organic standards. USDA was inundated with letters concerned about problems with the standards, so it isn't that we aren't out there, or have the same beliefs. We just aren't organized; hear the lion roar . . .

Policy Issues and Sustainable Agriculture

As far as policy, I've always been somewhat amazed at how easily the sustainable ag people can work within the existing policy. Just my own farm, converting to organics, I didn't run into any problem to speak of with government payments. I've run into a few problems with my organic pumpkins and the limits on the vegetable acres, taking away from base acres. When you get to bulk commodities and insurance, some companies that sell crop insurance categorize organic farms with less coverage. That could be a barrier eventually, a disadvantage.

Right now, we're in the driver's seat in many ways, like with the national animal ID program. If that comes into play, the source-identified food in specialty markets will be less important than it would be if there weren't an ID program. Right now, organic food is identified and traceable, and nothing else is.

Is that one of the advantages of organic right now?

The more people that enter that program could be a threat to sustainable ag, or to what we've built up to this point. Then it gets back to, you better have been developing a relationship between the producer and consumer to keep the farmer on the farm. As the animal ID system evolved, as more corporate farms get into sustainable ag, you are going to see some new creature arise from the depths that will be above organic.

What do think about the future of Ag? Is there a future in Ag?

It depends on what your definition of ag is. There are two ironies out there right now the way I see it. Farmers are falling by the wayside right and left, but all of the land is being farmed. What's been tilled is still being tilled, and even though there are one-quarter the dairy farmers, there is the same number of cows in Kansas. Until land starts not to be farmed, or until people start to go hungry, these trends will go on. I don't know that when they run out of water in western Kansas, and all that land that was valued as irrigated, becomes valued as dryland, and the banks call in the notes because farmers don't have the equity—

this whole buffalo commons things could actually happen. A depression in western Kansas will turn it back into a wasteland. It will happen first on the High Plains, where it is hardest to make a living, and then work east.

What do you think about agriculture as a way to make a living, for kids coming out of college?

It depends on what your goals in life are. If it's just the economic base, of course not, you have to want it. I was told when I graduated from high school in 1973, that with population growth, we need to feed them all, its going to be a golden age of ag, and we've been hearing that that ever since. I think that's what is really scary to me, like the ground water contamination in the county—at what point do we contaminate it with a product that's so horrible that we take production acres out of the system, land out of production. The land is mass sprayed now with various chemicals. If that were ever the case, that one of the chemicals caused permanent damage, it would hit massive amount of acres.

Is there policy work with Farmers Union that you are excited bout?

Most of the time, we are just throwing water on the fire rather than making creative or long-term, creative, innovative changes. I'm in Washington D.C. tomorrow morning; we're now fighting CAFTA[5]. We should have learned our lesson with NAFTA. There is still the same influence of industry every time I'm in Washington D.C. The control of our government structure and policy by private industry is so obviously evident at every step; it's just sickening. That is our biggest threat in sustainable agriculture. If we ever get successful enough to take a significant niche out of the economy, the structure of the status quo will not embrace that warmly. What I feel good about, is that I think the old, old successes such as developing the co-ops of the '20s, '30s and '40s is going to start coming into play again. We need a cooperative marketing structure in sustainable ag to have success. That particular skill in Farmers Union could become more beneficial all the time.

Other Farmers Union Accomplishments

In state, we hosted a food systems conference in Lindsborg, I'm very proud of that. We had great results. We had a future of farming policy forum at University of Kansas (KU) that went really well. The irony is that at K-State only five people came, while at KU we had more than 100 people. This tells you what direction you should be working. This topic was very much more embraced by faculty and staff at KU, which I'm sure doesn't surprise you.

I actually have some tools within Farmers Union to allow me to expand myself and embrace some things like that, and try to do some creative things. It's taken me a few years to feel comfortable enough to expand its horizons a little bit beyond what they had before. I hope we can have some success at that.

It may just be nothing more than pouring water on the fire in Topeka, but it seems like when I'm down there I generally feel like I'm a conscience man. You know it is wrong; I'm just the one that will tell you it is wrong. Once in a while you'll have a small success. Industry is pretty much in bed with the political system.

Chapter 8—Donn Teske

I don't know how it will ever be straightened out, I honestly don't. I think if you want to go back in to the defining moment was when they backed up on the corporation laws. Thomas Jefferson was extremely anti-corporation. It was explicit in the first constitution; there would be no corporations in the United States of America. Then Delaware was allowed to do their thing….and from then on, once you have a corporate structure, you loose your conscience, corporations have no soul. You get into corporation, and doing the right thing is not the goal of the corporation. The corporate goal it is to make a profits for the investors. That's two different things.

That would take some undoing? (laughter)…

I don't know if sustainable ag could tackle that one. We're getting awful far out, but I really don't know if sustainable ag will make a niche in politics in Washington, D.C. I don't think you can do it that way. The only way it will happen in society is when the consumer just flat-ass demands it. If the food just gets so bad, and there is another source out there they are willing to buy, that might make a difference. Right now it is so controlled that society is being taught that an inferior food product is a superior food product for you just because it can go through a machine. No ones like a hard tomato, but now we are taught that hard tomatoes are good, but they need the mechanical picker. All this mass junk on TV that teaches us what we are supposed to like and not like…..that gets old.

Big Picture

What about campaign finance reform? Would that help?
I think that only if it was serious, that would make a difference, but I don't think it will get to the point of being sincere enough and not just for show. They'll always put enough loopholes in it so they can make their bucks. It's a corrupted system from the word go. It didn't used to be that if you went into politics you came home wealthy. It used to be a vow of poverty to go into politics. Now nobody comes back from DC without being a millionaire. I don't think that I can conceive of the thought of truly campaign finance reform really happening.

What do you think about WTO, [6] NAFTA, etc global agriculture? I hate to see local farmers' prices undercut by China, but farmers in African and China need to make a living too?
I'm very pro-protectionist. Tariffs have been within society since the beginning of man. That's how one country maintains a standard of living different than another country. We funded our entire country on tariff income until the civil war. The 1890's were when we started income tax except for paying for the civil war. Look at the trade deficit we have now. It's obvious we can't keep doing what we are doing.

I'm not saying its fair to the other countries, but you can't expect a farmer who has to spend $40,000 per year to live here in the social structure of this system to compete against a farmer who can operate with $1000 per year. You can't correlate the two.

Farming in the Dark

At the same time, I tell people that modern technology has not been good for farmers. It may have been good for faming, or for consumers, and that is debatable....but not farmers. We're dying out at a fast rate. The over supply produced by artificial fertilizers put on, put in our ground water, has kept us in poverty and is keeping the world in poverty, so how can that be good?
Individual corporations are making a massive amount of profit, and the rest of the world is suffering because it is producing an over supply.
I understand what we are doing to the cotton farmers in other countries with our cotton subsidies here...and I think that is wrong. I'm not so sure that tighter borders at both ends couldn't help that both ways. Why are other countries allowing our cheap cotton to flood their markets?
So why don't they control their border? Control imports?
That's the way I would do it. Then if you are over producing within the county, you can use production controls. That way you have a buffer during famine years. I'm sure its broader than that, but I'm kind of a simple guy, I just look at the nuts and bolts of it. This world economy we talk about <u>can't</u> be good for everybody. We're different societies and different civilizations.

Other Things the Public Should Know About

From your point of view, what are other important things people should know about the topics we've discussed?

I think an awful lot of people have a misconception of what farming is. One of the Kansas legislators on the ag committee considers herself a farmer, because her family owns farms. Her husband is a doctor, and they live in town, so they think they farm, but my comment is that until you are out there pulling a calf, and you're out there watching a cow die and it's not just an economic loss but you feel bad for that animal as a soul, then you don't know what farming is. Just putting dollars and sense on farming does not do it justice. When you are a steward of nature, you want to do the best thing for it. It isn't necessarily that you are mandated to do it by law—maybe you'll never be truly compensated for what it is worth. My goal or effort as an individual is to try to keep a system, or reinstall a system that will allow farmers to keep making a living doing what they are good at doing. Once we've lost this generation, I don't know where agriculture is going to find the expertise to come back.

A couple of things happening on the ag scene that I think spells more doom for the future of agriculture than anything we've talked about to this point: 1) modern technology has allowed farmers to farm longer to an older age, 80- and 90- year old farmers still farming the same amount they farmed when they were 40. So you've lost a generation that has had to go off the farm to make a living that otherwise would have taken over that farm operation and 2) land values have escalated beyond what you can pay for with production. People talk about baby boomers, agriculture and a transition of land. As those estates come into play, there are generally four or five siblings; one has been farming, the others haven't. If the goal is to keep that one farming, you have to keep that intact.

Chapter 8—Donn Teske

How do you pay for that at market price? You can't. The one farming can't afford to buy out the others. So, do the siblings sacrifice their share of the inheritance to allow their brother to continue to make a below standard of living? He has to keep farming in poverty. It used to be that your share of the inheritance was enough to buy the others out, even pay more than market prices, pay open market prices but get part of it back. That no longer pertains. Pasture that can't pay for itself at $300 per acre is bringing $900 per acre now. It looks like there no end in sight. I think we'll see farms broken up and fragmented at a faster rate than ever.

The other downside of that is high land values are allowing farmers to farm longer than they should, based on cash flow and other criteria. The high land value allows them to get the loans because they have collateral. The only way to realize that collateral is to liquidate, but no farmer wants to liquidate his or her land. But if he doesn't have an operation that can cash flow to pay that debt back, he keeps building and building. Then he has to sell real estate to bring the ratios back into quotas (standard financial operating debt to asset ratios). It's a never-ending cycle. It isn't fair to the soul—it eats on people. High land values are not a benefit to ag, any way you look at it. It's going to mess up generational changes.

In the 1980s, our bank in Onaga went bankrupt probably because of loans based on high land prices. When land prices dropped, they didn't have the security to secure the bank, so they went bankrupt as a bank. That happened to quite a few banks across Kansas. With my financial consulting, I've seen some banks have made more loans with land values again, just to keep people in the community, to have them as part of society. If we have another crash, the security isn't there. I'm scared for the future.

How much do you think farmers understand about the forces, like the ones we've talked about, that affect them?

Actually, very little.

Do they know their break-even price? Do they know when they are making a profit?

Not very many. They all say they do, but they really don't. Every chemical company and fertilizer dealer out there has a spreadsheet that they do for them to show what they need to make money on this, all the input costs, but it's not bringing in the cost of machinery, family living, secondary expenses. It only shows the variable costs, not the net. The spreadsheets only show them how they can spend money, so they get lulled into this sense that they do know their profitability and break even, but they're not looking at the whole picture.

I got into a big argument back when I was milking. I called Washington, D.C. and talked to a congressional aid. He was a local person from south of Wamego. I was telling him my break-even price for dairy. He said that sounds high compared to all the figures I've heard, and I said, "That's my real break-even, that's what I need to support my family at home, make a living, service debts… If it's wrong, it's wrong, but that's what I need, and that's not where you're at!"

Farming in the Dark

We've lost control of our markets, and so we're not being paid according to society. It's a parity issue. [parity is a comparison of current market prices to the price received in the 1920's, relative to inflation and increased cost of living], We get into arguments within Farmers Union whether we should put parity in our publication every time, on our second page. It came to a vote one time, and we voted to keep it.

I think one thing you should do for a little bit of your data is to look into the Farm Management guidebooks, [7] where they break down profitability of example crop and livestock budget spreadsheets. Almost none of them come out positive. It gives you all kind of fodder to work with.

I looked at some of those when I first moved back to Kansas, and was surprised how almost none of them come out positive, without any net income.

And if you aren't making any money, you better buy a bigger farm! [Laughter.] I run into farmers all the time, they say, "I want to do this." It's a plan for expansion. I say, "Are you sure you want to do that?" I do the analysis using K-State's software and figures and print it out for them, and then they say, "Whoa—this is K-State saying this? Whoa . . . !"

Closing Comments

Are there any other things that are important? What did I forget to ask?

One thing I keep going on about is the society we are evolving into where we have low rural population and high urban population. Is that good for society in the long run? If you look at the demographics during WWII, those farm boys did a lot for WWII. It seems like they bled off rural America for a lot of the work force. They have been the most dependable work force, they're not afraid to work; they have a strong work ethic. What does that bode for the next generation? I think there is a sense of pride, ownership and responsibility to society that is indoctrinated into you when you come from a rural environment. There is nobody else to blame. If you are living in town you can blame just about anybody for just about anything, but in the country there is no one to blame except yourself; you have to look at yourself first.

Some people reading this may ask why we even need rural communities? What is the point of trying to keep them alive?

There is something in our souls that yearn for that, even in urban areas you see fragmentation into some kind of social structure that is smaller. In our small towns, in our part of the country, there are fewer empty houses. Small towns are being repopulated. People are retiring back into these towns. They seem to have a need for what small towns have to offer. If you take that a step further, what kind of society would we need to re-establish rural communities every seven miles—$10 dollar gas? What needs could be supplied by a rural community?

What would happen if we get a Ted Turner that would all at once turn into a sustainable ag fanatic? That sounds weird or patronizing, but we are probably not that far from it.

Well, we have Paul Newman organic salad dressing now.

Chapter 8—Donn Teske

What would happen if all at once this wealthy part of society said that it needed to be this way? It could change! Right now the dollar speaks. We say that right now, but it has always been that way. Wealth speaks, that's the way it is. I come from peasant stock, and I think I'm still there.

Notes

[1]Farmers Union – Founded nationally in 1902, and in Kansas in 1907, to help the family farmer address profitability issues and monopolistic practices during the American industrial revolution. With chapters in 26 states and national membership of 250,000, continues its mission to protect and enhance the economic well-being and quality of life for farmers and ranchers. This includes working with consumers to promote a quality domestic supply of safe food. http://www.kansasfu.org/

[2]The Kansas Rural Center is a non-profit organization that promotes the long-term health of the land and its people through research, education, and advocacy. The River Friendly Farm Program is a whole farm planning and assessment tool developed jointly between KSU and the Kansas Rural Center, promoted through the Clean Water Farms program by the KRC. See http://www.kansasruralcenter.org.

[3]The Kansas Center for Sustainable Agriculture at K-State, (KCSAAC) was established by Senate Bill 534, passed by the 2000 State Legislature, out of concern for the survival of small farms in Kansas. The Center works in partnership with state and federal agencies, nonprofit organizations, environmental groups and producer organizations to assist family farmers and ranchers to boost farm profitability, protect natural resources and enhance rural communities. http://www.kansassustainableag.org.

[4]USDA/SARE [4]The USDA SARE (Sustainable Agriculture Research and Education) program funds research and education through university, non-profit, and farmer competitive grants programs. See www.sare.org

[5]CAFTA (Central America Free Trade Agreement). Passed in 2005, follows the model of its predecessor, NAFTA (North America Free Trade Agreement – passed in 1993) which linked the U.S., Canada, and Mexico. Removes trade barriers for U.S. companies selling products to the countries in the agreement, but also removes some trade barriers (tariffs) that up until now had provided some protection to U.S. farmers and farm prices. http://www.ustr.gov/Trade_Agreements/Bilateral/CAFTA/Briefing_Book/Sectio n_Index.html

[6]WTO World Trade Organization ‾ delegates meet periodically to negotiate removal of tariffs and other non-tariff barriers to trade between countries. This includes both agricultural and non-agricultural products.

[7]K-State Farm Management – includes twenty agricultural economists who are faculty members in the department of ag economics. They work cooperatively with farm families to provide financial management information to members for use in decision making. The association also publishes updated fact sheets with current production costs and returns for some of the major crop and livestock systems in Kansas. http://www.agecon.ksu.edu/kfma/

Discussion Questions:

1. Do you think parity price comparisons are of value? Why or why not? What would be a better way to determine the worth of agricultural products in a changing economy?
2. Were you surprised at how blatant the dirty politics can be according to Donn's experience? Is there a way to be active in politics and not have to get "down in the mud?"
3. Donn mentioned a downside of sustainable ag and niche marketing as having people good at growing but not at marketing. How could this dilemma be resolved? Would more local markets for farm products help this situation?
4. In this interview, the future of sustainable ag appears to be threatened by a bleak future of agriculture in general. Donn suggest that the next generation will be worse off than this one – including concerns about food security, food safety, and the high price of land. Do you agree with this scenario? What could be done to reverse or change these trends?
5. The issue of ethics came up several times in this interview, and Donn suggests that the ability to create a corporation, which are legal entities without a "soul" are part of the problem. Do you agree or disagree? Why or why not?

9

Jackie Keller
Mto-sa-qua Farm, Topeka, Kansas

"I do love being out there, being on the tractor . . . seeing the soybeans get chest-high. . . .the gratification to be able to do that without chemicals."

Jackie and I share several things. We are about the same age; we both grew up in Kansas and we both left for several years to pursue our education before retuning to the state. We also re-discovered our farming roots and passion for farming. Differences are that Jackie's farm is full time grain, while mine is part time vegetable; she is farming the same farm that her parents' farmed. Her farm is named "Mto-sa-qua," the Native American woman's name on the land deed when she traced the history of the farm.

We both were exposed to new ideas while living out of state, she on the west coast, and myself on the east. We learned about sustainable agriculture through our coursework, but also through interactions with the organizations that have been leading the way. I worked for Rodale in the East, and she worked for organizations like Food First and Global Exchange in the West.

This afternoon we are sitting on the back patio of her mother's home, next door to Jackie's house. We just toured the vegetable garden. On another visit, I had a chance to see her crop fields, nicely laid out in strips on 108 acres of cropland, and nine acres of grass enrolled in the CRP (Conservation Reserve Program) on a horseshoe bend tucked in next to Mission Creek, on very good river-bottom soil. She has one field that is difficult to access, in hay right now, but the others are literally right outside her back door. There are sheds for storing equipment, and there was a tractor, rotary hoe and a grain truck to start with when she returned to Kansas, but she also had to scout around for other pieces of equipment for her crop rotation. As she mentions in the interview, she also lacks grain storage, an essential item for organic farms that must document how and where grain is stored, and keep the quality high to obtain the premium prices.

Besides farming, Jackie is active in her local OCIA (Organic Crop Improvement Association) chapter, and is familiar with and supportive of the work of the Kansas Rural Center and the Land Institute in Kansas. She and Nancy Vogelsberg-Busch have also joined me on a canoe trip or two on the nearby Kansas River when we can find a day when all three of us can get away at the same time. It is late August; the geese are already starting to gather. We hear them a few times overhead in the background.

Farming in the Dark

How did you get involved in sustainable agriculture?

I was in California going to San Francisco State University, in the international relations graduate program. I just really gravitated towards world politics. Through studying Cuba's sustainable agriculture movement, it circled me around to my roots, of growing up on the farm here in Topeka, Kansas. I decided to write my thesis[1] on sustainable agriculture, and really delved into the problems of industrial agriculture. Just being in California, around the people I chose to learn from, were sustainable ag minded. The organic farming movement there in general was thriving. I decided that was the only way to go. I moved back here after completing my degree in 2000. Just very gradually started to convert or transition my parents 108 acres of crop ground to organic. I used the Kansas Rural Center's[2] Clean Water whole farm program to help me get started, along with a $3,500 cost share to implement the legume cover crops. That really showed my parents at the time that there were people out there willing to work with me. Since I hadn't farmed, period, at all, except helping my dad very little growing up, that was a tremendous resource to have Ed Reznicek come out and walk the farm, and give me a rotation plan. That was a good tool, filling out the River Friendly Farm notebook[3]. I was able to get half the farm certified last year, 2004, for alfalfa and wheat. This year the whole farm will be certified, the milo and soybeans added to the hay, red clover and wheat.

Back to your California days, was there any one person or place that was a particular influence?

There were several. I took an undergrad agricultural class and we toured several farms. One rice farmer was converting nine percent of the farm to organic. We went to a dairy and saw how the milk cows couldn't walk; they were falling down trying to walk because their udders were so big. That was a conventional dairy. They were consolidating, getting bigger, bigger, bigger, because of the milk subsidies. We went to Harris ranch, where at that time there were 100,000 cows in a very small area.

I was learning about the water subsidies that California has for huge conventional growers and cotton. I went to the Asilomar Eco-Farm[4] conference in the fall of 1999, and that December to protest the WTO in Seattle and to see the people that I really enjoy speaking, such as Kevin Danaher from Global Exchange.[5] I did some work for Global Exchange too, while I was in San Francisco. That was very educational. Even though they aren't specifically agriculture, they try to help peasant farmers in Third world countries. I did an internship at Food First[6]. I dealt with USDA's research. I got on the Internet and found out just how many tariffs and quotas in Africa were hurting African countries with exports and imports. I really liked Anuradha Mittal[7] and Vandana Shiva[8], seeing her speak, and being exposed to her books, really extremely eye opening. Oh, and I can't forget John Jeavons[9] Bio-Intensive Mini-farming conference that I attended right before I moved back. I went up to Willis to their farm and toured it. I did a school garden project in the city, tried to help the Cesar Chavez elementary school start gardens there, did worm bins.

So you were living in the city while you were learning about agriculture?

Yes. And Pesticide Watch[10], I worked for them. When I worked for the Department of ag it was on integrated pest management. Fortunately, it then moved to the Department of the Environment. It was termed pesticide reduction at that point. I also went to Santa Barbara to a Pesticide Watch conference. We heard Michael Ableman[11] and Mark Ritchie[12] speak about the trade issues, about Monsanto being such an evil company. It was very educational, great exposure.

Given your history and exposure to sustainable ag and now practicing it on your farm, what do you consider some of the major achievements or milestones?

Well, I think that just making the organic food industry more mainstream, more popular, so now people in the Midwest don't scoff at the word organic. I think that California is the leader and that makes a big difference. I definitely feel isolated in many respects, being an organic farmer in Kansas, in my immediate area. As far as marketing our organic grain, I think the California people have helped tremendously, educating consumers. There is a very secure market for organic products.

Do you think it is because of California Certified Organic Farmers[13] (CCOF) or Organic Trade Association[14] (OTA) or the whole movement?

I'm not sure, but I think CCOF had a lot to do with it, being the granddaddy certifier. It's not good enough to just say, "Oh yeah, it's organic," but having to have that paper trail that lends certified organic such credibility and product differentiation. That's key.

Have implementing and having the NOP (National Organic Program) standards[15] had a positive effect, adding that next layer of authorization at the national level?

I think it has definitely been a mixed bag. Now there are twice as many certifying agencies as there were before NOP. It just makes me wonder how credible they are going to be and how strictly the standards are going to be adhered to. As far as getting it out there, making it more popular, giving it more exposure, is of course, a big help. Getting the federal reimbursement money for certification, I think that's really a help. I've been able to apply and for the next two years, the money is there. We get up to $500, or 75 percent of certification fees paid. I think that's a big step.

Have any of the other federal farm programs[16] helped you in any way?

My parents have always been recipients of USDA's farm program. I went ahead and kept that process going. I get somewhere around $1,300 a year, plus maybe $700 for CRP, it is not a huge chunk. [Laughing] I'm a little ways off the cap, the $250 grand!

Are you in a watershed eligible for the CSP (Conservation Security Program)?

Not yet, and that is another issue, Jim French[17] has said that it is a good thing not to be yet. He says that organic farmers have been being slammed for tillage. That's another frustration, trying to educate people on the detrimental effects of no-till. I look at my field after four inches of rain, with no water standing in it. I drive down the road to my conventional neighbor, with huge, huge equipment, and see water just standing, or running off into the ditch, with chemicals along with it. Being penalized for either losing carbon or the soil erosion, which yes, can be a problem. I have CRP grass all around, buffer strips all around my fields to keep anything from going into the creek. I just really try to keep the ground covered as much as possible. If it is healthy soil, I think the water tends to soak in, as I saw yesterday.

Just looking at your garden plots, the water has soaked in where you disked, but where you hadn't there was water standing, the asparagus area.

It does help infiltration, feeding those microorganisms is key.

One thing, my sister in San Francisco said they just couldn't find enough organic meat. The supply of organic beef is just not there. That makes it hopeful to me to have a market for my grain, or better yet, my hay; and the organic dairy in Texas that contracts with Organic Valley[18] is buying our hay. I think that is a big step. There is one in Dallas, which is not that far away. I see that milk in my Dillons grocery store down the road. I think that is a huge, huge step, that and seeing the little organic areas in the store. I'd rather buy local, personally, than food shipped hundreds of miles, but it does depend on what it is. Milk, I can't buy locally, organic anyway, yet. Then the growth of the farmers' market is a big success. The fact that the organic industry is still growing at 20 percent per year, and it just seems that the organic farmers that are my friends, they're selling their products!

What do you see as failures or shortcomings of sustainable ag or of the movement?

The thing that pops in my head first, very specific is helping with more technical assistance to beginning farmers like myself, or transitioning farmers, to make it not so difficult and frustrating. I think my situation is so different, since I didn't farm before; I have a lot more hardship as far as learning the specific details, the day-to-day operation part with the machinery and so forth. I'd say the biggest of course is the fact that we have the same problem as with conventional farmers. The small farmer is squeezed out now that we have huge vegetable or produce corporations getting into the organic industry. We already saw what happened with soybeans, when they were at $20 per bushel, everybody jumped in. I just signed a contract for $13 per bushel for soybeans, for feed grade. The price for food grade is $16, but because I don't have storage, I sell them out of the field as feed. Anyway, when everybody jumps in, lowering the price, the same thing with beef, we just have to get together and stay together on pricing and stick to it, and not allow big corporations to lower the market price.

Have some organizations been more helpful to you than others, like the Kansas Organic Producers[19] (KOP) and the Organic Crop Improvement Association[20] (OCIA)?

It's just more word of mouth. Organic farmer friends or others that are in the OCIA chapter, I talk to them. In terms of selling, KOP is really my only resource. This is my first year to have row crops certified. I'm pretty new at the whole marketing thing. Something might come from advertising in the OCIA newsletter. There have been conferences at K-State. I'm just gathering information from here and there. I've just been marketing through KOP, although I did put my hay on a website hay exchange, and I've gotten many calls from that, but only one of 10 was interested in organic, but no sales yet. That's the only marketing, except for just talking to other farmers on the phone, getting a few leads that way.

One thing on my list of things we could have done better is the economic side. Can people really make a living at this? Is sustainable agriculture sustainable for the people that are out there doing it? I know you are just starting, and there is a kind of ramping up process, but if you were producing certified crops on most of your acres, and KOP was able to get you markets, are prices where they need to be for you to make a living at this, or do we still have a long ways to go?

I think we have a long ways to go. To me, it's just educating the consumer about why they should spend more money on their food. Cancer is just such a big industry in this country. They are fighting, brainwashing every day with advertisements for things that are making them sick, and they don't know it. They don't care because they don't know. They are so brainwashed into thinking the only way our country can keep growing economically is the cheap food policy.

Is there a way for sustainable ag to contribute to consumer education?

Yes, I think so. If there could just be more information at the local, mainstream grocery stores, more articles in the paper, definitely it could be done. A problem with that is farmers are just busy farming, so they don't really see how they can do it themselves, and they aren't experienced at doing that, if an organization could do that. There are websites out there, but there is this big gap between people who are sustainably minded and who are going to go after that information. Then there are the other people who have not a clue about what sustainable ag is or why they should spend so much more money on organic. The information is not going to find them if they are not going to go after it.

Are there other things organic should address?

Another thing is the difference between the natural and organic beef, why certified organic is more expensive. I still see this lumping together of natural and organic. The word natural, what does that mean, what is allowed on a label? Other than organic, getting something on a label is pretty easy. When you reach certified organic, it is very, very difficult and people don't know that. The Merc[21] in Lawrence, for example, could educate their consumers on what the difference is. They put a lot of emphasis on organic produce, and a lot of other organic products, but when they get to the meat, they seem to slack off on educating why organic meat is so much more expensive than natural. Consumers don't even know the big difference between organic and natural. What is natural? It could be anything. No one goes and inspects. Sure violations can occur with organic as well, but it is a lot less likely when everything has to be documented; everything that cow touches is documented.

A colleague at K-State recently sent me a question asking how many farmers cheat, and call their stuff organic when it really isn't. She sent an article about farmers in England who were saying that their neighbors were spraying at night. My experience with OCIA and the certification process shows that it would be very difficult to do that here. People really want organic to mean something. You are the OCIA Chapter Administrator, what would be your comment to a consumer who wants to know if it is really organic?

A certified inspector has to be approved through OCIA, he or she has to go through their certification course. They come to the farm once a year, it's not very much, but they can get a good idea of what is transpiring on that farm. They come in July or August, so of course that's when weeds are growing their best. The crops are too. My milo has a bug eating it. That's the main thing, along with the documentation. It's not just the paperwork required at the time of inspection, but all year long. A calendar has to be kept when a crop is cultivated, the storage situation, so it's a continual process. Educated people are going to know when someone doesn't understand a bin log, or a bill of lading or a transaction certificate that OCIA requires.

When I was on the OCIA review committee, I remember that split, or parallel production[22] wasn't allowed.

It is allowed now, but it is just a paperwork nightmare. With OCIA, it is very, very tedious. When I was on the committee, we only allowed it if there was product differentiation, for example black hylum vs. clear hylum soybeans, or two different varieties of tomato, but it could be the same crop.

That reminds me, with OCIA there are chapters. I have 22 producers in my Eastern Kansas chapter. That serves as a farmer support group. We have a certification review committee that meets before the inspector comes. We go over every application. If there are any red flags, we can help that producer address the issue before the inspection. Then the applications are reviewed again after the inspection, before they go up to International, the office that gives final approval. It really helps that farmer, to keep the farmer certified, to keep any problems from coming up, any surprises, or things that would be too late for him or her to address. Within the chapter, if anybody has reason to believe someone is in violation, we can have "integrity visits," where a farmer is surprised with a visit; someone from OCIA International comes down.

It's not a peer visit, but someone from the International office.

That's the way I understand it. I think that is a good buffer.

That's a nice euphemism, an integrity visit! [Laughter] Have you had to do that within your chapter?

No, not at all. We have a good group of farmers. I heard of a group of soybean growers in Missouri where it was clear they were in violation. I don't know if they were spraying, but it seemed to be a result of the $20 bushel price, people just trying to see what they could get away with.

When I was on the review board, the only issue that came up was that some vegetable growers interpreted the soil building guidelines as including compost, and didn't have a rotation with cover crops. The inspector that year was big on cover crops and rotations, so we had to go out, re-inspect those farms and look at their soil-building plan. Also, some soybean growers thought that a rotation was beans after beans, and that clearly wasn't acceptable.

Chapter 9—Jackie Keller

We have rotation guidelines that the chapter has set up. Continuous cropping is not allowed except in a drought situation. In western Kansas, you might have to do wheat on wheat. Once again, the paperwork is just so tedious. Every fourth year a 17-page application has to be filled out; there are just so many detailed questions.

What do you see as the biggest challenges that we face now, and into the next 25 years?

I think the chemical companies are my biggest fear; they are so powerful. If we keep growing and growing, what they would do to put us out of business? Like the canola farmer in Canada, that Monsanto[23] has gone after—getting to the point where we make agri-business nervous, or beyond nervous, and mad, them preventing us from educating the consumer—and the chemical companies that have so much to do with the cancer industry in this country, creating the chemicals and the cancer medications at the same time. And what about the pharmaceutical crops and the genetic engineering—releasing those things into the environment and our seed supply? It's big business for them. Of course, they aren't going to want people eating right, and doing things to prevent them from getting cancer.

I don't have storage yet. With my last $5,000, do I buy grain bins? Do I pay people to weed? What's the best way to use that money? Without being able to save my own seed, I'm at the mercy of NC+ organic, or other the seed companies, or maybe some other farmers too. Another concern is the round-up ready alfalfa being released. These things could be cross-pollinated, it is bee pollinated. It has been developed, but it hasn't been released yet. So everyone who saves seed could be the next Round-up Percy Shemeizers?[23] I think they are holding back right now from releasing the GMO wheat. I did hear though that there are 200 test plots of pharmaceutical corn in Kansas and no one except the company knows where they are, even the secretary of agriculture. That seems risky to me; it doesn't seem democratic.

Are there any other elephants in the living room we should discuss?

Fuel. That is hurting everybody. It's hurting conventional farmers, hindering them to do their work. It's certainly hurting organic. I have to go out there and do twice as many passes as they do. But you'll save fossil fuel costs by not applying nitrogen fertilizer.

What else do you want to say about organic farming for you personally, as a woman, returning to Kansas? What do you like, not like, what are the joyful things, hard things?

I'm definitely passionate about it, and you have to be that way to do it. There are so many frustrations. I had to replant 27 acres of milo, and that was difficult. Frustrations? Making mistakes, not having the experience, maybe somebody in a different situation than me, who had been doing this for a lot longer would not make those same mistakes.

I do love being out there, being on the tractor. Seeing the soybeans get chest-high, to get that tall. Just the gratification to be able to do that without chemicals. Farming, as everybody knows is difficult, and to be able to do it without chemicals. You have to have the time to go out there during those very specific windows of time, in between the rains and so forth. Being able to do that is satisfying for me, to have a job situation where I can do that. One thing that disappoints me is not having livestock, not having that experience or background and trying to do it without livestock, without transferring nutrients like they need to be for a true organic farm operation. It seems that other people are doing it, so I think it can be done with soil and weather conditions.

Is that something you see moving into in the future?

Yeah, I think so in the future, but I'm just not ready yet. I'm still struggling with the crop farming, getting the basics. I mentioned the grain bins, some kind of storage, just keeping my equipment up to snuff, getting things sold, some cash flow. I do have a pasture, a quarter section in the Flint Hills, so it really is something to think about. I got a call the other day, somebody wanting organic native grass hay, so I had better think about certifying that. Right now, we just rent it out

Anything else you want to say about farming?

I guess one thing, as you mentioned; you were academia, now you are doing hands-on. I'm in the same boat. I studied it first, but I think that is what really motivates me to do it, studying all those problems, learning about problems with industrial Ag. That's what motivates me to get out there and do it.

I find doing it a much bigger challenge than reading about it. It's much harder than it sounds. We both laughed and agreed on that point.

We spent the next part of the interviewing discussing what was going on with the WTO, and the policy ramifications for the United States, especially policy suggestions by various sustainable ag groups doing lobbying in Washington, D.C. Since this is Jackie's area of expertise, I took the opportunity to update my list of books to read on the subject. Organizations with current information on the topic include the IATP, Global Exchange and Oxfam America[24]. Neither of us had the latest information from the most recent round of negotiations, but discussed how the overall result seems to be a power shift from nation-states controlling their borders to multi-national grain and other food companies having the bulk of the power. We also felt like the shift from "blue-box" payments to "green box" payments, though it seems as though it might limit the U.S. commodity payment programs in some future farm bill, are unlikely to have too much influence in the short term, as the large commodity groups and political inertia are likely to keep these in place for the present.

We wrapped up with one final question: do you think sustainable ag has done a good enough job of dealing with the international trade issues?

How can they, when it is done behind our backs? We can go to Seattle or whereever and protest, but they just continue to make their agreements. I don't know how we can at this point.

I responded by saying that the harm that cotton dumping is doing to Africa reminded me of the PL 480[25] grain dumping in the '70s; deja vu all over again. After lamenting that and other depressing topics like the rising rate of diabetes

in the United States, lack of awareness of the American consumers, we decided to turn our attention to something we could do something about.

"Would you look at my soil tests?"

Jackie had done what many organic farmers do—send her soil tests to two different labs so she can compare the results and the recommendations. "Sure. That is an area where I might actually be useful."

Notes

[1]Jackie's thesis, "The Genesis of Unsustainable Agricultural Processes," for Master of Arts in International Relations, San Francisco State University, May 2000.

[2]The Kansas Rural Center is a non-profit organization that promotes the long-term health of the land and its people through research, education and advocacy. http://www.kansasruralcenter.org

[3]River Friendly Farm Program—a whole farm planning and assessment tool developed jointly between Kansas State University and the Kansas Rural Center. The Kansas Rural Center is still being used in the Clean Water Farms program. See www.oznet.ksu.edu/rff.

[4]Asilomar Eco-Farm Conference, annual conference for more than 20 years covers a wide range of topics to sustainable advocates, students and farmers. http://www.eco-farm.org/efc_07/info.html

[5]Global Exchange—a membership-based, international human rights organization dedicated to promoting social, economic and environmental justice around the world. www.globalexchange.org

[6]Food First—The Institute for Food and Development Policy/Food First shapes how people think by analyzing the root causes of global hunger, poverty, and ecological degradation and developing solutions in partnership with movements working for social change.http://www.foodfirst.org/

[7]Anuradha Mittal, author of several books such as America Needs Human Rights, 1999, Views from the South: The Effects of Globalization and the WTO on Thirds, and Sowing Resistance: The Third World Speaks Out on Genetic Engineering. 2004.

[8]Vandana Shiva—is a physicist, ecofeminist, environmental activist, and author. Her books include (too many to list all) Stolen Harvest: The Hijacking of the Global Food Supply, South End Press, Cambridge Massachusetts, 1999; Water Wars; Privatization, Pollution, and Profit, South End Press, Cambridge Massachusetts, 2002; Globalization's New Wars: Seed, Water and Life Forms Women Unlimited, New Delhi 2005; Earth Democracy; Justice, Sustainability,

and Peace, South End Press, 2005; and Manifestos on the Future of Food and Seed, editor, South End Press 2007.

[9]John Jeavons—has been the Director of the GROW BIOINTENSIVE Mini-Farming program for Ecology Action since 1972. He is the author of How to Grow More Vegetables and Fruits, Nuts, Berries, Grains, and Other Crops Than You Ever Thought Possible On Less Land Than You Can Imagine, 10 Speed Press, 7th Ed, 2006. Frequently teaches workshops in the United States and internationally. http://www.johnjeavons.info/

[10]Pesticide Watch—Pesticide Watch and Pesticide Watch Education Fund provide California communities with the tools they need to protect themselves and the environment from the hazards of pesticides. http://www.pesticidewatch.org/

[11]Michael Ableman—Farmer and author, most recently of Fields of Plenty: A Farmer's Journey in Search of Real Food and the People Who Grow It, Chronicle Books, 2005, From the Good Earth: A Celebration of Growing Food Around the World, HNA Books, 1993, and On Good Land: The Autobiography of an Urban Farmer, (date unknown).

[12]Mark Ritchie—founded the Institute of Agriculture and Trade Policy and served as Executive Director and President of the organization. Mission of IATP is to create environmentally and economically sustainable rural communities and regions through sound agriculture and trade policy, http://www.iatp.org/

[13]CCOF California Certified Organic Farmers. Founded in 1973, was one of the first organizations to provide organic certification. Continues in this role today, along with education, outreach and networking. http://www.ccof.org/

[14]OTA—The Organic Trade Association is a membership-based business association that focuses on the organic business community in North America. OTA's mission is to promote and protect the growth of organic trade to benefit the environment, farmers, the public and the economy. http://www.ota.com/index.html

[15]National Organic Standards, adopted in 2002, see http://www.ams.usda.gov/nosb/index.htm for details.

[16]Federal Farm program payments (such as CRP, commodity payments, etc.) are available to farms of all sizes. A 1999 GAO report documents that farms with gross receipts of more than $250,000 make up 7 percent of farms, but they received 45 percent of payments. Sustainable agriculture advocates and others have proposed that there be some form of payment limitation or income caps to limit the ability of wealthy non-farmers to "farm the government." For data see http://www.taxpayer.net/agriculture/learnmore/govreports/gao.pdf

[17]Jim French, Kansas farmer, has been lobbying on behalf of sustainable ag for Oxfam American, also former employee of Kansas Rural Center. http://www.oxfamamerica.org/

[18]Organic Valley, producer owned cooperative, started out in the upper Midwest, The purpose of the Cooperative Regions of Organic Producer Pools is to create and operate a marketing cooperative that promotes regional farm diversity and economic stability by the means of organic agricultural methods and the sale of certified organic products. http://www.organicvalley.coop/

[19]KOP The Kansas Organic Producers Association (KOP) is a marketing/bargaining cooperative for about 60 organic grain and livestock farmers located primarily in Kansas, with some members also in bordering states. KOP's purpose is to help build markets for organic grain and livestock and to represent it s members in negotiating sales and coordinating deliveries of organic products. KOP was first organized in 1974 as an education association for organic farmers to promote organic agriculture, develop organic certification standards, and establish organic markets. http://www.kansasruralcenter.org/kop.htm

[20]OCIA – Organic Crop Improvement Association, provides certification services in the U.S. and internationally. See http://www.ocia.org/

[21]The Merc (Community Mercantile Food Co-op) in Lawrence, http://communitymercantile.com/

[22]Parallel production – when the same crop is allowed to be grown on the same farm under both organic and conventional conditions, when a farm is in transition for example, and some acres are certified and others aren't yet. The risk is the temptation to supplement organic shipments of product with non-organic; thus the tedious record keeping to track supply coming from each field, bin, etc.

[23] Monsanto—Percy Schmeiser case. See CBS article at http://www.cbc.ca/-news/viewpoint/vp_omalley/20040521.html for more information, or the video, "The Future of Food." http://www.thefutureoffood.com/

[24]PL 480, or Public law 480 allowed for "dumping" of U.S. grains (wheat) in Third world countries, undercutting their local farmers prices and farmers' ability to grow and sell, all the while disguising this as "food aid." This program continues, euphemistically called "food for peace." http://www.usaid.gov/pubs/cp2000/pl480ffp.html

Discussion Questions:

1. What would be good ways for beginning farmers like Jackie to get the help and advice they need to start out in organic farming? Should the Universities be responsible for this, NGO's, neighboring farmers, or all three?

2. What do you see as the pros and cons of "natural" vs. "certified organic" beef for the consumer? Is it better to have a certification system for both, or are there marketing situations where the word of the farmer may be enough?

3. Do you think organic farmers should be penalized in federal farm programs for tillage? Do you think no-till farmers should perhaps be penalized for additional compaction? Or maybe both should be asked to demonstrate their infiltration rates, since these can be farm and field specific?

4. After watching "The Future of Food" or reading material on the connection between chemical companies and pharmaceutical companies, do you think the same industry should be allowed to release chemicals into the environment that cause cancer, and also be allowed to profit from chemo-therapy medications?

5. Do you think an organization like the WTO should have the power to limit the control of nation-states over their borders and farm prices? How would you suggest sustainable ag getting a representative, or a "seat at the table" for the next round of discussions?

10

Kirk Cusick
Whispering Cottonwood Farm, Salina, Kansas

"The way I look at it, I'm a grower of food.
Food is something that everybody has to have, needs to live."

I visited Kirk's farm in Salina in the fall of 2005. We started out sitting in the yard protected by the shade of a row of cottonwood trees lining the south edge of his yard next to the road, and later moved to the front porch as a light rain began to fall. Kirk has been selling produce using the model of a CSA (Community Supported Agriculture[1]). I've known Kirk for several years through his activities selling local food in Salina, working with the Kansas Rural Center in the "Land to Hand Alliance." He was the mentor to Jane Drake, one of my CSA partners when we had a CSA in Manhattan. Now he is teaching farming to a much younger set in Salina, as he explains later in this interview.

Kirk's house is 104 years old, and it was built by Patrick Henry Dolan. Dolan was born in Ireland in 1833, moved to Kansas in 1878, purchased the farm, and served two terms in the Kansas Legislature as a representative from Saline County as a member of the Populist Party.[1] According to a packet of historic papers that Kirk let me borrow, Mr. Dolan was also a delegate to the Farmers National Congress in 1905. The essence of the movement can be summed up in their motto, printed in the New Cambria Times in 1893 (a town near Salina); "Equal rights to all and special privileges to none." Kirk hopes to preserve these historic documents and create a display in his house for visitors and the students he hosts for farm activities.

Kirk's spirit of caring and sharing with his community would probably make Patrick Henry Dolan proud if he were alive today.

About the farm

What was it like starting the CSA?

Probably the biggest joy for me was breaking out on my own; I really think that was the thing. I had a job where I was answering to the boss man, and it was nice to break away from that and to have a sense, depending on how kind nature was to me, that I was in control of my own destiny, where I was going to go from that point.

Farming in the Dark

The other thing about it, well, I was scared. You are always scared, because you don't know for sure if you are going to have enough people to be able to support the farm. You are scared when you first get into it, and I certainly felt that. I also always have felt—I don't want to be grandiose about it—not a calling, but that people do it because they deeply want to do it. Unfortunately, they want to do it so bad, they'll overlook many things that aren't right just so they can continue to do it. The way I look at it, I'm a grower of food. Food is something that everybody has to have, needs to live. If I don't grow food, people won't have food to eat. Farmers have never made that connection; they become a cog in a larger machine. Doing direct marketing, you get a feel for that, that you are right there. People like you, want to talk to you, they want to hear about your product. It just felt good, being a part of that. I've always seen myself as a professional, and I have skills that other people don't have. I utilize those skills in order to provide food, and it felt good being able to utilize these skills.

How did you initially gain those skills?

I've lived here all my life, and by all my life, I mean literally pretty much all my life. I left to go to school at the University of Kansas, and then I came back. We moved here in 1965, when I was three or four years old, so I've been here on the farm 30-some years. It was actually my uncle's house, passed on to my dad, his sister, then to me. It has gone through the generations. My uncle was kind of a gentleman farmer; he worked in town and rented the farm out. It was kind of strange to live on a farm and work for a guy who rented it. Doris Winslow was his name. He was a gentleman that was born to farm, he loved to farm, every hour that he was on the tractor he was thinking, "How could I do this different, do this better?" He gave me a feel for farming, he was really a neat farmer, to love what you are doing, that was really a nice piece of it. That is how I got started.

What did you produce, traditional crops?

Yes it was. I was running the disks and the combine, and the trucks, and that kind of thing. It was wheat. Actually, it was all wheat, maybe a little bit of milo and soybeans. But it also was a different time. It was a time when he didn't use fertilizer; he didn't buy into that idea, still farming naturally. Let me put it this way: sparingly he would use 2,4-D if there were a patch of bindweed, but very sparingly. He farmed old style, even though he was in a new era, still had set-aside, still a steward, even though he was in that much larger commodity era. I learned the commodity side, but you still have to care for the ground, it's not automatic, when you do things in what I consider the "right way."

How did you make that transition to horticultural crops?

I've always gardened. Way back in the corner of the farm, there is a foundation where there used to be a barn. My uncle sold the barn, somebody came, took the barn down and built it somewhere else and so we had a garden where the barn used to be. Things grew like crazy, my dad always had a huge garden and I helped him with that. That is how I got into the horticultural crops. I really got started after college when I moved back here. I started with three people at work; I worked with adults with disabilities at the Occupational Center of Central Kansas. I asked three people, fellow employees. I asked, "You know what, if I grow vegetables organically, with no chemicals, and compost, would you buy them from me?" They said yes, we'd do that. I started with those three people, delivered to them that first year. The second year I went to 15, most still people that I worked with, or people they knew. I held at 15 for or a couple of years then went to 30, then to 70, then to 100. I maxed at a hundred, that's where I held it. Then with the other things I was doing, I started to scale that back. This year I was in the 40 range. I brought it back down because I was doing so much other stuff. That is where I'm at right now.

Are you still doing deliveries or do they pick it up?

I deliver to Prairieland Market, and they pick it up there. It is a storefront that used to be a buying club. It is three years old, and we just hired a manager. It's just a tiny thing, but we are doing well, covering our bills. We sell a lot of local food through the Market; that was the focus of it when it started.

Are you still doing your bills with a weekly statement and a monthly bill?

Yes, it is working out well for me, and I've got people accustomed to that, so I probably won't change. If I was giving advice to someone now, I'd suggest doing the flat fee at the beginning of the season. Doing it monthly is a lot of paperwork, and you have to collect, there are those people that don't pay, that make it kind of difficult. You avoid that with the up-front fee. The up-front fee commits people for the whole season. I also think that it never hurts to have money at the beginning of the season. In my system, you are paying per product, the price of the potatoes for example, you get the price of each item and at the end of the month, you get a statement. I charged a $50 startup fee, non-refundable, you need paper, ink, you spend one of those fee amounts just buying one ink cartridge. I wanted it to be a large enough amount that they'd stick to it. Then the month's produce total averages $40 to $45 in a poor month, and $60 to $70 in a good month. Each person is paying this amount.

So you started as organic?

Yes, I did. We gardened that way when I was growing up, you just pulled off any bugs you saw.

Are you certified organic?

I'm in the process, in year number three of transition.

Which group are you going to certify with?

Probably OCIA[3]. Other people are encouraging me to certify with Indiana Certified Organic[4]. So I'll be getting certified starting next year. I own 88 acres. The reason I waited so long, this is my tenth or eleventh season of doing horticultural crops, but now that I have all the grains and grasses. It makes sense to do the whole farm now. This will be the first year I'm doing it, so maybe I'm growing the grains, and maybe I'm not! So we'll see.

Does that require different equipment?

Yeah, it does, but I'm doing it as low budget as I can. I have a big enough tractor with a bucket for doing compost. Many times, I'm driving past someone's junk pile, stuff that I can use. Many times, you can have it for $10 and as long as you can get it off the pile. It's not that expensive to get it. I'm doing this piece-meal. I'm planting wheat this year but I don't have a combine yet!

Farming in the Dark

What is your advice to someone starting a CSA? Should a person start with 20 people, or even smaller?

What I say to people starting out is start with 30 people. I think you are smarter starting with 20 or smaller. I tell people 30, because I could have jumped in at 30. When I was doing 30 here, it was my third or fourth year, the hardest part for me was getting the 30 people, at least in my mind, but in reality, once you put the word out, people start flocking in, the phone just starts ringing off the wall

I related some of my CSA experiences with Jane and Darrel, my two farming partners. Jane has now moved out of state due to her husband's job. We had 20 customers, and one skip week for the customers, and a skip week for us after the spring crops, but before the summer crops started coming in. It had surprised us not to have enough food to sell in late spring, with so many crops in the ground.

That time period, and anyone going in to this needs to realize, that the gap in late spring is the most difficult time. You need to start out early. That has happened to me every year, and I fill it in with root crops, beets, carrots, potatoes because they go forever, and turnips. Many people were overjoyed to see so many turnips in the bag! That is a good point. I don't think people realize that. It was worse this year than any year because we had that late frost, April 20-something. That meant you tried to beat the season by planting early and you were burned. Then the ground cooled; you had to wait to put stuff out, it was literally a month, then it warmed up. There were no spring crops! I was pretty desperate; I went and got some of Darrel's corn, luckily he had some of his corn ready, turnips were in the bag. I also do many real hardy Asian greens. They're amazing in the heat; they are so hardy. I put them under row cover and out in the open—the mustard greens, pak choi, all of those. They don't have the huge leaves you get in the spring, and they start to get rough, get some bug damage, but it saves me during that time.

You've been doing this for several years now, and some people burn out because of the commitment of having something for your customers every week. Have you felt that?

I'll tell you what I think has occurred for me. There are two pieces of this. You are an example of that. There is a movement; you try to build that, working on that with other people. Then there are the growers. Probably for six or seven years I was a grower, that was it, that was all that I did. I would give tours, maybe mentor, but didn't do anything else. Then I got involved with the Kansas Rural Center, and saw all that was being done. I started to work on the Salina Food Project[5] and the Salina Food Policy Council. You have to make a decision, at a certain point because it takes a ton of energy to run a vegetable farm, people have to know that going in. It's doable, but you need to be ready to commit all your time and all your energy (laugh); at a certain point, you just commit it all. There is the growing side, and then there is the building the movement side, the political side. Maybe fortunately, maybe unfortunately, but I'm being drawn more to the political side. Even though you are up against a bunch of knuckleheads, the political side is easier, physically easier—sitting in a coffee shop or meeting talking to people. Sometimes when you are in that marathon meeting, and you are thinking that you wish you were in the field; the physical side is easier and not as demanding. The mental side is hard, but I don't know if I would even say that, the farming side is mental too, thinking about what to plant, when, etc.

I like how the two sides complement each other, in meetings during the heat of the day, work in the field in the evening.

I do too. If we can coin a new phrase, "therapy farming." It is therapy for what you really do. Particularly when you have for so long, when you have mentored, lived and breathed it, really believed in it. And I really do believe in it! When you make that transition, you are breaking away from something. I've always maintained that if you are willing to work hard enough, if you can get the ground for a reasonable price, and you aren't going to use high-end equipment, you can make it on the farm. I've always maintained that. For the time when I had the 100 customers, I was doing that. I really believed that. But when I started doing the political side of things, I kind of gave up on that. It is where I said, the farm is no longer my sole source of support, but I'm going to put the two things together.

I see many people doing that, especially at the Rural Center. Which leads me to ask the question, is sustainable agriculture economically sustainable? It is hard to find a farmer who farms full time and says it is a good living and one they would recommend to others. It is a lifestyle choice, but is it really a job option?

That is a wonderful question. I think everybody, people who are farming, people who have farmed, people who want to farm, are really tossing that around right now. Here is what I'd say about that. First, I'd say yes, I believe it is doable, bringing in an income from farming as the only thing you do. I would add to that, you couldn't live the kind of lifestyle that people are trying to live today. You can't drive the newest vehicle, live in the nicest house on the block or in the county. You have to use old equipment, repair that equipment, and you have to market incredibly well. You've got to value add. And if you do all those things, I believe you can make it. The key to that is, deciding what is making it. My house needs a coat of paint. The reason it needs a coat of paint is that I have to wait for the proper time, to get the money together to do that. In our society, it is hard for people to make those kinds of decisions; they don't want to do that.

Even more fundamentally, could you afford health insurance through farming? I know many farmers have off-farms jobs for the insurance.

I have it through my wife's job, and that is a great thing. This is advice for me. I don't want to give this to other people. If I were alone, not married, no kids, I wouldn't even be concerned about health insurance. It is a scam. I hate to say it but this is what it has come to. All farmers, when you get a windfall (during a good year), what I would do is save that money as a security blanket. That's not good advice, but the system is set against the small farmer. You can't afford insurance today. We are paying $600 per month, and we hardly ever go to the doctor. It is so hard every month to do that.

It always seems they hold the risk of something serious over your head.

Farming in the Dark

A piece of that is true, because for people with land, you could loose your livelihood, everything. If you are paying $10 per month towards a medical bill, they have to accept that. It is not a great way, but I don't know what else to tell people.

Typically, people doing this kind of work are healthy. That's not to say everybody is, but typically. The other thing you could do is just have some sort of major medical policy, for serious illnesses. That might be a better way to do it. I wouldn't tell someone going in to this that, "Yeah, you can cover your insurance." What you have to do is be smart about health insurance.

This idea needs to be looked at: bartering for your health coverage. The largest percentage of my customers is medical professionals, and doctors are the main ones. Why not do a bartering thing, your vegetables are free but that is my medical account. You have to go further than that. If you want your farmer to farm for you, ask what could be set up so it would be covered.

For example, what about diagnostics like lab tests, MRIs, etc?

How much writing off can a doctor do, how many connections do they have?

This brings up the larger question of what are some possibilities for the ag community to negotiate with the medical community.

We could also do this for attorneys, not for huge things, but to do the paperwork to set up a corporation or something like that. I bartered with the person that did my farm logo. My sister did the original, but it needed to be modernized for the computer, so someone did it for me and we bartered for the produce, figured out what the bill was and how long it would take me to pay for it.

You need that kind of thinking, outside the box, if this is the life you really want to live. There are probably many other things I haven't even thought about.

I've heard of church communities setting up an account, where everybody puts in money, like a savings account in case someone falls ill they can draw on the account, sort of like insurance, but within a community.

If you think about it, that setup is what Blue Cross and Blue Shield were supposed to be. Why and when did they go on the stock market? When did providing medical coverage become a commodity that could be bought and sold to make stockholders rich? The whole thing is kind of weird if you think about it. That would be a great role for the churches; they could do that as not-for-profits. You know, perhaps NGOs (non-governmental organizations, also called 501(c)3 non-profits) could do it too?

How many employees do you have?

I say it takes a person and a half for the 40-person CSA, if one person is really putting all of their time . . . *If that person is you?* Yes, exactly [laughter] and you are willing not to sleep, etc. I have a student, the son of a customer, 18 years old, a great worker; I had him half days five days a week. I really needed him for picking day, which was always tough, getting all that picked, help wash it up, sack it up, and for putting it all together. He and I did it. There again, it is about what you can afford. That's what I could afford. If I could afford more, I would have more. Sometimes you have to sacrifice things you know you should be doing to get the main things done, in order to get out the bag.

One summer I had an intern part time, and her heart was in it, she was really dedicated, but I think she was surprised by how physical the work was, how hard some of the jobs are.

My employee's name was Casey. He worked hard, he really did, and some jobs I would ask him to do are just so miserable, it is hot. I felt bad, so I would go out with him, or do a job equally as miserable. I really tried to stay close. He never complained, always did what I asked, and he was sharp. I was lucky. Many people who bring on farm help aren't lucky. It depends on what you can pay, and on your pool of applicants.

We are hoping to start a student farm at K-State soon that could do some small research projects, in addition to production. What are some experiments that you'd like to see?

One would be soil work, where you could take extremes, like acidic soil and alkaline soil, and bring those back to neutral. What are the best, quickest, organic ways to bring those back to neutral. When I hear about people's problems, the soil is out of whack. That would be hugely beneficial to me, because I have an alkaline soil.

The other thing would be to look at would be the ecosystem and see how it is caring for itself. Too often, a farmer has a pest problem for one year, and then begins to believe it will be a continuous problem. That becomes their picture of what the farm is like, but probably the problem is cyclical if they knew how to watch. The next year the predator would arrive. We don't know how to watch. Setting that whole thing up, to look for what naturally preys on them, have that information. Then do what I would call an ecological survey, where you go out and look, and see how many of the insects are there. Push that, understand the natural ecology of the farm, and let it begin to work. Other than that, I don't know, I'm sure there are a million things if I thought about it.

How would you use the results in your production and planning?

If you look at the soil first, everything I do is in some way or another to acidify the soil, to bring it closer to neutral. For example, I have soil blocks, 2 x 2 inches; it is a block of soil. I use gypsum, bone meal, chicken manure and compost. I have a mix of all those things together. When I go to plant things in the soil, I spread that mix. In the greenhouse, I plant into those blocks. That's what I do for the soil[6].

As far as the ecology, I note things that are occurring the year before. Last year, the ladybugs, did you have them in your house and everywhere? What I looked for were the early season aphids. I'll get them in my greenhouse. The ladybugs were there instantly. Not even a second, there would be an aphid and then a ladybug to attack it. I had a sense that was going to occur that way. I didn't have to worry about the aphids on my sprouts.

By diligently walking through the garden, I see what is happening. Every year, I get the tobacco hornworms on my tomatoes. I usually just pick them off. One year, one was all brown and sticky on the outside, it was jerking around, different directions, a technid fly was laying eggs on it, backing off, coming back, the worm was using its own mouth to try to get the eggs off, that was why it was sticky. I left that tobacco hornworm; I didn't take it off. Chances were good that the technid fly would get an egg in there, and the worm would be a host. You can see a tomato plant and not really see it. If I hadn't seen that worm doing that thing, I would have missed it and destroyed a crop of beneficial insects. Now I have to watch, see if one is sticky, leave it, it is a host for something really beneficial. In the fall, I look for lacewings. With anything green you are mowing or anything, they fly out in the hundreds, to give me an idea of what their population will be, what they are working at destroying. It's not me doing anything, just recording what I hope is an ecosystem coming back to life. That is how I would look at doing it.

Do you keep a log on your crop, known pests, diseases and potential fixes?

Yes, but I don't keep my own log. I have a library of information. What I do then is read up on how to keep the plant as healthy as possible, and the diseases and insects that I need to be prepared for. That's what I'm doing in the winter, reading up and preparing. I look at what happened last year; I'll make notations of that, and connect in with this book or that book. For my soils, the first thing I did was to get a soil test and start working with it. I have a general journal or log. The thing about the technid fly was in there, but it's not really about improvements from the previous year. I take the problems, plug them in my library and see what I can do to make it better. Of course, I use extension a lot for those kinds of things; they can be helpful with information. I do have a journal here, but it's more of a picture journal, the farm starting, moments as they occur.

For production planning, how are you determining what you are going to plant at this time or another?

I have charts. I start with every crop you can grow in Kansas, and the chart is marked out by month. Then I go through the crops by month, when to plant in the heated greenhouse, when to put it in the ground, then draw a line to when you would harvest. Or it may be to plant in heated greenhouse, then move to cold frame or tunnel, harvest, then remove the tunnel. I look at the whole year. There's something growing every month. January and February—I'm maybe 25 to 30 percent successful for growing all year long, Sometimes having a crop grow all year long, and having it be harvestable all year long are two different things, the leaves may be too small, but it is growing.

I stick with this chart like glue. That is why I had the beets and carrots during that typical down time in late spring, because I did plan. Even if you plant root crops early, they will stay in the ground forever. If you do succession plantings, you are going to have them longer. Probably the biggest problem is following the charts. Things don't work the way they do on paper. It rains, or you have a late frost, sometimes you aren't prepared, but I stick with the chart. One thing I got from K-State extension is a small pamphlet showing how much a person would grow for a garden to produce enough for a season, then I multiplied that times 100 people. From that, I can figure out row feet for 100 people, and then come up with my seed needs. It's a system, it all comes together and it's perfect on paper, but never in the ground [laughter].

Even with the chart, isn't it a challenge with 40 or so customers to keep enough planted to harvest at each point in time? To keep the timing where one crop overlaps with another; how has that worked out over time?

It has worked very well. A lot of it has become just the know-how of how to do it. My rule of thumb is to be planting all the time no matter what the weather is doing, no matter what they say is going to happen, constantly plant. If you do it that way, you're never going to go wrong; something will be there. What it also does is it means that you'll have the crops coming every few weeks.

In my multiple plantings, sometimes my second planting catches up with the first. Then I wonder, why bother to space them out?

You need to look at many things when one catches up with the other. Sometimes one will be infested with something, and the other one won't, just because it is at a different stage. Other times, even when one catches up with the other, if it gets hot, the one that was planted later has adjusted to that, becomes acclimated to the heat, the other was used to the cool and it died. There are more reasons than just having the crop available; it is also the strength of that crop, other benefits, the younger ones are more tender, that kind of thing.

What about variety trials, would those be useful to you from a student research farm?

Yes, it would, but only heirloom varieties for me. I use hybrid varieties, but I really want to move over to just doing heirlooms. Heirloom trials I would really like. I'm just not that familiar with very many of them. When you read in the catalogs, they are all beautiful, but they are from Vermont or from Maine. Now that you talk about that, maybe even provide a seed source. There are no heirloom seed sources in Kansas, where it is grown here.

I recently found an heirloom seed catalog in Missouri.

That would be good. Now I get my seed from Johnnys, or from Vermont, or Harris, those three companies, but it would be so much better if the seeds that I grow where acclimated to our climate.

Another idea we are thinking about would be some small scale vegetable breeding and/or selection, because we don't have many university vegetable breeders any more, it is all being done by the seed companies. As gardeners and farmers maybe we need to take that back?

I think that is a great idea. That would be very useful.

Do you have any other advice for starting our student farm, or any CSA?

Ok, initially I think it is incredibly important to make sure that your customers, who ever that ends up being, are totally invested. People are going to assume, "This is K-State," and think they are going to get a beautiful bag of vegetables every week. Not that you should lower expectations, you're a player in this whole thing. The thing about a CSA is that it is an agreement. Play your part in this, but this is a relationship . . . Have hundreds and hundreds of recipes, for things that aren't that well known. You want to grow things that aren't well known, people like these things, but they don't know how to use them, help people know how to use the product, very few people know how to cook Swiss chard or okra. I think that is really important, and that is where your buy-in comes from.

Do you still do a newsletter for your customers?

I call it my bi-monthly newsletter that goes out, if I'm lucky, annually. It is a good idea. I just backed off because I tried to make it journalistic. I went too far; I should just scribble stuff down, put it in the computer and put it in the bag. I did a thing one time where I had the customers send me recipes. It would come out in the newsletter, they liked seeing their name in the newsletter, and getting the recipes.

Political Linkages and Implications

Are there some political linkages, activities for a student farm that you know of?

Right away, I think about the farm to university programs, and the Kansas food policy council[7] wants to look at those kinds of issues. A big part of food policy council is also food security, to make sure that there is food in the future, that there are farmers, so mentoring, teaching young farmers, giving them experience is of interest. The USDA is also concerned about this, that there aren't young farmers, there aren't new farmers coming in. You should not look at the statewide food policy council for financial support; they're more about working with the bureaucracy, more about policies and procedures...

You mean if we wanted our city to pass something, for local entities to buy a certain percentage of food from local sources.

This (your student farm) could be the birthplace of a local food policy council, grow it right out of this garden.

When we talk about movements and the political side of this, there are a few people doing a lot of work. That isn't healthy. It would be better to have a huge number of people doing a small number of things that are very effective. This fits, and it's almost eerie, but I actually just have a grant from the EPA to use my farm, to bring kids on the farm, I'll be training the kids on the farm. Kindergarten through twelfth grade. It has three focuses: ecology, looking at the ecosystem; health and nutrition, we're working with all these vegetables, and yet obesity is out of control. We'll be working with kids on healthy meals, snacks, what is healthy. Outreach, the kids will go back to their cafeterias, lunchrooms and display information for other kids in the school. It starts in a couple of weeks. I'm having an open house and ribbon cutting here at the farm, but it's tied in with a larger event. The art center here in Salina[8] did a project about art and food. One piece we are going to tour is a thing called "edible estates." An artist went into somebody's yard and made it all edible. We are going to look at that, and see how that works, see the new sorghum mill in New Cambria, then come here to the farm, do lunch, and tour the farm.

What a great way to celebrate the whole food idea.

We are just lucky because we had an incredible director here at the art center, reaching out in to the community. We did the food thing the whole month. They are also doing book discussions at the library, *Epitaph for a Peach*. Before that, they did *Fast Food Nation*; people are discussing these.

The local food policy council is responsible for setting up the book discussions. It is just a matter of working with your local library; anybody could do it. We just happened to have a good art center and food policy council.

I think I read an article about how for another exhibit related to food; she made people's favorite meals, then shrink-wrapped them and made a quilt out of it.

Yes, it is in the window of the art center (laugh) and had the kids do the same thing. She went around to the schools, talked to the kids; they made little quilts. I didn't see it, so I don't know what it looked like. It's all part of that.

Anyhow, so I'm setting that up right now, with the grant from the EPA.

How did you go about getting the grant?

There is a community food system project here in Salina that the Kansas Rural Center started with Claire Homitzsky. After she left, I took over for Claire and then I was able to get it funded by EPA. It also funded the River Friendly Farm[9] part of my work; I got farmers together to do the river friendly farm planning. There was a second round of funding for environmental education. It was through that funding that I got the kids program funded. They are looking at it as funding the ecological part and funding the farm part since it is an organic farm, looking at the impact of agriculture on the ecosystem. That is a lot of what we are going to do, as farming decisions have to be made from year to year.

The root reason that I started this is my belief that kids are so disconnected from nature[10] that nature is no longer part of their decision-making. It's not even there. They are being taught in two ways, seeing and hearing. These are the only ways we are working with our kids now. Nature teaches using all the senses. Its automatic or you don't pick it up. I want to work with children. I just hope that these children one day are politicians, developers, teachers as well as people who understand nature.

Kirk then relates a story that illustrates the current disconnect he sees in adults.

I'll keep this short, but with the Salina Food Policy Council, I had to go in front of the school board to talk about removing soda and unhealthy snacks from vending machines. Well, it was at a time when they wanted to get new turf, fake turf, for the stadium. Central High School has won the state football tournament for the last five years. That's what we were up against; Pepsi was going to put up a chunk of the money. We went and talked about health issues. It was like pulling teeth to get the board to see beyond the turf, that our kids are fat. FAT. Drinking five sodas a day, it is disgusting, eating snickers . . . It isn't that bad occasionally, as a treat; but they are doing it all the time. They couldn't see past that.

It is ironic that this vending machine money funds sports fields, turf.

Isn't that weird? We actually succeeded. They now have a policy for their vending machines, for snack machines and their al a carte. During the day no soda; they can get water, 100 percent fruit and vegetable juice, and two percent milk. And the items in the snack machines have to be below a certain percent of sugar and fat. The dieticians figured that out, and the same for al a carte. The point of my story is that it shouldn't be so hard. The kids in my program will be sitting on school boards, and they will "get it." If we keep going the direction, we're going, with kids having less and less exposure to nature, who cares if we pump oil out of the Arctic National Wildlife Reserve? If more and more kids go through this, who knows, maybe it is too late, but it is still worth it. Right now, this is funded under the Rural Center, but I'm forming a corporation called the "Whispering Cottonwood Educational Center." To do this, I'll have a huge liability insurance policy, be a separate corporation and continue to work with the Rural Center.

Comments about Sustainable Agriculture

When you look at the movement over the last 20 years, locally or nationally, what do you consider the accomplishments?

The biggest thing is that it is recognized now as a movement. I think too that it has gained an incredible amount of importance, people don't laugh anymore, [laughter] people don't think you are fringe, or way out there. Overall, it has had an impact on big commodity agriculture. The reason I say that, is when the USDA looked at organic certification[11] and how they were going to set it up, they got more responses to that than any other issue. That says to me that there is a group of people who care about this, who are saying, "Don't try to water it down, don't try to benefit big ag." That, to me, is excellent, the growth of a network that is very strong.

If you want to take it broader, to the idea of organic, look at the fact that it is the fastest growing market. I was reading that a certain percentage of food that ends up on the table will be organic, that is a step in the right direction. The growing number of farmers' markets is wonderful! It really is. It is a sign of where people want to be, what they want to see occur, coming back to community. Bringing it down to a local level, what our Salina producers have learned is that we are strong only if we work to be strong. We need to work with consumers, and the commodity folks don't get that. Those farmers aren't set up to do that and the system doesn't want them to do that. We know that, we have learned that.

The movement, not just producers, but also the consumers, they get it. They got in contact with the farmers market, a CSA, they read *Orion* or they read *Newsweek*. We've got that nailed. Now we need to move that to a smaller level. To tell you the truth, the level of the local community, communities need to start saying, "I want to be fed by my community," And that can be a region, not just your small town, "I want to be fed by my region. I don't want to bring things in unless I absolutely can't get them anywhere else." To me, that is the reason to go to the local food policy councils, local marketing groups, local vendors. Some have done wonderfully, some not so well.

Chapter 10—Kirk Cusick

The Importance of Local

The reason I say that it is important, is that now the big boys are getting into organic. At the national food policy council conference, and I don't know if this is a good thing or a bad thing, Sysco was there, talking about how they are using sustainable products. I didn't buy that idea. I'm not buying it, but maybe there is something I'm not seeing? The reason I say local is important, it may come down to the fact that the only thing the small sustainable producer has going for them is their community. Sysco can't be your community; the supermarket that buys from all over the world can't be your community. A local community can make it possible for your regional farmers, your regional people are producing and buying food, nobody else can claim that. I think that all of us are getting that idea. I think that is because the whole thing is gaining strength. Where people are looking for local and wanting it. Overall, that is my picture of what I see happening.

There is a flip side. The movement itself made locally grown food important, but I don't know if it made locally grown farmers important. They still want it cheap. The thing we haven't communicated is that if you want this food, and if you want your farmer to have any kind of mental health, to be able to take a vacation, to drive a dependable vehicle, to have insurance, you have to pay for it. If you really want this, you have to shell out the money to do it. We haven't gotten that point across. People are willing to pay more, but I still don't know if we are asking for the amount, we really need. We need to attack that gap between what people are currently paying, and what people need to be paying locally, to keep that local farmer there and thriving, and to bring more of them in; to make the ground more valuable as farm ground than for a development. People in town think, "Oh, what a beautiful housing development they are going to put up wherever, what a nice thing for the community." Do they see that in the farmer, these vegetables and fruits, being nice for the community? We haven't got to where I want to be, where the farmer is respected, I mean really respected. That the customers pay that price because they are paying someone with a set of skills that they don't have, and probably will never have, and maybe can never acquire. "Those farmers are putting up their land, their equipment, all of that up, in order to put food on my table. I need to pay for that." We really need to look at the real price of food.

I get into trouble when I make blatant statements, but I think the current form of commodity agriculture is literally destroying our environment. I mean literally. People need to realize that. This is what we get for the cheap price of food. Water isn't drinkable; rivers aren't swimmable or fishable. They have "blue baby alerts" in Des Moines, Iowa because the nitrate levels in the drinking water from the river are so high! What kind of society do we live in where it is ok for blue baby alerts to come over the air? What is happening that we aren't standing up; we are just letting it happen?

I've heard that the bottled water companies are making a profit. This is an example of our society's mindset, why we are allowing this to happen. Then it becomes a marketing tool, "So now that they are having blue baby alerts, we need this much water on the market." It is the mindset of our society.

Where we are, we haven't done a very good job at what we need to do. I hate to see these things as always in opposition, but when will the sustainable movement be strong enough to tell the commodity groups, "You got to stop." Tell people you have to stop allowing this kind of agriculture to continue. You are talking about the big folks, they have to stop, Farm Bureau, Monsanto, but you have to educate. I go back to when I bring kids on the farm; we'll discuss issues like who has the right to own a seed. Where did the seed come from? Do we want to make sure this seed will be available a million years from now, or whatever? They have to think about that. The environment is at red alert. There is no more waiting. We're done waiting. Things have to start to change. I hate to say this, but there is a cynical energy, where one side of me has accepted that the world is the way it is, but the other side, has to say, even so, I'm going to die saying that I respected myself and my land and my community. That's what is important when it really comes down to it. I have to struggle with that, chip away as best I can. We're closer, to having people say that we have to do it sustainably across the board, but we are still pretty far away.

One of my frustrations, discouragements and reason for the book, is that I can't show through statistics that sustainable ag has accomplished anything—looking at soil or water improvement for example.

Particularly when you came into this believing that is what it was going to do.

Some of the practices like crop rotation, cover crops, make so much sense, but I don't see anybody really adopting them.

I don't either. I feel that same kind of frustration, same kind of thing. We have this network in place now, it is a pretty strong network, I think we just need to crank it up a little bit. If we can bring that much pressure on USDA when it wanted to make the National Organic Program decision, why can't we do it in a political campaign? I don't want a PAC (political action committee), I hate that term, but why don't we have a group that says, "We have this many people, you aren't getting our vote unless you support sustainable agriculture." Is it strong enough to do it now? I think we should throw it out there, see what happens, tell the next person running for governor or at the state or local level, put it into action.

One person that I interviewed said we aren't a "movement," only a group of interested individuals. A movement is a group of politically empowered people. The only example where we behaved like a movement was the write-in comments on NOSB.[11] She wants to see it happen, create a think tank, see more attention paid to "framing" so that the public understands, so they get it, so they are on our side. She wants to see money put into doing this. We are ready to be a movement, and I think I hear you saying the same thing?

Yes, I think that is a wonderful way to put it, and I think she is right about the whole think tank idea. I have, for lack of a better way of putting it, written off the "older" generation. I want to start with the kids; in a way start with a fresh group, see what we can do. These kids will influence their parents; will influence each other. I'm creating a "little think tank," [laughter] and working with kids to do it.

What would happen if we had a million micro-think tank? If everyone had a teaching farm, we could do this.

When I go to Rural Center staff meetings, a lot of the talk is about the political. Mary Fund is involved in sending out alerts, which is what I see as the system, and the system is getting good at getting stuff out, but are we doing things that have political clout?

The Future

So my third question is what should we do for the next, the next 25 years? Where should we put our efforts?

I think that if you look at the broader picture, what we need to do is show on paper that this movement is reducing the use of pesticides, antibiotics, pollution. We have to show that it is making the world a better place. If it is not, it is all for naught anyhow. I don't want to sound cynical, because I think we can show it. If it's not there, if it's not occurring, then we have to make it happen. We need to make cleaning up the earth a number one priority. Through our practices, we have to be a political entity that tells people, not that I want a boss machine, but if you want to campaign in this area, you have to please this group of people for example, have this as a plank in your platform?

We have to make people understand what the real cost of food is. If you want small-scale farmers, their job is to not only produce your food, but also to take care of your water, air and soil, that is also their job. If you want your water, air and soil to be taken care of, you need to know your farmer, and pay the right price for their product and keep them alive and doing well, not just subsisting, but doing well. People need to know what that cost is. Let them know that, if I charge $1.25 per pound for potatoes, you are getting closer to a fair price for farmers to make a living with what they are doing. If you pay 30 cents per pound, as you do in the store, you really are using your dollar to do some destruction to the earth. You really are, because you are saying that it's ok for large commodity-type agriculture, or single cropping type agriculture to be the dominant force. That dominant force has polluted our air, our water, our soil. Was your savings of 95 cents worth not being able to drink, or breathe, or grow anything? Come on, people. We need to get that point across, where people start to hurt, to be disgusted by what they've allowed to occur. They won't know what the real cost of food is until they understand that. That is very important.

One of the things I'd like to add to this book is some actual budgets. I can get some from K-State, some general estimates, especially for commodity crops. It is harder to find the numbers for vegetable crops.

That is something we need to look at in the next few years. What are our budgets being made from, where are we pulling the numbers come from? Notoriously, this particular group is not too good at saying, "It really did cost me this much, and I've GOT to sell it for more than this," and even that is not much profit. Maybe that's why you can't find numbers. We say, "Well, $2 should be fine."

Sometimes we go to another local market, see their price, charge the same! [Laughter]. We do some calculations, but for small-scale vegetables, it is sometimes hard to track the labor for each crop, except for certain ones.

Yes, look at peas, for example. The work that goes into peas is just phenomenal. I don't know if any amount of money for a pound of peas is worth what you put in to them; you have to weigh that in. I don't even know what they sell for at the store, maybe they don't sell them fresh. I sell them for $5 or maybe $6 per lb. I'm just not going to short myself on this. If you want to complain about it, well, come and pick a row with me. There is that pricing thing. How do we figure what our prices are? Maybe we are the ones shorting ourselves.

In some communities there seems to be an upper limit to what people would pay. They can charge more in Kansas City than in Wamego, for example.

I don't ever want farmers to have a chokehold on people with food. I never want to see that. Maybe there is room for a kind of payment plan for farmers, if we want to keep food prices low. Something like a payment plan, not about the products, but the practices. The policy makers are leaning that direction now, not very well, but some ideas have come out of it. In the future, if we get a farm bill that says it wants to increase the respect for farmers, keep food reasonably priced, and keep the earth clean and environmentally sound. Maybe there is room for a payment, maybe we can never say that payments are gone, that we do away with them. Or, maybe it is a tax thing; not a payment anymore, but you get to claim something because you are a farmer and you do it sustainably. Maybe that will always have to be there to keep food at a reasonable price, not as cheap as it is now, but somewhere in-between.

Are there any examples of something like this that you can think of in agriculture, or in other professions? We have public schools, and pay teachers with tax dollars, for example.

We have a friend in New Zealand. I'm not sure what their system is, and I know it is much smaller, but the farmers there are very well respected. One of the most respected professions you can have. Why should it be any different here? Where did the respect go? Did the farmers give it away? The populist movement tried to bring it back, tried to grab a hold of it again but didn't have enough clout at that point, so there was a big switch.

Could you say a little bit about what drew you personally into sustainable agriculture?

For me, it was definitely the Rural Center that linked me to the whole idea of sustainable agriculture. To tell you the truth, I had farmed and gardened a certain way all my life, but I didn't call it anything, I just did it. Then I realized there was a whole group of people farming the same way. I have always disliked chemicals, disliked that kind of thing.

I'm trying to think about how K-State is a part of it. You'll appreciate this, it probably has been 10 or 11 years ago, I went to a K-State seminar here in

Salina about how to start a small business, a small agriculture business. We got to talking; folks were doing neat little gardening things. At one part of the session I said, "What do you think about organic vegetables, selling organic? The person from K-State said, "Forget organic, you'll never make it with organic, don't go down that path!" [Laugh] They gave me some of the best business stuff, to help me get started, but the worst advice about organic. I threw away the organic advice, and kept the business advice.

Then I went to Dan Nagengast's (director of the Kansas Rural Center)[12] farm. The reason I know about Dan's farm is that I subscribed to *Growing for Market*.[13] When I went there, Lynn Bryczynski gave me a tour, Dan came later. I asked, "Is it doable, can a person live off it?" Dan said, "Yeah, here is how we are doing it." Then Claire Homitzky, when she came to Salina, she knew I had talked to Dan, she contacted me and we started the Land to Hand Alliance. There is a neat group of people here, doing some neat stuff, and there is a movement of sorts. Then I started doing work with the Rural Center. That's how I got here today. I'm very happy I got involved in this.

2007 Postscript: Kirk is no longer producing for individual customers but taking all his produce to the Prairieland Market where it is sold off the shelf, not as shares. Many of his old customers visit the store to buy the produce. His work in sustainable agriculture has resulted in obtaining non profit 501(c)3 status for the Whispering Cottonwood Farm Educational Center. He works with the local schools to develop on-campus gardens and also has an after school program where children and young adults come to the farm and its diverse habitats to connect to nature. The Center is also focusing on farm to school in order to get local produce into local schools. He believes if we connect our children to nature they will be good stewards of the Earth. As good stewards they will eat local food close to home and they will be willing to support the producers who are good stewards of air, soil, and water.

Notes

[1]CSA Community supported agriculture, provides bag or box of weekly produce for members or subscribers. For publication see http://www.nal.usda.gov/afsic/pubs/csa/csa.shtml and to find a CSA near you, see directory at http://www.localharvest.org/csa/

[2]Populist agrarian movement in the late 1800's and early 1900's who's platforms were eventually adopted by other political parties, included anti-trust legislation against railroads, meat-packers, and others. For history see for example *The Populist Moment: A Short History of the Agrarian Revolt in America*. By Lawrence Goodwyn, 1978. New York and London: Oxford University Press

[3]OCIA – Organic Crop Improvement Association, provides certification services in the U.S. and internationally. See http://www.ocia.org/

[4]Indiana Certified Organic, offers organic certification throughout the U.S. and the Virgin Islands. See http://www.indianacertifiedorganic.com/

[5]Salina Food Project and the Land to Hand Alliance were local food projects sponsored and organized by the Kansas Rural Center to promote local food and sound local food policies in this central Kansas community.

[6]The soil blocks use peat moss, bone meal, aged chicken manure, and composted leaves which all are acidic to some degree. Another very important aspect of this mix is that lime is never used.

[7]Kansas Food Policy Council – objectives are to bring together a diverse group of public and private sector stakeholders to examine food systems in the state. The KFPC makes policy recommendations regarding ways in which the food system and related practices can be improved to enhance the health of the Kansas population, strengthen local economies and market opportunities, improve coordination and efficiency, protect the environment, and reduce hunger and food insecurity. See http://www.kansasruralcenter.org/kfpc.html

[8]Art Center in Salina exhibit; "Eating – Exploring what, how and why we eat." See http://www.salinaartcenter.org/Pages/Eating.html

[9]River Friendly Farm Program – a whole farm planning and assessment tool developed jointly between KSU and the Kansas Rural Center. It is still being used in the Clean Water Farms program by the KRC. See www.oznet.ksu.edu/rff.

[10]See "Last Child in the Woods" by Richard Louv, Algonquin Books of Chapel Hill, 2005.

[11]NOSB National Organic Standards Board, see http://www.ams.usda.gov/nosb/index.htm

[12]Kansas Rural Center is a non-profit organization that promotes the long-term health of the land and its people through research, education, and advocacy. http://www.kansasruralcenter.org/

[13]Growing for Market, Monthly newsletter for market gardeners and farmers. See http://www.growingformarket.com/

Chapter 10—Kirk Cusick

Discussion Questions:

1. What do you see as the pros and cons of working for a boss vs. being on your own (in farming, for example)? Psychologically? Financially?
2. What are the best ways to gain the skills to begin farming? Are specialized courses or college necessary, or is experience more valuable? Does one need to grow up on a farm to be a good farmer?
3. Over the years, Kirk has had a small CSA, a large CSA, and now sells vegetables through the Praireland Cooperative, rather than directly to his customers. What do you see as the pros and cons of those and other marketing methods?
4. What do you see as the pros and cons of farming full time, vs. combining farming with a career working for a non-profit, or writing grants to create your own non-profit?
5. Kirk mentioned watching the ecology of his farm coming back to life. What would you consider evidence of the ecology of an agricultural system coming back to life?

11

Paul D. Johnson
East Stonehouse Creek Farm, Perry, Kansas

"Visiting these farms in Sweden, seeing that lifestyle, the serenity, convinced me
that was the better choice."

*Paul farms near Lawrence, Kansas, and sells his produce through the Rolling
Prairie CSA with five other farmers. He is a board member of the Kansas Rural
Center. In 2000, through his capacity as a lobbyist for the Kansas Catholic
Conference, Paul helped pass/introduce legislation in Topeka to create the
Center for Sustainable Agriculture and Alternative Crops at Kansas State
University. In the past, he also has served as an active member of the Midwest
Sustainable Agriculture Working Group, which helps craft sustainable
agriculture pieces for the federal farm bill every five years, and also helps the
appropriations process necessary to get funding.*

 *Paul and I first met in 1994 prior to my moving back to Kansas at the
"Sustainable Agriculture Dialog." This was a rather large gathering of
"sustainable ag types" that came together to try to reach consensus on our
priorities for the upcoming 1995 farm bill. Our goal at that meeting was to come
up with five or six priority items. If I remember correctly, we ended up with
something like 23! It makes sense that Paul and I continue to collaborate and
communicate on policy-related issues, including what K-State could and should
be doing in the area of sustainable agriculture.*

 *We found time to visit about the future of sustainable ag in August 2006.
Though I'd known Paul for several years, I had never asked how he got started
in farming or in sustainable agriculture, so I asked him that, before getting more
into the policy discussion. I have visited his farm a couple of times, once when
he was a collaborator on our soil and water-testing program. Paul's farm lays
in picturesque valley north of Lawrence, with his home on a hillside, vegetables
and berries in the rich bottom-soil near the creek, and bits of pasture here and
there for his small flock of sheep. I started the interview by asking Paul how he
got involved in sustainable agriculture.*

It is probably a good question, how I got started. As I look back on my
involvement with gardening, it was minimal. My folks were avid gardeners, and
ate a good portion of their food supply out of their back yard. They did a lot of
canning: potatoes, beets . . . in McPherson. They were the first generation off the
farm. Gardening was just really in their soul, but it was the furthest thing from
my mind growing up. I spent little, if any time, helping with the canning; doing
any of that.

Farming in the Dark

I started at the University of Kansas in 1969, and met people there who did a lot on nutritional awareness, understanding the food. I'd worked for a couple of years at a fast food joint in McPherson when I was in high school. I think I ate myself out of the fast food industry in some ways.

The back to the land thing was evolving—the hippy culture—around Lawrence. That intrigued me; it was tied to a much stronger awareness of whole foods. In the early '70s we had a small cooperative, distributed raw milk around Lawrence. We would go to a dairy, get 50 gallons, take it around to various households. Then the company who was buying most of it told the dairyman that they would take all of it or none of it. That was a rude awakening. I watched the evolution and development of the food co-op in Lawrence,

The break point for me was in 1980. I went back to the home country, to Sweden, with my aunt and my parents. I was at that point in life where I was either going to get real serious about politics and move to DC, make a career of it, or do something with land. Visiting these farms of relatives in Sweden, seeing that lifestyle, the serenity, convinced me that was the better choice.

I came back, and started looking for land. One week I was out in the country with some friends who I'd occasionally go running with and saw this for sale sign. The next week I bought 40 acres. This was in September of 1980. Then I started thinking about what I could do with it. The people before had raised some strawberries, raspberries, various things on this acreage. What it first evolved into was a u-pick strawberry operation. That had its impractical side. I sold at the farmers market, which went ok, but I could see it was something I enjoyed. Then in 1993 and '94, Dan Nagengast moved back to Lawrence from Topeka, with seed money from a Kellogg grant to the Kansas Rural Center, and we started the Rolling Prairie (CSA)[1].

So you've been part of the CSA ever since?

Most of my sales go through that. I started with berries. I've diversified a lot—growing a lot of different stuff now. If my parents were alive today, I think they would be bewildered seeing the child they probably couldn't have gotten into the garden in my teen years, although they did have a chance to visit my farm several times.

How did you know you wanted to do sustainable/organic?

I was influenced by some people I saw in Lawrence. I was introduced to the health food, natural foods angle. I lived in a house where an artist used to make tofu in the basement. He rigged up a bicycle and chain for the soybean grinder. Tom Leonard, another friend, came by and stored miso in vats in one of the outbuildings. I went to a book discussion in Lawrence about tofu. It was an oversized book, called *The Book of Tofu*, by William Shurtleff. They did a whole analysis of how to use tofu. I had many friends getting into that, meat substitutes.

You've seen sustainable agriculture evolve over all these year. What do you count as the milestones or success? These can be big or little things.

A lot of that I wasn't aware of at first; it was in the mid-nineties when I started to get into national and farm bill issues. I went to the national dialog meeting in

Chapter 11—Paul Johnson

1994 and represented the Kansas Rural Center at the Midwest Sustainable Agriculture Working Group meetings from 1995 through 2003. In the back of my mind, I remembered my folks were the only siblings that left the farm. My aunts and uncles stayed with farming. Then I just watched this trickle, no, deluge of people having to leave the farm, seeing it in people's faces, living it out, what that consolidation meant. I remember printing out pictures of my uncle who died right before I was born, plowing with horses.

So, milestones . . . I think nutritional awareness is something that has slowly but surely, managed to catch hold. There is so much groundwork laid for people understanding the quality of food; that really underscores a certain drive for more fruits and vegetables, for whole grains. That has come from a lot of different sources. The growth of the food co-op movement in Kansas has been erratic, but I've stayed with that ever since college. Today, certain stores are seeing the writing on the wall. They have to offer organic items. We are seeing organic milk as the first major introduction—the rise of Organic Valley[2], how they have put together a network; that is pretty admirable.

But it is fair to say, that is in the context of many other losses. Kansas, since the 1980s, has lost 90 percent of our dairies, over 90 percent of hog farms. By the consolidation of farming into larger farms, there is less flexibility, more monocropping, boxing farmers into what their economic choices are. Another loss is their lack of any political ability to try to stem that tide, and to try re-balancing the economic players and interests in this game. Farmers have been their own worst enemy about trying to work together, to collectively bargain or bid. Other losses are the federal government's virtual abandonment of any anti-trust enforcement, the consolidation of the meat packing industry, seeing supermarkets like Wal-Mart and Costco have entered the fray from no involvement 10 years ago to being the biggest supermarket chains.

We had so much more of a diversified, decentralized food system 50 years ago. And it was more nutritious. In its diversity, it helped a certain food security aspect that we've allowed to disappear. We've gone from independent, small business entrepreneurs of various sizes out on the land to more and more contractual corporate servitude farming systems. There are perils on many levels there. I don't think our farmers have the diversity or ability to respond to what we are going to see in terms of certain trends. When we move animals off the land into confinement situations, we've lost a certain biological diversity.

It also seems like we've lost some of that infrastructure on the land too; people bulldoze their barns and fences. Absolutely. There are many challenges there. It sounds like maybe some of the losses are bigger than the gains, in spite of our best efforts. I do think that, I absolutely think that. Part of it is rebuilding the infrastructure you talk about. A lot of it is making new. I do think we have some technological tools and informational devices and skills that can start a new food system more readily. Given the internet and our ability to understand information across the world, we can access small farm technology that we see in Europe and other places: drip systems, handling water, dynamic gardening techniques. It can sprout in many ways; I think there is the potential out there. I think those can be drawn upon. Then the question becomes, who is going to do it, how do we integrate the new generation, into this discussion? What role do educational institutions play in providing some of those kinds of skills? How do we integrate them into our small business development centers, small colleges, Vo-techs? How do we start selling the idea that horticultural enterprises of various scales and sizes can be part of one's household income stream? That is what I would sell to folks. If both domestic partners have to be working, why not free one of them to try some of their own small business horticultural ventures?

So don't give up your day job?

Oh no, absolutely not. We need to integrate farmers into that service economy. Obviously, where the food comes from <u>has</u> to be as valuable as getting your car fixed, your plumbing, building skills. If you play it in terms of the food you consume being the most important thing you can do for your health, and the most important medicine you take every day, you put it in a different value structure. So if you are eating properly, you have lifestyle choices too. What is that real value to you? If you integrate that around better insurance rates for people who are eating better, we are going to have to have a different mindset. What we have now is a fast food, cheap food, cheap quality food mentality. Our movers and shakers in Topeka are looking at the obesity crisis, and they are damn scared. They are absolutely taking it seriously. They are looking at billions of dollars being spent, starting to graph it out, where these elementary kids will be . . . Here are some sobering statistics: nationally, one out of six kids is either obese or at risk of obesity, and 25 percent of Medicare in 2020 will be spent on obesity related disease.

I'm not an expert on this, but my sense is that we still have a dismal connection in medical schools and nurses training about nutrition. What they call it is still sterile. *You mean like only covering vitamins and minerals?* On many levels. *We only now understand some of the connection with the other phytochemicals; some are still in the research literature.* Things like the "South Beach Diet" and "Atkins," for all their shortcomings, really have shed a light on the dangers of white bread and white rice, especially the almost nutrition-less carbohydrates that turn to sugar so readily, so it is the mind change. When are we going to see a dean of ag at K-State stand up and say that Wonder bread and whole wheat are not the same? What does it take? What kind of scientific analysis?

Would you put that on your list of things to pursue for the next 25 years?— integration of nutrition with other fields?

Absolutely. I would do that with curriculums across the board, and especially with our medical providers. I would also have a more concerted effort for business skill training, especially for horticultural business planning; expand that. In places like Wisconsin, they have short courses on grass-based dairy, things like that. Others place much more work on perennial agriculture, nut tree type of efforts, obviously some fruit trees. Maybe try to build more of that into the land use planning that is slowly starting to take hold. As we get more into conservation easements and land trusts, building sustainable agriculture opportunities into that open space, green space mentality, that more people talk about and think about.

Do you mean that we should do these things at the educational or policy level or both?

Both. I guess I would try both. I think that at the policy level we have to have a much more fundamental discussion about what our farm policy, farm bills, have done to the structure of agriculture. I would ask that a fair share of Land Grant University dollars go towards support of local foods research and extension. In addition, we have to have a major re-writing of the inherent subsidies that we've given to certain styles of agriculture. We have to plan or fashion phasing out the commodity payments. We are not going to do this overnight. Similar to what we saw in the 1996 Freedom to Farm[3] bill, that the direct payments were going to be phased out over seven years, a glide path, and exports were going carry us, there were some fallacies there. But, I would revisit that. A seven-year phase out or phase down of the direct payments and ramp up the green payments. These would be primarily for the working farms, being as progressive as we can be with cropping patterns, encouraging perennial agriculture, etc. Build that as the income base for a much broader number of farms in the country. Make these payments tied to the land as opposed to that particular farmer, so if it turns over—is sold, or rented—we can get away from those changes.

I think we can overcome some of those barriers. And then I think we have to get our fair share of rural development loans and grants and micro-development. These were promised in farm bills and not fully delivered. For example, the value-added producer grant program was listed as $40 million per year in the 1996 – 2002 farm bills, and we barely got to half of that. I think the SARE program was promised higher levels. We've fought for that repeatedly. We also need to do a lot more with beginning farmer programs of various kinds, credit, mentoring programs, land, special tax credits for sales of land to beginning farmers. I think we need to come up with a whole package of how we can help another generation get in to land.

Some of the work Pete[4] is starting out on—this networking aspect—someone with the technical skills, between the growers, and the purchasers in various contexts, from various institutions to supermarkets, etc. That piece is absolutely critical.

Should that kind of activity be institutionalized? Made more formal?

We should. It seems our food system is at least as important as our highway system. We need that 10 to 20 year vision, and we need to sit down and grind out the numbers, like your slide show[5]—where we are now, what we are growing, what we are consuming. Look at targets; how would we get 20 percent of our fruits and vegetables being produced locally, using a mix of private and public resources. We need a food system plan for Kansas that is good for economic development and good for a nutritious product.

Is anyone doing that now? In Kansas or nationally?

Connecticut is maybe one of the leaders there. It started out of Hartford[6]. They had a hunger element to it, with a portion of their production going to food banks. Then they started thinking about the city, food planning around that, and now they have a full-blown state of Connecticut food plan. Ken Meter is doing a similar thing in Minnesota. And the Leopold Center is one of those core places that are trying to take the next steps, put pencil to paper. Obviously, Dan Nagengast[7] is starting to do some of that now. I think Governor Kathleen Sebelius could understand that. I think she could see a bigger picture, and has worked with Dan to start a state food-policy council.

Right now, I'm hoping she'll do that with energy. Her natural resources advisor was there 20 years ago, as head of Health and Environment and helped lay the groundwork to write the Kansas water plan. He brought the players to the table with the legislators, and had two weeks of solid, full day sessions on what we should do about water. Out of that came the water plan and the Kansas Water Office. I'm hoping now he'll get excited about doing a similar thing with energy. The governor has an energy council.

Have the food policy councils started to meet, to lay some groundwork?

Sure. I'm not sure where it is. They have started a state employees farmers' market near the capital building in Topeka, and are also making EBT (machines for accepting food stamp cards) available at several farmers' markets in Kansas. The Kansas Health Institute and the Sunflower Foundation are both more interested in the obesity question and nutritional issues. Dan thought the discussion we had earlier about the food programs tied in with the farm bill, bring in those players, hunger advocates in with the farm policy discussion might make some sense.

There are lots of barriers here with the mythology about where we really are with ag. We have organizations that say we are going to save the family farm if we kill the estate tax. There are just so many misunderstandings . . . That we have the most efficient food system in the world. Obviously, fossil fuel is the base of what we have developed. The most efficient thing we have is the labor and the quality of our food system is what has declined. We can't seem to have a very open debate about what it means to feed kids "Wonder bread."

We don't have anything close to a free market in agriculture any more; the grains are somewhat still open, but all poultry is grown under contract, about two-thirds of hogs, and 80 percent of beef moves by captive supply. Maybe the public will never care about the nuances, as long as it shows up at the supermarket and they don't get sick from it. They never think about the subsidy from their tax dollar. If you add it all together, would we still have the cheapest food system? Of the farm bill, commodity payments are make up 75 percent, and conservation payments only 15 percent, and we are still losing soil at unsustainable levels.

So do you think these barriers, these myths, are the main things holding sustainable ag back? I'm personally frustrated by our lack of progress or ability to document progress. The data are not there, and I don't see many new farmers adopting these practices. If it is such a great idea, why is no one doing it?

Some of it is "show me the money" mentality. When most of the farm bill spending is commodity payments of various types and only 15 percent is conservation, then commodities are driving it. And, 80 percent-plus of it goes to five crops . . . I think you'd have to say that the conservation reserve program (CRP) is a success, and even with a reluctant USDA relative to the CSP (Conservation Security Program), we have had far more farmers actively want to sign up, eager to get involved, and no-where near the money we were promised.

I think environmental concerns cannot be as readily dismissed today, as they were 30 years ago. Farm Bureau, and the major commodity groups at least have to drop it in the discussion, they can't write it off as rhetoric. They play the game, mention sustainable 50 times, mention it in work plans; the level of consciousness is that much higher than before. But boy, we have to change substantially the economics of where farmers are boxed in today. Some of it is their own doing, but a lot of it was system change that was set up to make CAFOs (confined animal feeding operations) a lot more profitable. Growing subsidized grain for particular processors, that is where we have to attack this. Now we have the pressure of ethanol on the corn price, and more studies on grass-based systems.

Another thing to consider is the generational age that we are looking at. The average age of farmers is 55. We have lost half or two-thirds of farmers under 35. In 20 years, we are going to see tremendous changeover in land ownership patterns in this county. I have seen some statistics of land rented versus owned land and who is farming it. In terms of who is farming it, that is a structural element. 'Aunt Jane' in the nursing home in Hoxie and probably a few siblings are living off the income of certain sections of land. As that starts to drift away, and we loose that, what can we do, practically, about those land ownership questions?

Do you think there is hope? Can we reinvigorate rural America? Create a rural renaissance? Or, do we just keep doing damage control? [Laughter.]

That is a tough one, I don't know if it is an either—or question. Many people talk about whether you have the right accoutrements, if you have a good lake, good fishing, a mountain, vistas, those amenities help certain places. Kansas will struggle with that. (More laughter) *But, we have such nice weather . . . and it can be healthful too!*

I think we need a more practical approach in Kansas. With heightened communication efforts—the internet—you have the ability to do business in larger ways. We need to be more practical about what core food-production systems will make sense. I think you can grow larger venues of winter squash across the state. I don't know if you can process tomatoes and make that go. I think asparagus is one of the special niche crops that has the viability to move in quantity with a high enough value. I think we need to think through, pick and choose, certain . . . I think 20 years from now we are going to see 50 dairies selling their own-labeled milk in Kansas. I think grass-based, or a minimally fed, natural meat is another thing. Once again, we need to come up with better processing opportunities or options.

I would be a lot more pragmatic about it, built around houses and businesses that are state-of-the-art energy-efficient, use those practical things. I think wind farms and wind resources will continue to grow and expand. I don't think we fully appreciate that as a value.

I think pasture based meat systems are only going to keep growing. I think the demand for a better chicken is definitely out there. Those are the trends. The commerce dept, they pick their trend of the year, and they're on wine now, they bend over backwards to help with ethanol. I hope we go for better, more practical choices with long-term impacts. Wine might be ok.

We seem to be the last state to get on that bandwagon; every state has wineries and tours

We could do pumpkins. Have we done enough, have we really figured out the best vegetables to grow in the sandy soils of the Kaw River bottom? Leopold Center in Iowa[8] is going down to an island in the southeast corner of Iowa that was renowned for a certain kind of melon; people grew vast amounts of them. The 'Muscatine' muskmelon, historically was grown on the sandy soil near the Mississippi River south of Muscatine, Iowa They are trying to re-discover what was special about it, what the varieties were, and then how could they play that. Could we do that with new potatoes, or sweet potatoes? What research would it take to identify some of that? When we updated those figures for the Cornucopia Project [9] for the rural papers, I think 1,200 acres of asparagus would do it in Kansas.

I'm starting to put together some statistics . . . current Kansas production from the 2002 census of ag compared to historical figures and current consumption. It is pitiful; we only had 54 acres of asparagus. The bottom line is that we need between 37,000 and 68,000 acres of vegetables to meet Kansas needs.

We had 65,000 acres of fruits and vegetables in 1930, although 39,000 of those acres were in potatoes. But, we have nine million acres of wheat! And we currently have less than 3,000 acres of fruits and vegetables, and we need more than a 10-fold increase. We are only supplying around seven percent of our fresh fruit and vegetable needs. Then multiply that, compare the profitability of an acre of asparagus versus an acre of corn. What would it mean to shoot for 20 percent of our food needs, and then include the multiplier effect of a dollar spent at a farmers market versus a dollar going to California for their carrots or whatever? That data is wonderful . . .

Well, it shows how far behind we are . . . and how food 'insecure' we are, how we don't really have much local supply . . . .

Some people think we'll just import it. At least we'll never run out of Wonder bread!

What else should we cover? What message do you want to leave for the youth of America?

This is my bias, but I think farming is a more fulfilling career. People's lives will change, they can do it for a while, come back to it or whatever. I'm struck by how fundamental this is, growing and supplying food—and how nourishing it is for the soul as well as for the body—that connection to certain cycles of life that our market economy struggles with, to value. There is such wisdom in it;

there is a mental health therapy aspect to it that we should not underplay. Clearly, there have been horticultural therapy programs evolving. So how do we measure that? For kids that are "consumered out," if you know what I mean, they are longing for something more in-depth, something a little more meaningful.

Do you see your neighbors growing more stuff?

No, but I'm in a kind of bedroom community expansion. I'm a couple of miles away from Buck Creek Valley. Lance Burr and Roger Johnson, they put their land under conservation easements[10] down there, so that is what I'm starting to think about; set my land up under that. I guess I want to believe that kids are hungry, not literally, but desirous of some other challenges and that this can be rewarding, in such practical ways, what you can produce. I also think that as opposed to "working for the man" this country has thrived on its entrepreneurial, "have my own business" mentality. We can play off that, encouraging people in these endeavors. I think we have to be practical, realistic about people making a full-time salary off the top. We don't want to give illusions to folks, but we need them to understand, to show them how to build into it. I'm not personally convinced that there isn't a supply of willing folks out there, but they want to do it working for their own business. That's what we have to keep; we have to keep that option open for them.

Is there anything else you want to add about national agriculture policy—do you want to comment on successes and failures, challenges for the future?

I said some of that earlier about restructuring the commodity programs, moving towards conservation payments, getting value added development grant programs expanded, micro-enterprises endeavors funded. The Small Business Administration does a little of that. The Center for Rural Affairs has been on the cutting edge, trying to develop and spur that on.

We are back to the same problems we have at K-State, when they receive block grant funds, where do they go? How does it get to isolated priorities? There are worse challenges at the federal level, where we try to earmark money for sustainable ag, for organic, and we get a pittance compared to the big picture. It is a constant battle.

Do you think we should try to get more earmarks for sustainable Ag?

(Long pause, sigh) I think we are probably going to have to. Some of that is how the game is played. As I understand it, the Land Grant Universities just came up with a sizeable increase in their basic grant. I think Ferd Hoefner[11] said, over the last 10 years, this is re-compensating for some lost ground. Then the question is how does it get used? When it gets here, how does the meat grinder work? Are we are going to lose that to the turf grass?

What I see is more funds moving from block grants to competitive grants, and then we spend most of our time writing grants so others can do research. People sell it as an accountability issue. But another problem is that you can only do research with a 2-3 year planning horizon, not 10 years, as would be preferable for sustainable ag.

The narrowness of it, what is out there, what is available? That is a challenge. I have yet to get a good fiscal read-out on how it is spent out here when I go before the subcommittees at the state legislature. I can't get it broken down by category, how much is spent on wheat, etc.

Even the researchers in Topeka can't break it out?

They hand you a 40-page book with research projects listed, and total head counts, but any more detail than that is . . . the question is if we give you 50 million dollars, can we expect that $10 or $5 million go for basic research?

Part of it is knowing what we need to ask for. I've said that to you before. What would be on your short list for K-State? Crop breeding? Long term alternative cropping? Is it better soil testing? If you can help me to promote sustainable agriculture or organic agriculture, what needs to happen here? If you can help me fine-tune that, I can help carry that message. The ranking democrat on the senate ag committee, Marci Frannciso, is one of my vegetable customers. She asked Fred Cholik, K-State Dean of Agriculture, some of these same questions—can you give me some numbers? Which she didn't get. She is starting her third year, and she is better grounded. Another representative, Josh Svaty, is from Ellsworth. His folks do some truck gardening. He has been involved and he understands the issues. He's young, charismatic and has quite a political trajectory in front of him. He was the one that dropped the bill in, to say that before genetically modified wheat can be commercially sold in Kansas, it has to be approved by the secretary of ag. Boy, the feed and grain folks, and the Kansas wheat growers, they came unglued. They hated it. In fact, they hated that we were even holding a hearing on this issue; it was sending negative messages to the major seed companies of the world, that we are an inhospitable environment in Kansas. Literally, that was the tone.

You mean even holding an open debate on this is considered off-limits? It seems like too many topics in the political realm are like that.

Paul and I then wrapped up the interview by pulling out some files, and looking more closely at where vegetables were grown in the Kansas River Valley historically, where they are grown now, and plotting a strategy for bringing them back again, since the acres have dropped to about 1/10th of their historical averages.

Notes

[1]Rolling Prairie CSA (Community Supported Agriculture) started in 1993 using the subscription farming model. Members sign up for a bag of groceries per week during the growing season, and pay all or part of the cost up front. http://www.rollingprairie.net/

[2]Organic Valley (milk producers cooperative). Originally founded in 1988 in Wisconsin as a vegetable marketing cooperative, continues to be farmer-owned with profit sharing arrangements among the farmers, employees, and the community. For more information see: http://www.organicvalley.coop/

Chapter 11—Paul Johnson

[3]Freedom to Farm was a policy first implemented in the 1996 farm bill that decoupled program, or commodity payments from requirements to plant or set aside acreage of certain "base crops." The commodity-based payments were then to be phased out over the next 7 years if/when prices for the crops rose in the marketplace. This did not happen, and subsidy payments continue in current farm bill programs. For more information, see article by Brian DeVore, 1997 for the Committee for Sustainable Farm Publishing at http://www.ibiblio.org/farming-connection/farmpoli/features/freedom.htm.

[4]Pete Garfinkel is currently the 'local food liaison' for the Kansas River Valley food initiative, a joint project between the Kansas Rural Center and Kansas State University. For more information see the marketing website at www.kansasrivervalley.com, and for background see http://www.hfrr.k-state.edu/DesktopDefault.aspx?tabid=735

[5]See websites above for tables of per capita consumption of fruits and vegetables, estimated quantities required for Kansas' population, current, and historical acres of these crops planted.

[6]The Hartford CT food plan. For more information see their website at http://www.hartfordfood.org/, see more information about Ken Meter's work in Minnesota at http://www.crcworks.org/rural.html or the Leopold Center in Iowa at http://www.leopold.iastate.edu/

[7]Dan Nagengast at the Kansas Rural Center has helped develop several local food councils and a statewide coordinated effort. For more information, see summaries at http://www.kansasruralcenter.org/kfpc.html

[8]Iowa Place-based foods, see website: http://www.iowaartscouncil.org/programs/folk-and-traditionalarts/place_based_foods/assets/pbfmast.jpg

[9]Cornucopia Report – a project begun in the 1970's by the Rodale Institute, provided funding for each state to research and write a report about the status of their farming systems and food supply relative to their population and sustainability criteria. Kansas' report was published in 1981, and a follow-up report, by the original author, Kelly Kindscher, professor at University of Kansas, was published in 'The Rural Papers' in 2003 (ref here) Unfortunately, both reports are out of print, but can be found in various libraries and archives.

[10]Conservation easements – "A conservation easement is a voluntary agreement that allows a landowner to limit the type or amount of development on their property while retaining private ownership of the land." For more information, see http://www.landtrust.org/ProtectingLand/EasementInfo.htm or for the Kansas Land Trust see: http://www.klt.org/

[11]Ferd Hoefner is the long-time Washington staff member for the Sustainable ag Coalition and the Midwest Sustainable Agriculture Working Group. For more information about these organizations see: http://www.msawg.org/ and for a directory see http://www.kansasruralcenter.org/publications/msawgguide.pdf

Discussion Questions:

1. Do we need a new food system? Why or why not? What would it look like? Physically? Nutritionally? Business-wise?
2. What infrastructure and/or policy changes might need to take place for this system to become established?
3. What would our economy look like if people were healthier?
4. How would the insurance business be impacted if they were required to give lower rates to people who ate better quality food?
5. Do you think the public should have a role when there is a debate about whether a new genetically modified crop species is released? Do you think farmers should have a say in the final decision? How could a process be set up that would make these decisions be more democratic?

12

Jon Cherniss and Michelle Wander
Blue Moon Farm, Urbana, Illinois

"I fell in love with the food...that's a strong motivation for me, the quality, something to get excited about, those tomatoes, eating them" . . . – Jon, farmer

"The soil remembers itself, it embodies a consciousness." – Michelle, soil scientist

It's a steamy morning (August 2005) in corn country, or the "corn jail" as Michelle comments about their location on a 20-acre square of land boxed in not by trees or houses, but by corn. The view for as far as one can see through the fog is flat, only broken by corn, soybeans and the occasional farmstead. The sun rises orange, the smell of mown grass field borders is strong, and Drummer, the three-legged dog barks notice as the three or four farm employees arrive at 6:30 a.m. to begin their work on Blue Moon Farm. Drummer, a black dog, was named after the state soil of Illinois, since Michelle is a soil scientist, and Jon raises vegetables full time. Strong coffee is brewing in their large country kitchen, and locally baked sour dough bread topped with butter and a thick slice of fresh golden tomato start their day.

My husband and I are staying with Jon and Michelle while we are in Illinois to attend the Midwest Sustainable Agriculture Working Group meeting to discuss possible changes to the farm bill. Though it is still two years away (this is 2005, and the farm bill will come up in congress in 2007), the sustainable agriculture groups, and I'm sure the conventional farm groups are already preparing proposals and strategies for changes they'd like to see. I've known Michelle for more than 20 years, since my days at Rodale, and her days at Ohio State University, exploring the mysteries of different types of organic matter that up until then were not documented in organic cropping systems[1]. Our long term cropping systems trial at Rodale was a good proving ground for her theories about the effect of organic matter on soil physical and biological properties, and how conventional and organic practices affected them. Michelle is now a professor at the University of Illinois, continuing her research on soils in addition to teaching. Jon has been farming organically for a couple of decades, first in Ohio, then Georgia where Michelle had a post-doc position, and now in central Illinois. You will hear the rest of their story in this interview, which started after breakfast one morning on their farm.

We started out by visiting with Jon at a picnic table in the front yard, under the shade of several large trees.

Farming in the Dark

How did you get involved in sustainable and organic agriculture?

I'm certainly not a great thinker or big picture thinker in sustainable agriculture. It's funny; I haven't read much of Wendell Berry or Wes Jackson[2], it's not how I came to farming. I came to farming more via Nancy Creamer[3] and Michelle. I finished my masters in international relations and needed a summer job. I swore I'd never raise livestock because I grew up on an egg farm. I knew it was 24 – 7 and I had no real interest in livestock. I took a summer job on a farm, and loved the food. I worked on Stone Free Farm in California, a certified organic farm owned by Stuart Dickson. He was really on the cutting edge of the salad mix wave and heirloom tomatoes. I fell in love with the food. I was never motivated to make money for money's sake, but I grew up an entrepreneur, had an egg route. Our family had a medium sized egg farm at the time, 100,000 chickens, caged. In the 70s dad started processing and selling directly to stores and restaurants. You could no longer just raise eggs. It's the "story of ag." The processors were holding prices down, and direct marketing was the way that dad dealt with this for four or five years. Then he ultimately sold out. That size farm wouldn't exist at all today. Anyway, that's how I got started. I guess it wasn't really so direct, but it seemed like a good thing to farm, I loved food, I could be an entrepreneur.

I had an interest in machines, the systems; I wasn't a plant person. Some people are avid gardeners and scale up, but through Nancy, I met Michelle. Nancy had land in Ohio and we all started a farm. Then Michelle started her PhD, and Nancy had a baby and had a full time job, Nancy's husband had a full time job. I was the only one full time, and by the second year, it was really just me. Probably my personality drove them away too. [laughter]

That's how I started farming. We were certified organic the first year. That was the challenge. For me, it wasn't, and still isn't, just about producing vegetables, it was about, "lets do it this way, and show that it can work." It's about the food too, that's always been a strong motivator for me.

When you say about the food, what do you mean exactly?

Taste, quality, something to get excited about. I was exposed to all those tomatoes, eating them, as you work. That, and people are so appreciative. When you deal directly with the consumer, you get a good sense of what they like, no one really ever complains. I really enjoy working with customers and scheming at how to do things better on the farm.

Where did you get your information to start farming?

Stuart's farm in California was a seven-acre intensive vegetable farm, some equipment, not much, just the main tractor. He grew diverse vegetables, a lot of salad greens and a lot of tomatoes. That's really, what I'm still doing. Some of the information came via academia and Nancy and Michelle, they were both involved in sustainable agriculture. *New Farm*[4] was an early source; ATTRA[5] was a great resource, and conferences to a certain degree. I think Elliot Coleman's books[6] although I'm critical of what he sees as true, were helpful. I still use his method for planting green onions. I was always more concerned about the mechanical side of things, the mechanisms of what to put after what in a crop rotation etc.

Chapter 12—Jon Cherniss and Michelle Wander

Do you go to conferences?

Not as much anymore. Periodically I go. I generally can't use the information I have now, so when I go to a conference it just ads to my "to do" list. It actually makes me nervous. The hardest thing is that we grow 40 different crops; we do many succession plantings. Mother Nature seems to have the biggest impact on how our crops do. If I get too much information, I can't do anything with it.

What would be an example of something that you heard about at a meeting that wasn't useful, or you couldn't implement?

That's hard to say, all information is useful but it seems there is too much presented without economics. Knowing how to predict a pest cycle is not that useful if I don't also know if the remedy (for example, a biological control or botanical pesticide) works and if it works well enough to cover the costs of application. This type of research is expensive, it takes years to do and it can be site specific; but it really is what is needed.

When you go to sustainable ag meetings, do you get something out of the camaraderie, out of interacting with like-minded people?

I'm not the kind of person that can easily strike up a conversation with someone I don't know. In this region, we don't really have any farmer groups that get together. In the last few years, there has been an Illinois Organic Conference but it is just getting started. At the farmers market, we have a lot of give and take, in terms of information exchange.

Could you talk a little bit more about marketing, the challenges, what has worked for you, what hasn't?

Marketing has been the easiest thing for me. I was already comfortable working with restaurants. I think I have a good sense of quality and I think I've always had an attitude that the customer is always right. I also knew that if I wanted to sell to a restaurant I'd have to call regularly. I've met growers that said they tried to sell to a particular restaurant and delivered once, and then complained that the restaurant never called them. I have worked with the same restaurants for 10 years and if I want to make a sale, I call them. I don't even leave a message for them to call me back. Out of 12 restaurants I bet I can count on only three or four calling me back..

I know you marketed through a CSA for a while. How did you decide to do it, and why aren't you doing that now?

We had a CSA for five years. We were fortunate when we came to this community. A farmer who had just started the CSA had decided he wanted to go back to school. He rented a farm and had a stand at the farmers' market for about 10 years. The CSA was only a year old. We bought most of his equipment, rented the same land, took over the CSA and sold at the Farmers Market. I wasn't sure about the CSA at the time, so we got another farmer involved. We had 30 or 40 customers at first, than we got up to 130 families. We split it between us. He would handle some crops and we'd handle others.

We had a very good core group that served as a board. We sponsored shares for needy families and tried to do a lot of education in the community. I was being pulled everywhere—farmers market, restaurants and CSA were too much.

Farming in the Dark

After five years of deciding what everybody was going to eat that week, I decided something had to go. Having the money upfront was not worth the extra pressure. I wanted to be able to till in a crop of weedy carrots and not feel that I had to save them for the CSA. It was hard to meet everyone's needs. One member wants more beets; one does not want any beets. Every year we tried different ways to manage these things. It's not that members were negative or unhappy, but they were telling you what they liked and did not like. In fact, we surveyed them every year to find out how we could do better.

After five years, things like farm field-days and work parties started to become chores themselves. We like people, but we also didn't want to have people out to the farm all the time. We went as far as having someone manage CSA members who wanted to work on the farm. One time nobody showed up at the designated time, the person who was going to manage the workers went home, then just as Michelle and I were sitting down to dinner, somebody showed up. I had to go and weed with them for an hour. I once heard a speaker telling a group of farmers, "You are not selling your crops, you are selling an experience." I have never been fond of "entertainment farming." That approach was not for me, I prefer selling tomatoes.

What happened after you decided not to do the CSA any more?

The other farmer continued the CSA and our gross sales from the farmers' market and restaurants grew 15 percent the next year. We never noticed a loss of income. Sometimes I think I am going to start a CSA again. I still think it is a great idea. If I did it though, I would stop selling to the other markets.

What do you think sustainable ag as a movement should do next, and also what is your perspective living and farming in several states; Ohio, Georgia, Illinois, and also California? Are there support networks in place for farmers like you?

I'm mostly interested in some old-fashioned variety trials, product trials, breeding and especially cover crop breeding. I am not that interested in all the marketing work that is being done. Certainly, marketing helps organic by increasing demand and this helps new farmers, but given that I am already farming and have markets, I guess I am more interested in the basics. I am also a little concerned that we are putting the cart before the horse. Demand is going to have to keep up with supply.

What are your thoughts about on-farm research? For a while, that was a big push in sustainable ag, but there is still the limitation of time and money. Have you been involved in any, and/or do you feel there is still a role for the university to do the product testing and other types of research?

We have had help from researchers on a couple of things at the farm. Last year we looked at cover crops for controlling Canadian thistle and this year we have looked at mulch in tomatoes. The problem is that research seldom gets funded for more than a couple of years. I think the sustainable ag movement really needs to break the mold and start funding research for 6 to 10 years at a minimum. You can't look at cover crops for two years and find out what covers will work best in controlling Canadian thistle. How can farming be sustainable if research funding is not?

Chapter 12—Jon Cherniss and Michelle Wander

How do you calculate your economics? At the end of the year? Or by crop? Do you know what your break even costs are, for example?

We keep track in varying degrees. We track sales by crop and record harvest and pack out times. In the past we used to track all activity times weeding, tillage, planting etc. too. It became a nightmare getting people to remember to write down their times and another chore to get the information into a spreadsheet.

What is your break-even cost on something like a tomato?

I can't tell you off the top of my head. My labor costs have ranged between 35 or 45 percent of gross. If just harvest costs get above 35 percent of gross, I know I have a problem. I can't spend my entire labor budget just on harvesting. How will I pay for machinery, insurance, seed, planting and weeding? When harvest costs get too high I either have to raise prices, figure out how to harvest the crop faster or stop growing the crop. We used to harvest bunch carrots at 30 bunches an hour and sell them for a dollar a bunch. Customers love our fall carrots because they are so sweet, but I was essentially growing them as a loss leader. Fortunately, we raised our prices to $1.50 a pound, bought a barrel washer and started selling them in bulk. Now we can harvest and wash carrots at 65 pounds an hour.

Another way we look at individual crop budgets is by spreading all our costs except labor by bed. Other than labor costs, our other variable costs and fixed costs have been stable throughout the last five years. This allows us to determine what our minimum bed costs are. I also add my salary and a yearly capitalization cost to these bed costs. If a crop does not make enough to meet this minimum than we have to take a closer look and make sure it does not require a lot of labor or management. With 40 different crops, I think it would cost too much to look any closer. My grandfather, who also had an egg ranch, used to say if he had more money in the bank at the end of year then at the beginning, then it was a good year. I think that's the whole-farm planning approach and may be more suitable than individual crop budgets if you grow forty different crops

Farming like this, can farmers afford to buy land in Illinois and start a farm at your scale? If a recent graduate from the University of Illinois came to you and said, "I want you to be my mentor," can they start out with zero, or a loan from the bank and make it work?

And pay him/herself? If they really knew what they were doing, maybe, but I doubt it. This is where I think we are lacking information. What year does a new small vegetable farm become profitable? What is the average capitalization cost per acre for a small farm? Is there an average profit per acre for market farmers? There are so many types of small farms, these numbers would be very hard to come by but they would be very useful. We would have to separate between full time farmers from part time farmers and farmers that have income from a spouse. Decision-making is significantly different on each of those farms. I do think you would see some patterns and it would be very useful for farmers and people wanting to start a farm. It wasn't until year nine on this farm that I started to feel like I was actually making enough money to make it worth it to me. We never had a problem with making the farm cash flow. My initial farm investment was about $15,000, for equipment etc., to get started and we borrowed money for the land. Michelle has a good job so I could work free and reinvest the farm profits each year. We don't borrow money for the farm input expenses or equipment. This has meant the farm has grown slowly but it has taken 10 years on this piece of ground to have enough taxable income that I can call it a living. Each year we do better and I hope that all the investment we have made will start to pay back, especially if we cut our capitalization costs.

It's almost like we are creating this parallel reality then, of sustainable ag farms?

Maybe we should. I think small farms are products of peoples' personalities more than anything is. Most growers I know don't want to hire that many employees, if any, and I think that puts a real cap on what a farm can make. I think only the top 1 percent of growers could have taxable income of more than 30K per year if they don't hire outside labor and most people would not even call that a living. So small farms become a question of lifestyle and we have to change our definition of "making a living." That's where a good very critical economic look at these various farms would be beneficial. [7] It certainly would be beneficial to me.

There was a slight pause in the conversation as an employee came up to ask Jon how to hook up the hydraulics on a tractor, and brought us a fresh apple from their tree. After enjoying the apple and talking about the risks of late frost in a continental climate, we continued.

Did the implementation of the National Organic Standards[8] help you in any way?

I think it's a work in progress. Many growers in this area chose not to become certified and we have seen a lot of new terms used to describe growing practices. It's almost worse than before. It used to be that the growers that used the word organic were for the most part organic. Now these growers that are organic but not certified are using terms like beyond organic or going back to terms like natural or sustainable. Customers really have to know what questions to ask farmers. One grower at our market tells customers that they don't use any dry synthetic fertilizers (they use liquid 28) and don't spray. Next thing I know this farm is on a list of "Organic Farms" being toured in the summer. The tour was put together by our local Extension. Customers see this and then I am being told by them that this farm is organic. So where does that leave us, as a certified organic farm? We're feeling the need to do more consumer education than we did before USDA got involved.

How do you feel about the "big box" stores and large chains starting to sell organic, mostly imported, and at discounted prices?

If you want a community, you have to support it.

****

At that point it was mid-morning, and Jon needed to start making restaurant calls. We just had time for a quick walk around the farm.

Chapter 12—Jon Cherniss and Michelle Wander

An historic barn serves as a combination farm office, complete with phone and computer for record keeping, washing tanks for the vegetables, and boxes for packing out the produce. A new metal shed next door keeps the equipment clean and dry and provides a place to work on equipment out of the elements.

We saw the well-kept fields, blocked in a way that matched the rotation diagrams.

To meet the crop rotation requirement of the organic certification guidelines, Jon has the farm parceled into 12 bed blocks, 300 ft. long. Every other block is planted to 3 years of annual crops, and 3 years of alfalfa/clover perennial cover crops. These legumes feed the soil with nitrogen, and enrich it with organic matter. Jon's soil fertility program includes three years of cover crops supplemented by soybean meal as a nitrogen source for heavy feeding crops. He doesn't apply compost directly to the field, since his phosphorus levels are so high that he could upset the micronutrient balance.

Jon's rotation is blocked by land preparation, spring crops, summer crops, fall crops and by what goes around them. His rotation is based on timing, with the individual crops planted in succession into ground managed and prepared for the "early- late- spring, summer and fall plantings that have similar management needs.

Lettuce and tomatoes make up almost half of Jon's total production system His customers depend on him for these crops. The other 40 crops include five types of sweet basil, potatoes, summer squash, winter squash, peppers, eggplant, onions, leeks, green onions, garlic, etc. Summer cover crops include buckwheat, Sudan grass and occasionally cowpeas.

Weed control is always difficult in a not-very-competitive crop like lettuce, so bed preparation prior to planting is important. Jon does this by tilling and shaping the bed and then pre-germinating weeds using a row cover fabric and watering. Then the germinated weeds are flamed, using a propane torch on a backpack flamer or a three- point whole bed flamer. Then lettuce is seeded, and the row cover is put on again to encourage germination of the lettuce. Once up, hand weeding is required until the crop is ready to harves, in about four weeks. Beds that are waiting to be planted are covered with black landscape fabric to keep the weeds at bay until they are intentionally germinated. Lettuce is planted in succession, or weekly in the field between April and November, and year-round in the high tunnels or poly houses.

Jon's tomato production system is equally complicated. Early tomatoes are seeded in February and transplanted in March into high tunnels. High tunnels are small, temporary unheated greenhouses with galvanized pipe uprights and hoops over the top, covered by plastic sheeting (about 12,000 square feet). The earliest tomatoes are varieties that will tempt the hungry consumer at the beginning of the tomato season, before the market is flooded with garden tomatoes, undercutting local prices. His mid-season tomatoes feature special heirloom varieties. These come in a variety of shapes, sizes, colors and flavors, increasing the diversity on the dinner plate, as well as on the farm.

Walking around the farm, I see a lot of lettuce, but a lot of everything else too, including greenhouse tomatoes. A large bunch of transplants was waiting to go into the field. As is typical of most vegetable farms, there is a point in the season when everything needs to be done at once.

He explained that in terms of this growing season, it had been very dry, then the remnants of hurricane Dennis came this far inland, they got 6 inches of rain about two weeks ago, and the fields were just starting to dry out from those rains, so they are behind on the cultivation. This is also typical of farming in general, that you need the rain for the crops, but certain things can only happen when the soil is a little drier, like tillage and cultivation, so every weather event can feel like a mixed blessing.

I commented on some purslane in one field, which happens to be an edible weed that is sold in some ethnic markets. Since we sell some weeds at our farmers' market, I asked Jon if he is selling the purslane, an edible weed recently popular because it contains omega-6 fatty acids.

We didn't have a problem with it when we were first here. I don't know if that was because of herbicide residual, but it seemed to all the sudden come on. I haven't been selling it, but we should, out of spite.

Jon's field season begins in late January, with the starting of lettuce and tomato seeds in flats in the greenhouse for later transplanting. His last market date is the Saturday prior to Thanksgiving.

****

After the tour, I sat down for another cup of coffee with Michelle, to get her view on things. I want to ask about your personal experiences as well. How did you get involved in sustainable agriculture?

My dad was an environmental consultant, and I was a suburban brat. I wasn't an ag person at all. I'm typical of a certain type of kid that was suburban, but who was taught to value rural culture. My parents had a rural background, my dad in particular. He loved animals, and he loved good food. His mother grew most of what they ate. I heard somewhere that children's culture is not just their culture, but is taken from where their parents grew up. The family values will be so infused. I would say a lot of my values came from an ag environment, from my dad. He grew up in Minnesota.

My mother is very political. You see young people who want to have vocational employment. They want to do well. They get that from their parents. I think my mom made me want to be a different person than one that just wants to make money. It's about how do you define success. Both my parents communicated the value of "do good for society."

When I went to college, I had a boyfriend in soil science. My dad was in waste and water management so I saw a connection, and I thought I would do environmental conservation. Then I read *Food First*[9] and got interested in food politics, South American politics, Nicaragua. The route to sustainable ag was partly through subsistence ag, and that beautiful Miguel Altierei[10] vision of agriculture where everything is perfect. When I was an undergrad, I was sucked in through social justice, small farmers. It is still parallel; the issues that are there don't go away. For educators, it would be helpful for us to view issues as common, but locally nuanced. That's how I was sucked in.

I might be dumber than the average person might be, or don't realize the assumptions, but sustainable ag has infused in it assumptions that are so central that we don't even see them. We all have very solid assumptions about what is good. The truth is, as natural scientists, we think that nature is our model, and that is good, with no rational basis for that. It is totally an assumption. In sustainable ag, we have our dogma. I think the dogma that I really went by, that has been driving me and continues to do so, is romantic. Of course, the restriction of doing my science has resulted in a split personality. My original assumption is that if you manage the soil well, and you care for the land, the centerpiece of good stewardship will be recorded in the soil. I love this quote; I got it from a retired NRCS (Natural Resource Conservation Service) scientist: "The soil remembers itself, it embodies a consciousness. It knows what happened to it." You can read the past, and you can, especially with organic, forecast the future very well. The soil is a great crystal ball for stewardship and for a piece of land. Many properties tell you what it has been done to it.

Even though the "soil improvement plan" is the central, initial paragraph for organic standards, there is not one requirement in the standards that says to do this or do that, and very little money to do research on it. If you can talk sideways to your certifier, you don't need one. That's still propelling me. I could spend all my time, and not have to think about world politics, which is a lot more difficult and having a greater impact on sustainability, than knowledge about soil.

Before I went to graduate school, I worked on biological nitrogen fixation at a research institution in Hawaii. This free nitrogen is still the centerpiece of much of sustainable agriculture. They were developing the key components of the technology, breeding and selecting, and knew the need to match rhizobial genetics with the plant. You can't inoculate into soils, due to the competitiveness of the native micro-flora. You had to localize selection for compatible strains.

It was a brilliant project, a nice model. I was working on the intercropping research.

I really wanted to think of soil stewardship with the combination of rotations and management. This is the Holy Grail for me, and a whole lot of others. Can we understand, and then intentionally make nutrient and resource efficient production systems? That doesn't include social equity and labor. It's one of the key dimensions, a production wedge I can contribute my technical training.

In my professional life, I try to keep the technical and social separate; that isn't my scholarship. My right to contribute and to weigh in on those issues comes as a citizen, not as an academic. Academics don't always obey those rules, nor should they. I've tried to do that. I've been more comfortable doing that. It works better for me.

What do you see as some of the successes of sustainable agriculture?

In agronomy, an issue we should count as a success is that we see research on a systems level coming along. That is a great accomplishment. Right now, we need to get serious about answering the question "how do we do systems research?" I think that sustainable ag birthed that question; brought it along, but now we are at risk of it being taken away from us.

There has been a lot of scholarship effort, we've done great things, but with no credit given. This is a huge accomplishment, but there is frustration that comes with it. We are really positioned, if we get it right, to take leadership in the area of sustainability research. We will probably always be doing leadership, because we feel the questions so acutely.

We, meaning agricultural scientists have been hostile to our own; we have done a lot of the legwork, important thinking, but not articulated or interpreted our findings forcefully. For example, there are bio-geochemists describing nutrient cycling in agriculture and being admitted into the National Academy of Science. We've been there, done that. Instead of congratulating our scholars, we have tried to suppress ideas that are challenging.

As to the university experience of sustainable ag, we've waged many of the fights. We spent a lot of time defining, and fighting. I don't think we have acknowledged that we were stuck in that place because of the worldview, which had nothing to do with the science. I think that these are remaining issues.

When we talk to Jon, and he talks about his research needs, what is best for him is back to the basics. That is attractive, right? That is easy; we can do that. It is close in. A challenge for us, and my own agronomy piece of the pie, is that we have a history of research, where we've been, what we might do. Some of the issues we were talking about with Jon, such as how do I do research, or think about rotations for intensive market vegetables versus those for corn? I struggle with that it is not helpful to Jon's immediate decision making. One way I've handled this, is to realize there are different issues and questions for Jon, for corn growers, and for academics. We all are grappling with systems, but with different questions about sustainability.

It's a good thing that sustainable ag raised these broad issues, so we have all these issues to consider, but we haven't yet figured out how to field them all. We talk about economics, but we use that as a proxy for yield, right or labor? Some scholars have gotten into the broader use of aesthetics, and we talk about community. There has been some good theory, but not a good way to get it into practice. We also need to be able to move forward, to be motivated to make this practical. If there was an invitation rolled out, "Hey, you guys, we are really going to do this," we could have a chance to unite our voice. I think we are too competitive.

Do you think we are competitive in sustainable agriculture in particular or agriculture in general?

I mean all humans are competing to have their ideas heard. I think now, our territory, and our chance to get our scientific ideas into practice is being challenged across the board, we've become more marginalized by skeptical public and vocal industries. For example climate change, critics including popular authors and politicians have tried to discredit climate change scientists.

Chapter 12—Jon Cherniss and Michelle Wander

There are several reasons scientists are not given serious respect. We were raised with the assumption we could identify and implement truths. For us, it would be easy if that were true. It is a changing frontier. We have to cope with the status of science, perceptions about rational thought, and ways to get information into play. Beginning in the 1980s, we assumed that we were arguing about sustainable ag versus. conventional agriculture practices. To address this, we also argued that you have to do this different kind of research, and not just collect narrow facts. The design of Rodale's Farming Systems Trial[11], which did not have little factorial treatments, was revolutionary in this sense. In the past 20 years, we have racked up proof that alternative research and farming methods perform well and discovered that there is more needed to get those practices on the ground. We could also count as successes the number of granting sources that support that kind of effort.

I think SARE[12] has raised the issues really well, and they've done a great job of that. In terms of funding useful applied work, I'd be less positive. There are some great projects, but SARE has different goals; their outcome focus is not satisfying to me because I don't see the results having lasting influence. We can look at some NRI, [13] and the new organic programs; it is notable that many are more focused on marketing than on ag practices. I think the ideas have come out. If you look at National Science Foundation and their emphasis on interdisciplinary and systems research I think you can argue we sustainable ag researchers, and agro-ecologists have educated them; it's encouraging. If you trace the origin of the idea, it comes from sustainable ag, and from systems ecology, our close cousins, the applied ecologists. I think that the main ingredient, the reason we've been leaders, is that we are applied scientists. We are different from engineers, who begin with a solution they want to reach before they attack problems.

Something that struck me recently - are we still asking whether sustainable ag is good, and whether small farms are good? Or, are we taking this as our starting point and assumption? This is something that is interesting because so many people are in the sustainable ag "movement," which presumably doesn't have anything to do with science. We need to give credit to advocates in the movement and accept contributions that are not purely science based. The importance that society places on agriculture is tied to our ideas about democracy. This is the beauty of the land grant university system, which is true in this culture, and tied to ideas in France and much of Europe. There is a lot of commonality about this, a central resonance in the human species about the importance of agriculture and land. This is very primitive. They say the story in Genesis is an agricultural allegory and is about stewardship. I think people recognize this, and are romantic about agriculture. Whether you are urban, suburban, or rural, you have some need to participate in agriculture, and you want it to be beautiful. We see this as an important opportunity; we also see it exploited and marketed. This is a scary point and we as citizens see green washing happening.

When the land grants were established, their doctrine was that of sustainable ag. They were talking about healthy agriculture for healthy people.

Some critics say that land grants never took any of the social dimensions seriously. You could look at this as a cynic, and assume that they were immediately corrupted and serving corporations. Or, you could accept that part of this challenge is the difficulty of doing research for highly individualized systems. For example, I could spend $10 million a year just on questions that Jon comes up with from our own farm.

We need to generalize. I look at organic as the model system for questions we ask about sustainable agriculture. Organic doesn't own sustainable ag, but it's the only ag production system legally requiring a sustainable system. It obligates itself. Now we see the local foods movement being energized explicitly because the organic standards are weak with regard to social issues. I think the organic standard lost power because it didn't commit itself fully to local, to community, to labor, so we see other standards and branding taking off[14].

In Europe we see more emphasis in the IFOAM[15] standards on the social dimension, and more interest in actually requiring those things - in the European Union standards, we see a more aggressive attempt to include them explicitly. The U.S. standards are generic and don't deal with the issue of local. You definitely need certification if you are going to export corn or other commodities particularly where they need to be processed, shipped and aggregated. At the farmers market, where people can broker the relationship directly, they are not going to include that at all. For us, what we need to do is to work together and simplify our vocabulary so we aren't spending 80 percent of our time educating about definitions. I don't think consumers care about the regulatory stuff; they care about quality. To satisfy them, we need to communicate effectively.

Somehow, in this discussion, we need to maintain local and global perspectives. Right now, the labor laws in Europe are going to let Turkish labor into the Netherlands, and this is already depressing returns to established organic farmers. At the same time, academics from the Netherlands are actively working in Eastern European countries that have few resources, to increase organic production. Wal-Mart has seen this fit, and so had focused on Turkey for organic cotton production. What are the implications of this for organic agriculture and its relationship to sustainability?

The truth is, agriculture, while traditional, is also one of the most rapidly changing industries, because we don't have to build multimillion-dollar plants to change our product. I can retool my production plant quickly with relatively very low investment. All of Eastern Europe could go organic tomorrow.

One of my motivations for this book is that it is about time to do a self-critique. We aren't going to get better unless we do one.

If I am going to do a self critique of the sustainable ag movement I want to first start by saying that the tiny group of people that deserve the most credit for successes have been advocates. Even though academics have played a role, they have not had the impact they might. The reasons for this include the usual list of gripes- including peer pressure, rewards, lack of courage, and competition. Everybody is so pressured to keep their science alive, keep people employed in their science and attract students.

Chapter 12—Jon Cherniss and Michelle Wander

Do you think we hit the right balance in sustainable agriculture in terms of promoting ourselves?

I agree we definitely have to promote and avoid the trap of competing among our selves and with others with aligned interests. The contribution that academics can make is to promote knowledge and observations in an open way, not as true believers, or evangelics. We have to define what we are, and be clear about who we are. Even though pure objectivity is a myth, we need to strive for it and do so while being passionate about pursuit of our questions. Social scientists call this "reflective objectivity."

Here is a question that I have; is there a unique role for non-profit organizations to play, outside of the university, or are they models for what we think sustainable ag programs should be? What should groups like Rodale and Michael Fields[16] do that we shouldn't do? This is a question, not an assumption. I think we can say at this point, this will all come together. We need to figure out how to compliment positive efforts.

Part of this is driven by social values. It is a lot easier when we were production agronomists, and could just say, "That's outside my box." (That's not my research area).

There is a hunger for applied learning and it needs to be satisfied. I don't want to be an advocate in my day job. Research on production practices or related economics needs to be respected on its own merits. We farm ideas and not crops. Folks coming to field days to see agriculture in practice want to see the courage and a realness of someone farming and not some academic person talk about principles. This has tempted us to do science from an advocacy point of view, and maybe this has a place, but if that is all that sustainable ag research and outreach is about we will let the public and consumers down.

I am surprised at the level of influence that advocacy groups have. For example, I asked a class of students what they thought was the most important aspect of sustainability and was amazed, because they all said "local production." This was a large group of students, and I was shocked that they said organic wasn't important. This is at a time with the science on organic, and the health impact of pesticides is known. To me it shows that sustainable ag is catering to the social movement of local. We want to be invited to the table. Science has to be strong enough on its own to evaluate social theories and, when necessary, to poke the sustainable ag movement in the eye when it is wrong. It may cost us our credibility if we do too much advocacy.

How exactly are you defining advocacy?

Advocates work to promote an idea to which they are committed. In science, we are supposed to withhold judgment but of course we investigate questions based on assumptions and these reveal how we think about the world. You could assert that because my research on the influence of organic practices on soil quality has proven that organic practices build soil, that it is advocacy. Playing by the rules of science, this is not so, because the test was designed in a way that allowed fair comparison. We should define how to collaborate well with others asking driving questions—this is how to do advocacy science. We need to also say, "this is our assumption" and be comfortable. We natural scientists know we should work side by side with someone else doing evaluative science, to scrutinize assumptions, to ask how components of the system are valued.

Then it comes back to assumptions, including assumptions connecting sustainable ag, sustainable society and community. These creep into my science—so is this advocacy? Agriculture has to take its place working with others, thinking about the role agricultural plays in society, and global ecology. It is just one of many of society's sectors. When it comes to implementation, systems science needs to serve community members. The issues of sustainable ag are local; issues can't be generalized without loosing the ability to optimize choice for each community. This is a great opportunity for participation and collaboration.

Agriculture is a model system that is so emotional for people. You have basic needs; shelter, your food. Folks support the notion of sustainable agriculture, but are they practicing or supporting it? We have the opportunity to help them practice or support it if we give them the tools to make good choices. This is what we academics tell ourselves anyway. McDonalds is seriously considering going to grass fed beef. Partly because of the health implications, that is increasingly supported by research and empirical evidence, and the humanitarian issues that concern many people. All those little girls who became vegetarians at age 12 because they saw beef slaughtered might be won back, right? I can ask, because I am one of these vegetarians challenged by the "Slow Food[17]" emphasis on meat. The way we will get sustainable culture is through agriculture.

Along these lines, I had a friend comment that she sometimes feels powerless to change many political things in the world right now, but at least food, and what she eats, is one thing she has control over, and can help change.

I think you are right about this—the tremendous number of efforts to build networks to help folks do this is amazing. Think about the Slow Food movement. All these people are coming together to think about food and food culture. It goes way beyond Epicureanism. It's an excellent vehicle to bring a new set of people, consumers, to the issues. There is a real interest in having a quality life. It is about consuming responsibly, and about locally produced food, and about sitting around a table. We see people responding to values. They go to the farmers' market not necessarily to get the heirloom tomato grown organically. They go to the farmers' market to see children, or to hear some child playing Suzuki violin over and over, those things that we are starving to death for, quality of life things.

****

The rest of the interview started with both Jon and Michelle at the table the next day. We discussed the book process in general, mutual friends we hadn't seen in a while. After some chitchat, Jon left to go work in his office in the barn. Michelle and I kept talking about world politics and sustainable ag.

Chapter 12—Jon Cherniss and Michelle Wander

We are very comfortable at the universities now with organic, because it's taken out all the values. But there is no soul in it any more, so we can just talk about upholding those rules. However, you take one step back, the rules have very little to do with science. I do think we have a lot to offer from science, but the thing is we have to have this discussion about pitting science against values. This relates to our conversation the other day, how do we engage in something that is value-driven with non-biased science? It is a challenge. Your question, have we been too quiet, or too strident? We have to figure out how to do all of these things, to say we are going to use science to its full effect for things that are socially valued. We absolutely have to do that. We can't get our scientist stripes removed by engaging something we think is really good.

We spent a little time enjoying the view of a humming bird in the garden. Then went back to the general topic of the book and its three central questions.

Your questions are of interest. You could spend a lot of time on what we've done wrong. Some other things are so discouraging, but this topic of what to do next is an area you can control, you can make progress. Look at opportunities we can see. We are getting better at describing what is going on and understanding this at different scales, with different connections. Now if we can figure out how to share and divide the duties, and be strategic with other organizations that are more motivated with much less ambitious goals. Sustainable ag is really is a sort of euphemism for sustainable culture, right? So I think when we get down on ourselves, we have to remember we are trying to do all this, and we are trying to do it with great respect and inclusion of one other, versus somebody with a really narrow motive; to make money, in this accounting year, by selling something. They don't necessarily care about telling the full truth, the whole truth, and nothing but the truth. They just want to knock our argument down and sell, and we want to do all this, which includes selling things as a component.

One person commented that this is our greatest strength and our weakness, that we try to cover all these issues, because they are all connected.

Do you consider sustainable agriculture a movement?

Calling it a movement is appropriate, because it is a movement promoting certain social values. I think for me, we can see the question really well defined for ag, while relevant to societal issues in general, which is the question about the role of scientific information. People with narrow motives—climate change or holocaust deniers—often succeeded in marginalizing informed voices. This is serious, deadly serious; it's more serious for some things than others are. We are going to run out of oil before we exhaust the soil, but we will accelerate the rate of soil depletion because we have denied climate change for too long. We need to take science literacy seriously or cede the field.

Farming in the Dark

What is your view of the control of corporations over universities? Some of this can be off the record if you want.

We are on the dole of course and this is getting worse because public support for education and research is drying up. That said, I think people's perceptions about the university, and how corporations influence what we do aren't exactly right. The public believes that access to funding entirely defines what we do. Money is one factor of course, but other things do as much or more to hold us back. The way that we academics are hamstrung is not that clear. We are mostly prisoners of our own fears and our assumptions about peer's judgments and the reward system. We are totally free, and can do whatever we want to do. When I came to the University of Illinois and did work on sustainable ag, I limited myself, and didn't work directly on organic. People who disagreed with my worldview, which was revealed by my work on soil quality already suspected me. I had nothing to lose and really should have been more aggressive about my views. That was my mistake. I censored myself.

Of course, there is the fear of never getting tenure. In general, I don't think folks don't get tenure because of their worldview assumptions, but having alternative views can undercut your productivity for a variety of reasons. This is like a handicap horse race where horses carry different weights in a race. We need to remove those weights. Nobody tells you that you can't work on this or that. Right now, the universities are afraid, the public is divesting in them, so we are losing our critically important thing—the land grant mission. If you look at the values in those original founding documents, they are for sustainable ag, and they define agriculture for culture. They include aesthetics; they include wildlife. It really is these ideals of people on the land, with knowledge, so they can be good citizens in a democracy. It's the full meal deal, everything we could want. We've never gone for it, have we? People talk about it, these are so important, we have to go back to our roots; or was the charge in that mission just a rhetorical flourish, and we've never been serious about the land grant mission?

Right now, we have this real problem of marginalization of the value of education, of the citizen. This isn't just ag corporations training work forces. People, citizens, we've agreed with industry that we should be training workers, instead of citizens. If the public isn't willing to pay for a broadly educated public, the liberal arts education will die.

Does the public really want to pay for the education of Cargill employees?

This is accelerating the public's divestiture. Unfortunately, our administrators have largely accepted the "life sciences revolution," which is another term for biotechnology, as the solution, and they see public-private partnerships as the key to a well-funded future. North American and European institutions have all bought into this as a means to solve their financial problems. In my view this is the fastest way to put land grants out of business. I think we have to articulate our opposition to this approach very loudly, and fight very hard for the land grants, or they will not, and should not be funded.

I think we are all our own prisoners, we are afraid to dress the dean down. Go do it, and the deans will actually listen. We're just too cowardly.

190

Chapter 12—Jon Cherniss and Michelle Wander

I know that I've let other people fight those battles for me, but it also seems more effective if farmers do it.

You are probably right. Again, were back to figuring out job assignments, and who and how to get ideas heard. That's one thing the farmers can do. There are things some farmers don't understand, or have time and motivation to articulate.

I don't mean to say that we, the sustainable ag research community does not have important knowledge to contribute to big corporations or the biotech industry, we do. If industry was smart, and our institutions backed us appropriately we could do the kinds of research that get out in front of environmental, health, equity related systems-type problems before they occur. This is almost completely impossible given the way research relationships between industry and public scientists are currently brokered.

For sustainable ag in the university, it's a question of how can science inform the movement, and work for the movement, while sometimes taking its case as a given, but also questioning it at the same time. That's what we have to figure out how to do. It would be great if the movement had more confidence in scientists and went to bat to support related research. I know that some of the non-profit groups try to do this. Research support from SARE and other groups is great, but these programs frequently specify what researchers can ask. Where is the room to ask the big, challenging questions about equity, scale and resource use that we need to tackle?

Notes

[1]Several publications resulted from Michelle's dissertation research at Rodale, including: Wander, M. M., Dudley, R. B., Traina, S. J., Daufman, D., Stinner, B. R., and G. K. Sims. 1996. "Acetate fate in organic and conventionally managed soils." *Soil Science of America Journal* 60(4):1110-1116. Wander, M., and S. J. Traina. 1996. "Organic matter fractions from organically and conventionally managed soils: I. Carbon and nitrogen distribution." *Soil Science Society of America Journal* 60(4):1081-1087. Wander M., and S. J. Traina. 1996. "Organic matter fractions from organically and conventionally managed soils: II. Characterization of composition." *Soil Science Society of America Journal* 60(4):1087-1094. Wander, M., Traina, S., Stinner, B., and S. Peters. 1994. "Organic and conventional management effects on biologically active soil organic matter pools." *Soil Science Society of America Journal* 58:1130-1139.

[2]Wes Jackson and Wendell Berry, authors and spokespersons, especially in the 1980's as the movement was just beginning. Jackson's early writings include *New Roots for Agriculture* (University of Nebraska Press, 1980) and Berry's most influential book was probably *The Unsettling of America – Culture and Agriculture.* (Sierra Club Books, 1978).

[3]Nancy Creamer – Was also in grad school at Ohio State University in the early 1990's, and now coordinates the Sustainable Agriculture Program as a professor at North Carolina State University. She also happens to be Jon's sister. http://www.cals.ncsu.edu/hort_sci/faculty/creamer.html

[4]New Farm Magazine, a Rodale Press/Institute publication from the early 1970s to the mid 1990s, now available on-line at www.newfarm.org.

[5]ATTRA (appropriate Technology Transfer for Rural Areas, funded by congress to provide alternative agriculture information, www.attra.org

[6]Elliot Coleman, author of several books including *The New Organic Grower*, Chelsea Green, 1995, and *Four Season Harvest*, with Barbara Damrosch, Chelsea Green, 1999.

[7]An economic analysis of Wisconsin organic vegetable farms at 3 different scales was published after this conversation with Jon, that partially addresses some of his questions. See http://www.cias.wisc.edu/pdf/grwr2grwr.pdf for the executive summary and full report.

[8]National Organic Standards, developed national standards beginning in 1990 (authorized in the farm bill) which took effect in 2002. See http://www.ams.usda.gov/nosb/index.htm

[9]*Food First* by Francis Moore Lappe, 1981.

[10]Miguel Altieri, author of several books and many research articles. See *Agroecology—The Science of Sustainable Agriculture*, Westview Press, Second edition, 1995.

[11]Rodale Institute—Farming Systems Trial, started in 1981 at the Rodale Institute in Pennsylvania to look compared three cropping systems; two organic and one conventional. Several papers are in refereed agricultural journals now about the study, which is still on going. For current information see http://www.newfarm.org/depts/NFfield_trials/0903/FST.shtml

[12]SARE—Sustainable Agriculture Research and Education, USDA program funds grants for researchers, farmers, students, and promotes education of agricultural professionals on the topic of sustainable agriculture. See www.sare.org

[13]NRI (National Research Initiative) funded by congress, often for more basic research, now includes several programs that lend themselves to funding sustainable agriculture, such as the Systems Research program, the Organic/Transitions program, and others. http://www.csrees.usda.gov/funding/nri/nri.html

[14]For some examples of other standards that take labor, wages, etc, into account see Consumer Reports Eco-Label website at http://www.greenerchoices.org/eco-labels

[15]IFOAM—International Federation of Organic Agriculture Movements, goal is to provide an umbrella organization within which to unite the various organizations within the organic movement. www.ifoam.org.

[16]Michael Fields Agricultural Institute, research and advocacy non-profit sustainable agriculture organization based in East Troy, Wisconsin, http://www.michaelfieldsaginst.org/

[17]Slow Food movement, see http://www.slowfood.com/ and http://www.slowfoodusa.org/

Discussion Questions:

1. Do you think that farming at the scale and intensity that Jon farms, with employees, financial risks, etc, should be considered a lifestyle choice or an economic choice?
2. If it is an economic choice, what would be a reasonable remuneration for someone with the skill set to manage six employees, several acres, machinery, create a marketing network, etc.?
3. What did Michelle mean when she said, "The soil remembers itself, it embodies a consciousness?"
4. How would you answer this question: "How do we engage in something that is value-driven with non-biased science?" How has the role of science changed in the last 25 years, with respect to society?
5. Do you think that sustainable agriculture could help bring about a more sustainable culture?

13

Terry and Sheila Holsapple
Greenup, Illinois

"And that's what I'm trying to say about sustainable agriculture. Farmers simplified it. The universities made it hard, and put it out of touch to the reality."

I first met Terry and Sheila when they became on-farm research cooperators with the Rodale Institute in the late 1980s. They live in a mixed terrain region of Illinois, mostly flat, but not all flat. Their farm is beautifully kept, and Terry was featured in The New Farm *magazine[1] more than once; first with his amazing hairy vetch crop, and later with his pumpkin patch and other marketing innovations. He was definitely on the leading edge of agro-tourism. This interview with Terry is particularly interesting, because of the on-farm research cooperators[2] that I interviewed for this project, Terry probably surprised me the most—he isn't farming any more!*

He doesn't have any regrets though, or at least any that he expressed to me. His views on farming, on working with universities, and working with groups like Rodale and others are insightful. He also has some great stories, like the time that Rodale asked him to go to Washington, D.C. to speak to legislators about sustainable agriculture, and the shuttle bus from the airport wouldn't stop at his hotel in DuPont Circle, so he had to jump off with his luggage while the bus slowed down. Or the time he was in Washington with George DeVault (editor of New Farm *magazine at that time), and a hotel employee drove George's truck around from the back parking lot with a flat tire, much to George's dismay.*

My husband and I stopped in Illinois to visit Terry in the fall of 2005 while driving to Washington, D.C. Sheila was at work that day, so we didn't get her take on things, but I always enjoyed my conversations with Sheila at our on-farm research cooperator meetings. We were also able to meet Terry and Sheila's daughter, Page, while we were there. Just to give a sense of time passing, she is in college now, and was four-years old the last time I was out at the farm!

How did you first get into farming, and interested in sustainable agriculture?

I raised livestock as a kid. After high school and after I got out of college with a two-year ag production degree, or something like that; I started out farming at a rather young age. My parents farmed, and had a business, so they did both. But it turned out that this was absolutely the wrong time to get into farming. I got out of college in 1975, got started, and within a few years interest rates went to

20 percent in the early '80s. Boy, that nailed me pretty good. Everything was working just fine but the interest rates, we were working on borrowed money. It pretty well took us down to nothing. We were subsistence as far as cash flow goes. There really was no money to be made. We just survived.

How did you get into hairy vetch and the other cover crops?

I'd read an article about it, a person just south of here 60 miles was raising some vetch. I found some seed and started growing it. I swear I can't remember why Mike Brusko, editor at the *New Farm* magazine called me up and wanted to come and take a picture one time. Shelia tells me I wrote a letter to the editor about some article or something. It just happened to be an outstanding spring that year. When he took that picture, the vetch was this high . . . I remember that picture. He always said that was the most asked-for picture they'd ever done. That really got things started. Not only individuals, but also I had colleges, professors called me up wanting to know about hairy vetch. Back then we didn't have the internet, so it took a lot of research to figure out how much nitrogen was in it, the potassium, Another farmer, over here by West Union, Mike Strohm, he was kind of interested in it too, so we worked together, figured out a lot of things about vetch, and promoted it.

Next thing, Ken McNamara, Rodale's on-farm research coordinator, came in for the on-farm research part of it. Once we were on-farm research cooperators, they published stories on everybody. We had field days, and then they got into policy, sustainable ag in Washington; they were flying some of us out there . . . went out there a couple of times. They gave my name out, and I had to go to Minnesota to give a talk to some clean water people or something. I'll never forget that because it was the first time I had ever really spoken in public, had a slide show and everything, and I didn't know how it would come out. But, it was all right. It worked out fine.

I'll tell you what got me interested in organic agriculture. My wife ordered a *New Farm* magazine from some school-kid that came out here selling them for band or something. The first one I just threw in the trash because I didn't want to have anything to do with it. The second one had Glen and Rex Spray's picture (organic farmers from Ohio) on the front, and said they were farming 600 acres without chemicals. Man, did that ever strike a note. I read every word of that, and every one after that. That's what got me started.

What was it about their not using chemicals that was appealing to you?

They just never used any. They were going against conventional wisdom, and they won the battle. A couple of years after that, I was on the same speaking tour in Ohio at an organic convention, and got to meet them.

When did the hay bale maze and pumpkin patch start?

That started about 1985 I believe. I'd never grown pumpkins before. I just took a little plot out here, grew an acre of them. People had never seen an acre of pumpkins. They'd come out and drive back and forth out here looking at those pumpkins. The second year, I did the same thing again, and that year people were driving by again, so we decided to sell a few of them. On the weekend, I'd cut up a few and make some Jack-o-lanterns out of them, so it looked like the field had Jack-o-lanterns in it. And people really got to driving by. We always had family dinners or suppers every Sunday; we'd go to somebody's house, and it was at my house in early October. These people were driving by looking at those pumpkins. My mom said, "I'm going to put a sheet over my head, and I'm going to go stand out there by the road in the weeds, and when they stop to look at those pumpkins, I'm going to scare those kids." Well she did. The haunted pumpkin patch—I mean, the word got out. The next weekend it was just cars going up and down the road looking for that ghost. That's how it got started. We started thinking that maybe there was something to that.

The next year, we had pumpkins for sale, and I thought I needed a little education on this. I went to the Illinois fruit and vegetable growers convention. They had a speaker there, flipping through slides, and he flipped through one and never even stopped, and said, "This is our hay maze" and just went on. I thought, "hmmm, a hay maze, I wonder what that was." I got to thinking about it, and how a person could do it, and I came home and built one. We had 700 bales, it wasn't very big, I didn't know what I was doing and this was with little square hay bales. It had to be straw, because if you had hay, the nitrogen content was too high, and it would rot. We used square bales for several years, and we got up to 3,000 per year! It was quite a job. We'd actually build it in one day, and tear it down in one day. I had as many as 15 kids helping build it. I would told them where to set the bales and they set them.

Then we went to big squares. That saved a lot of labor, and I built the whole thing on the seat of a tractor. We would use about 400 big squares, 3' by 3' by 8 or 7'. These were two bales high. When we built with small bales, they were five high. We went that high so nobody could see over them, but they were hard to keep stationary, we had to drive posts, tie them to the posts . . . people got rowdy in there you know. When we went to the big squares—I love technology! The big bales were fantastic for what we were doing. We could save the bales from year to year, as long as it didn't rain too much. That's how we got the pumpkin patch started and the hay maze was a big part of it because that was a big draw, and then everything else . . .

We tried to add one or maybe two things a year as we went along. It was very, very good to us. We had two "fun houses," we had a hayride; I made a train. It was actually pulled by a four-wheeler. We used wheelbarrow frames, and then put on feed tubs, mounted them. We just did weird things like that, but made it so people could get in and out of them easy. It was very popular. We put in the "tunnel of doom," which was a rotating cylinder that people walked through in the dark; it threw you off balance. We put in archery. Believe it or not, archery was very, very popular. We bought a 3-D bear, a deer, a groundhog, turkey, and then had just a regular target. We had a section of our shop out there with an overhang on it, and just made it into strictly archery. We put Styrofoam on the walls on the ends, decorated with corn stalks on the sides, little trees out there, made it realistic. People would just stand in line to do that.

Did people pay for each individual activity, or as a per-day rate?

As we built this up, we charged individual fees for everything. The last couple of years, we went to a one charge for everything, and increased our income 25 percent when we did that. It was great, really great the way everything worked out. I think the rate was $7.50 to come in and do everything you wanted to, and it worked out very well. We did things like the "rat racers."

What's a rat racer?

A friend of mine, he's just a natural at carpentry. He said, "I always thought we should make something people can get in to go out to their pasture and look at their cows." I said, what are you talking about? He said, "Something they can get in and just start walking, and it will roll with them, and they can go out and check their cows or something." I said that sounds like a good idea, but I have no idea how to make it. Two weeks later, he pulled in with two of them already built on his trailer. He already had the idea in his head. That was another very popular thing; unique, nobody had them. People came from all over. I kept it a close secret how to build them. I let a friend in North Dakota have a copy. He's also a natural carpenter, and he re-designed them a little bit.

I don't understand how this works?

They were 9 feet tall, 24 feet around, and 4 feet wide. They look like a giant roller. You just get in them and start walking. I rigged it up so that whenever they walked in it, it unrolled a rope off it, and when it came to the end of the rope, it would stop them. Then the rope would wind up when they walked back. So I had an automatic stop on it that way. For years, they would take them and go as hard as they could and run into a bale, and that was a little hard on equipment! So I finally put these stretch ropes on them, it would stretch out so far, and then stop them. That worked really well.

As time went on, we learned how to manipulate those things. We put in a basketball court, and did something called "Barnyard Olympics. They had to climb up over a couple of big bales, slide down a slide into some inner-tubes, get out of the inner-tubes, jump over a bale, go through a tunnel, swing on a rope, then get on a pedal a tricycle to a certain spot, get a golf club and try to knock a ball into a hole, that kind of thing. We tried to change it a little bit every year. That was a lot of fun. We did it all

Your customers came from how far away?

At then end, as word got out, they actually come from as far away as Chicago. We never ever really tried to tap that market. It just happened. Advertising for something like this is fickle. The very last year I finally figured out what to do and how to do it. Radio is really not very good, newspaper is too expensive; television is just too expensive. What we did, it was the election year, and I was driving the mail route, seeing all these political signs around. Some of them, you couldn't even read, they had the wrong colors, you know. I thought, an orange sign with black letters would really work well. I had some signs made up with our name, the location, all the information we needed on orange with black letters. I took off and went 300 miles one day, setting these signs all out, within a 60-mile radius. You talk about bringing them in. It worked. I put them at intersections, exits at Wal-Mart's, any place I could stick one and I thought it would stay a few days. Marked down where I put them, and went back and picked them up. Most of them, believe it or not, were still there. It was like an instantaneous hit. I only shared that with one other person, and I don't think they ever used it. If I ever do anything again, man, that's the best advertising I ever did.

So you were making money on the farm entertainment activities. Were the crops, the hairy vetch, any of that making money?

Well, crop farming is hard to figure out. My brother and I would sit and figure out the numbers, and when you were done, there wasn't anything there. You sit and watch some of these guys, and you wonder how they do it? Most of the farmers who are still farming around here, that are any good, started years and years ago, had land paid for, so they can actually take and subsidize a little bit. If you have land paid for, you can make it. If you are trying to buy it, you just can't do it. When my granddad was farming, they bought land back in 1947, planted soybeans that year and paid for the ground that year with two-thirds of the crop off that field. Then it got to the point where it would take an extra 20 acres to buy 40 acres, and then it took a 40 that's paid for to help pay for a 40 that you just bought. Then when I was doing it, it would take an 80 to buy a 40. Now it wouldn't pay for itself, in a lifetime. I haven't even figured it up lately.

What does land go for around here?

$3000 per acre, or more.

Are there still cash grain farms around here? Are they making money?

I'd say 90 percent of them are cash grain. As far as making money, they are still farming, so I guess they are. As long as the American people are willing to subsidize farming, we'll continue down the path that we're going now. That is the only thing that makes it worthwhile. That's like so many other things, as long as we are comfortable subsidizing different programs and things, it will continue. If the subsidies were to go away, and they had to do it on their own, they wouldn't last very long.

Do you see anybody having to sell off land to stay on the farm?

Not near as much now as you did 15 years ago. In fact, there is very little land changing hands.

What does that mean?

It finally just shook out the people that were weak financially. It's in good hands now. There is a lot of money out here on the farm. When land does come up for sale, a bigger farmer will just normally step in and buy it. They actually have the cash. They've had their land paid for 20 years, they're farming a lot of acres, taking a little off every acre. They're making it work, but they had to have that initial land paid for.

So a recent college graduate shouldn't plan to come out here and start farming?

For someone to step in here now and start farming, it would be a miserable life, as far as I am concerned.

What my neighbor did down here, he's approaching 60 years old, he went and got a college graduate that's farming with him. As they farm together, he's actually transferring over to this kid. The boy is buying him out as they go. That is the only way he could have ever done it. I see farmers here, they went together and they share machinery now. I'll be honest with you; farms are still too small. When you look at the technology that's available, these guys ought to be farming five and six thousand acres, not two and three.

You mean to make efficient use of the equipment?

Exactly. Some of these people have the equipment to farm that much, and they're only farming two and three thousand acres, and they should be farming five and six. It's probably going to narrow down even more. I know that's not what sustainable people like to hear, but facts are facts, and truth is truth, and as technology goes on . . .

You really want to know what allowed farms to get so big? The machinery has always been there to farm big, for the last several years. The one thing that has allowed farms to get big is Roundup (an herbicide, also known as glyphosate).

Along with the Roundup ready soybeans?

They can farm thousands and thousands of acres and all they have to do is spray the weeds once, maybe twice, and that's it.

Herbicide programs used to be a lot more complicated.

Absolutely—and you didn't know if they were going to work of not. Roundup is 100 percent. Roundup is the one technological product that has changed the face of agriculture.

So everyone around here is growing the GMO beans?

Absolutely, these river bottoms used to be infested with horseweed, giant ragweed, shatter cane; you name it. These bottom fields were just horrendous. Shoot, they are clean as a whistle now. Clean as a whistle. All Roundup.

Of course, that speeds up harvest if you don't have any weeds. It all worked hand in hand. Roundup is the one product that has changed the whole face of agriculture. People can argue whether that's good or whether that's bad.

It's sold as an herbicide that isn't as environmentally harmful as some of the others, but it is also simply harder to detect chemically in water supplies, streams, etc.

I would argue the point that some of that nasty stuff that we were using, is now put on the shelf, when we went to Roundup, and probably the environment is cleaned up dramatically. That's my view on it now. As I stand back and look at it, some stuff out there was just terrible, terrible.

Are you seeing any Roundup resistant weeds yet?

People speculate on it, but I don't see it here. With the high-energy price right now, you'll see a tremendous move to no-till. Roundup has made that possible.

What these guys do now, they go in the fall, put on very small amounts of a Sencor type product, another class of herbicide, to keep the winter annuals from coming up, like creeping Charlie, shepherd's purse. It keeps the fields 100 percent clean going into spring. The ground warms up quicker, they can go in there and no-till quicker, because the ground is warmer without all that weed cover. They are using small amounts in fall to take care of the winter annuals. In the spring, they go in there and plant, come in and spray their Roundup. It's simple.

Are people renting much ground?

Like us for example, we cash rent every acre we have to others who want to farm it. It works out for us; we're guaranteed income.

Does anyone share rent?

There are still shares out there. I know a person from the mail route who's doing shares. We talk every day, and I get all over him. Like this year, he'll probably lose money, and he could have locked in the rent. The crop insurance will pick up the loss to the farmer, but now he's taking a loss. That's ridiculous. He's just a small guy, he doesn't have crop insurance.

Back to sustainable agriculture. What do you think are some of the accomplishments that you've seen over the years?

The best thing that I've seen is intensive grazing. That is one thing where there is still money to be made. As far as crop production goes, we talk about sustainable ag, I went into the organic side, another step beyond sustainable, I guess you would say. When I was doing organic, many people said, "He's got weeds," but I made more money doing that than any other time I farmed. We were getting premium prices, $14 for soybeans, $0.60-$0.80 per pound for azuki beans.

You were raising cantaloupes too?

I raised watermelon, I raised tomato, I raised cantaloupe, sold it all locally. We did that in a period when cash flow was low, we needed cash and it got us through. We survived to fight another day. It can be done.

Did you get certified organic?[3]

Yeah, we had 80 acres certified with OCIA (Organic Crop Improvement Association). The produce part, I didn't advertise it much as organic, or if we did advertise, we didn't get a premium for it. We just got wholesale prices from the stores, or if we sold it on the street, we got retail. It kept us alive.

Would organic be on your list of accomplishments?

Absolutely. I had some reservations and still do about the organic people certifying and things like that. When they let the government into it, that was the biggest mistake they ever made. That's when I got out. We had an opportunity to sell the 80 acres, and get completely out of debt. I jumped at it. So we sold that 80 and got out of debt.

Did they keep it certified organic?

No, they just wanted the land.

As far as sustainable and ag crops, it can be done. I proved it can be done, but you have to operate on a smaller basis because of more weeds, you got to control the weeds, you really need a livestock enterprise to intermingle with that to get the nutrient base. It's just more work. In this day and age, people are looking to get out of work. It makes it tough.

Are there any other organic farmers here in this area? Anyone growing cantaloupe?

No.

At that point, my cell phone rang, and our house sitter was calling to find out where we set the coyote traps in the pasture, so she could go check them. We had lost several lambs that spring, and were just starting to try different methods of control. It turns out Terry has been quite a trapper.

I wish I was there. I'd trap them for you. That's one thing I always loved to do was trap.

He pulled out several magazines with articles about trapping, and mentioned that the best bait was a dead cat. He said they even sold a sort of bait paste that smelled (or was made out of?) dead cats or bobcats. My husband Raad was intrigued, as he's never liked our cats that much anyway, but I quickly tried to change the subject!

What are some things the sustainable ag movement hasn't done as well? How would you critique the last 20 years or so?

Well, there are so many organizations. There's Rodale, there are all the universities in each state, there are so many different groups that are probably not coordinating with each other. For instance in our state, I see people promoting sustainable ag, and actually what they were after was a position somewhere. They jumped on the bandwagon to get the position, and once they get the position, their focus goes somewhere else. I've seen that a lot.

You mean at the university?

Within the university system. Off in another direction they would go. Probably that turned me off more than anything did.

I can't say that it (sustainable Ag) really failed. It filled a void for many people looking for something different. In many cases, it helped some of them get through the rough times. There were some rough times back there. Is it filling a void today? It goes back to the old definition, what's sustainable?

So how would you define sustainable Ag?

As I said, for some reason, I jump straight to organic. To me, sustainable, you're trying to cut your input costs with sustainable. That is just so limited on what you can do. Why not take the next step, go organic and be done with it. That is always my take on it.

It goes back to the old definition of sustainable. I remember sitting in meetings, and people would argue for two hours about that; what's sustainable, what's not sustainable . . . I'd get so sick of it, you know. It meant something different to each person.

It seems like we spent a lot of time doing that in the early years.

It'd just drive you nuts! That would be a good complaint that I would have, during that whole episode, when it was really a hot thing. Everybody was interested in it, it was just so tied down to what was and wasn't sustainable and the arguments would break out. It took a lot away from what should have been done to move it forward. Once it got the definition, everything just went away, or the focus went away, I don't know. Or maybe I just went away? [Laughter] I'd had enough, or something.

For some people, the USDA/SARE program is a milestone. It offered grants to universities, and had the farmer grant program. That program seemed to help the people in the university, helped their credibility within the university hierarchy. Did it have that same effect outside the university?

Chapter 13 – Terry and Sheila Holsapple

You touched a hot button for me, when you talk about money flowing into the university system, the program and stuff. I know you are in the university system, and I'm not trying to diminish that. I have some personal experience with that. They were starting a sustainable ag deal here at Illinois. We all met at Allerton Park up at Montecello. We had all the different interests, including the mainstream corn growers, the soybean growers, then we had the sustainable people and the organic people—we hade them all. The Farm Bureau was directing this. They were trying to get a few million dollars so the universities could be funded to do on-farm type research. Well, what it came down to, the university, they wanted a block of money for the maintenance of the university, they wanted a block of money for the faculty and they wanted a block of money for the research. They had it broken down that way. I said no, it should not be done that way. It should be competitive grants only. The university is going to get tax money to run the university, to keep the lights on and everything. I said, you do not need part of this money going for that. And man, the war was on. They were there to get the money for their programs, to keep their buildings open, to keep their heat on. That was their concern. I fought against it; it didn't do any good because all these people outnumbered me

That is how the SARE[4] (Sustainable Agriculture Research and Education) grants are set up. There is no overhead for infrastructure. Other grants, we can get as much as 45 percent overhead, and in the SARE grants, there is zero. All the money goes to the research.

That's good. it ought to be that way; that's the way all of them ought to have been as far as I was concerned. In this state, that wasn't necessarily what happened. What I'm trying to get to is that I can go tell another farmer, "this vetch, it will put out 100 pounds of nitrogen to the foot, loosen up this much potassium in the soil to use next year," and he'll believe me. If you go to the university they say, "We need to do some research on that to see if this is true or not." The research has been done, but now they are trying to go back, update it, figure it all out. The numbers were already there. The others and I who have done on-farm research with vetch, we can go one to one with another guy and convince him, then they can do it themselves and find out on their own. All this money was being spent, going down a rat hole. You know what happens when it gets into academia; it's hard telling what's going to come out the pipe on the other end! [Laughter.] I always thought there was a big waste there.

Then they come out, and want us to do research on the farm. You know, farmers are not oriented . . . Dick Thompson[5] is the only one I ever knew who can do it the way the university did it, the way it is supposed to be done. There isn't one in 500,000 people who could do it like Dick Thompson does. I'm proud of what he's done, and he's done a lot! But I didn't want to do that. That was not my focus, not my interest. My interest was growing things, seeing how to grow it, sharing that. I wasn't interested in little bitty plots, pulling weeds, and weighing things. I'll tell you how a farmer thinks. I love to tell this story. I had a big field of vetch, here in my lot out here. McNamara, the Rodale on-farm research coordinator comes out, says we need to take and get some samples of this, test it, see how much protein it has in it, so we can convert it to nitrogen and all this. I said, how do we do that? He said, we 'd make this square, throw it out here randomly, clip a sample, dry this down, and weigh it. I said, I can probably do that. It's going to be a job, but I can do that. He shows up out here one day, and I said, what do we have to go through to do the calculations on this? He said, it's like this—he took a manila envelope, laid it out on the table, started out at the top, and when he got done, he had an algebraic formula, on the whole cotton-picking thing! Well, I hadn't had algebra since high school. I said, I have to do this on every single sample? He said, yeah, that's right. And I have to do how many samples? And he told me. I said, Oh my gosh. I don't even know what this is. Explain what this is . . . explain this . . . so we went down through it. Then I said, why didn't you just do it this way? I added, I subtracted, I multiplied, and divided and I had the same answers he did. He said, "you farmers, I hate you. You're always taking short-cuts." [Laughter]. I took his algebraic formula, and I simplified it.

That's what I'm trying to say about sustainable agriculture. Farmers simplified it. The universities made it hard, and put it out of touch with reality. Now, there are a few of them, well, nerdy farmers I'll call them, robo farmer, they like this stuff. Most of us didn't. We just wanted to grow things. To me, that is a good analogy. He said, "I don't know how you got that answer. I can't do it that way."

We went on to talk for a little while about how hard it is to even get a hairy vetch sample, with the vine nature of the crop.

I went on to suggest that having numbers, like nitrogen content, can be important too. Having a number is better than not having one, especially for field days.

Let me say this. As far as farmers go, Mike Strohm and I, we figured out that to relate to farmers, they'd ask how much nitrogen do you get out of this? Finally, we'd have to say, well, for every foot of vine length there is 100 pounds per acre. It worked out real close to that. They could relate to that, instead of some algebraic formula.

Do you have any other comments about various sustainable agriculture organizations or your experiences with the university? These can be off the record if you prefer.

There were some good people, no doubt about it. Some of them were very well entrenched. I'll never forget going to a university of Illinois in-service type meeting. The professors were there and they had somebody come in to speak. Lo and behold, who did they have in, the head of the Potash and Phosphate Institute. I was invited to come by someone from the university who said, "You need to come and sit in on this." I was sitting there, listening away, and my name comes up! On the stage! This guy was saying that we got people running around out here saying that they can grow crops without putting on potash (potassium).

Did he know you were in the audience?

No, he didn't have a clue, and I knew you could grow crops without potash! I was mad. I perked up and got real interested then. They had people with a vested interest funding their program, and saying, "You guys, you don't do this, as far as sustainable ag goes, because we're funding you." That was the intention. He didn't come right out and say it, but that was the intention. "You guys need to watch who you are walking down the street with." I found that very enlightening.

A friend of mine who was up for secretary of agriculture for the state of Illinois, we cornered Ted Peck at the University of Illinois on this potash thing. The soil test recommendations that they were issuing at the time; you take a soil test, this time of year etc. They were wrong. As far as I was concerned, they were wrong. If you use a test that shows the uptake of potassium into a plant, don't you want to take that test when it is actually available in the soil? Not when it is tied up? We jumped him on several different things. They got the Morrow plots up there, been there since 1876. He said, we couldn't understand it —the potash numbers coming up in one plot where we never ever put any on. He was agreeing with us that you probably could loosen up the potassium in the soil, but on the other hand . . .

I'll be he still wanted you to put some on?

We went around and around about that. We could prove it, raising vetch. I can't remember the percentages now, but if you could get 100 pounds of nitrogen to the foot, you could get something like 43 pounds of potassium that would be available to the plant the next year. That's more than enough to grow a good crop. You couldn't get those people . . . can't get them to do it.

What is your prediction for the land grants? Will we need them, or will people telecommute? Are they earning their keep?

Land grant universities? That's a good question. I hadn't thought about that for a long time. You still see them constructing buildings. At the University of Illinois, they just moved the sheep barn that's been there for 100 years, sold it to somebody to move so they could build a new complex of some kind. Out with the old, and in with the new. As far as needing them, boy, for farming and things like that, boy, I'd have to sit and think about that for a while.

Do you still have county agents out here?

They person we have is doing two counties. To be honest, he does more 4-H than being county extension person, that's what it has turned into; and they've disintegrated 4-H in my estimation. All the rules and regulations that they have, it's just ridiculous. It's more community service stuff. It's just insane. If I want to go out and do a community service project, I'll just go out and help someone. You don't make somebody do it to get the recognition to be a club. That's where they are. As far as the university on-line stuff, I'd have to think about that. I'm not a supper big university fan, as you well know, but they do have their place. That's where the technology is going to come from, to lead us on.

Some of the other organizations, they jump around where the money is. They may go over here, "Well the hog farmers have dire troubles," then they jump over there . . . They started out as coal miners' spokespersons here in Southern Illinois. Then they jump over here and got into sustainable ag because they see they can write grants for that. They made people aware, and they have somebody new running that organization every month it seems like. They just kind of jump around where they can get funding, that's my take on it.

Farming in the Dark

In terms of information sources, how would you compare non-profits, universities, and other sources? Where did you get information? Mostly from other farmers?

Some of it was. A lot of this research had already been done in the '30s, '40s and '50s. The folks at the university helped us find some of those numbers and some of that information from way back then. I'll give them some credit. One of the best people that I ever worked with in Illinois was Don Bullock at the University of Illinois. He's done a lot of good research. The way Don approached it, he always told me, "You know, you farmers have a second nature, you have an instinct to know that something probably is going on there. And your instincts are probably right. It is up to us to figure out if it really is, and what it is." We got along great. He did some great research.

I'll never forget, Dick Thomson was ridge tilling and seemed to be tying up his phosphorus. Remember that?

Oh yes, he called it his "phosphorus problem."

He was just going nuts. I told him what his problem was. I said, "You need to flip that ground over just one time and plow it. You'll loosen it up, and you'll be just fine again." Get the air, bacteria, and everything going. He did that. He came back to me and said, "Terry, you were right about that." He got on a system where he plowed his ground, what, every four years?

Yes, it seemed like he plowed it after the hay crop in his rotation.

It solved his phosphorus problem.

What do you see as the challenges coming up? Where do we go from here?

We are seeing a completely new change on the face . . . you probably remember, growing up, there were houses on every corner, or every section in Kansas. Here in Illinois, there were houses on every 80 acres. I see a lot of that happening again. Not necessarily on the high-production ground, we see a house, and then we see a fence, see cattle, see a horse. We see people doing that kind of thing. The sustainable movement will be catering those people more than to the conventional farmers. Just the last 10 years year on my mail route, I picked up 100 boxes on the same number of miles. A person who delivers mail on the other side of town, same thing, she has another 100 boxes in the last 10 years. We are seeing a re-population out here.

Why do you think people are moving here?

The same reason they moved to town to start with I guess. There is a lot of money out here. They want to get out of town, get their own place out in the country. They want to mow grass; they want to be with nature. There's nothing wrong with that! I think it's great. I think it makes society livable to do that. One thing I tell people is stay out of the way of technology; you'll be mowed over if you get in front of it. It's the same way with sustainable ag. Use it to your advantage instead of fighting it. I see that a lot with the organic people and stuff. I shouldn't say that. The old hippies maybe? They don't want to see technological change. I'm just the opposite. I love technology. It just makes my life so much easier.

Chapter 13 – Terry and Sheila Holsapple

I can go to Minnesota four weeks a year now. I couldn't do that before. I can coach basketball now. I couldn't do that before. When you are farming, you are totally focused on the farm; how you are going to make it work, how you are going to pay your bills. And you're a weatherman every second of the day—is it going to rain, is that cloud going to produce . . . ? I still look at the weather, but then I'll go weeks without looking at a weather map. I don't care. That is just a stress relief, if you can imagine that.

When I was on the farm, it was 24 hours a day, thinking about how you were going to make it, what kind of crop you are going to have. If you have a good crop, then if it rains in October, how are you going to get it out of the field? It was just constant stress.

Was there stress relief once you paid off the debt?

Yes, absolutely.

Was farming more fun?

It makes it so you, well, you are very careful then. It's a funny thing. It is harder to invest your money once you have money, and keep track of it, because you don't want to lose it. When you don't have it, everything is kind of a gamble, you know. It's kind of a hard thing to pin down. When we had the debt, we knew we had to go to the bank to pay the note. It was a given. Now you got money, and we're thinking, "We don't want to lose that." It makes you a lot more conservative when you get out of debt. Then you can save more and more, and do more things. The money part is a big thing. That was a real stress relief.

In my particular situation, at one time, I was driving the mail, running the pumpkin patch and farming all at the same time. There weren't any hours left except for sleep. Something had to give. I knew what it was going to be, it was going to be the farm. It wasn't too bad then, but we kind of were forced out when the barn burned down at the pumpkin patch, and what was left was just my job and basketball. It freed up a lot of time. Until you get away from farming, you don't realize there are other ways to make a living. I always thought I had to have these aces in the hole, had to have a few hogs, a few cows, something like that. I still have a few cows. You always have to have something to fall back on.

Do you think many other farmers feel that way? That they need to keep something going on at the farm. That there is no other way to make a living?

Absolutely. Even the big grain farmers. I know a guy, all he wants to do is farm. He wants to be a big, big farmer. There's no glory in that! But that's what he wants to do, you know. I don't think he'll ever make it because he's not going to have the chance. He loves sitting out on a tractor at night and driving. I hate it. He likes working on machinery. I don't. I did it because I had to do it.

I know we have it a lot easier today than our grandparents or their parents. I think about that a lot—what they went through to get us where we are today. For us to lie down and quit, that wouldn't be right either. Man, they went through a lot, and died young for it. Sitting on open cab tractors, sucking in dust, spraying DDT, and doing who knows what else, to get us to where we are today.

Are your kids interested in farming?

I had one boy that would have, but I wasn't going to allow that to happen. We had enough land base that he could have probably done it. He was a sit-on-the-tractor kind. Animal agriculture wouldn't have appealed to him. He's in business management. I tried to lead him away from it as much as I could. I see so many more opportunities out there.

I'll tell you what; I missed my two boys growing up. There are things I don't even remember. I was just trying to figure out how to survive, and how to get out of debt, 24 hours a day. At night, you close your eyes to go to sleep, and you have the computer going in your head, what you were going to do, and when you woke up, it was right back in that mode. I missed things I don't even remember. When they got older, I remember some of that stuff, but when they were little kids, I missed a lot of that. That was during the really hard times, then Page (their daughter) came along, and I remember that, it was easier.

But when you first started farming you enjoyed it, right?

Yes. Man, I was enthused.

Did you even consider another occupation?

No. But it beats you down. As I said, the people that were my age that got into farming at the same time, very few of them survived, because the interest rates knocked them out. Most that survived are the ones that were hooked up with, maybe they had an uncle that had a farm and they were farming his land. They weren't out spending a lot of money buying land. The ones that tried to go out and start from scratch, most of them didn't survive.

The ones that didn't survive, are they happier now, or do they still wish they were farming?

Oh, I don't know. The ones that I know, I know they are better off. I'm sure there's a few that regret it. A few will tell you that the best thing they ever did was quit farming. I'm sure there's a few that regret it. One thing I've seen a lot of, were sons that went back to the farm. Maybe the farm was 600 acres, it couldn't handle both and they ended up losing the whole thing. Maybe going into a livestock venture when they shouldn't. That wasn't good; it was hard on families. Even some of them went out and got good jobs.

I'll tell you, I do have an interest yet in farming. I would love to go to probably southern Illinois. Of course, land prices are too high right at the moment. Somewhere there is good grazing land. I really enjoyed that. The problem around here is getting the numbers to make it feasible. In the south, there are auction barns, and cattle everywhere. Up here, we're in the grain belt, so cattle have moved out of here. Although, one advantage of the cattle market here, in terms of buying feeders, is that it is one of the cheapest places in the nation to buy. Deep into the cattle belt, they get higher prices.

We wrapped up the interview with one more "Washington story," about the time he and another Rodale cooperator from Indiana tried to stop and see a congressperson on their way back to the airport, and the security screeners got mad because they had to check each and every suitcase, backing up the line of people waiting to go through. Then Terry treated us to lunch at the local café/restaurant in Greenup, a thriving small town in a countryside where some towns make it, and some don't.

Chapter 13 – Terry and Sheila Holsapple

Notes

[1]New Farm magazine, published from 1970's to the mid 1990's, now on the web at www.newfarm.org.

[2]Rodale's on-farm research network, active beginning in 1987 through the early '90's. Included a Midwest network, southern states network, and east coast network, each with a full-time on-farm research coordinator located in one of the states included in the network, technical backstopping from Rodale Institute and New Farm magazine staff. Goals included research and outreach, the farm data was published, and cooperator farms hosted field days and tours.

[3]Organic certification has been available since the late 1980's through several private organizations. National organic standards were adopted by USDA in 2002, see http://www.ams.usda.gov/nosb/index.htm for details. OCIA – Organic Crop Improvement Association provides certification services in the U.S. and internationally. See http://www.ocia.org/

[4]The USDA SARE (Sustainable Agriculture Research and Education) program funds research and education through university, non-profit, and farmer competitive grants programs. See www.sare.org

[5]Dick Thompson – one of the Rodale Midwest on-farm research cooperators, he also worked closely with collaborators from Iowa State University and USDA/ARS Tilth Lab scientists. Dick designed and conducted replicated on-farm research, usually in long, field-length plots rather than the much abbreviated plots common at university research stations. He had as many as 20 trials/experiments in any given year, analyzed the data, and presented it at his annual field day. He also founded the "Practical Farmers of Iowa" who used Thomson's model to continue to conduct on-farm research.

Discussion Questions:

1. When land prices and interest rates do not allow a new person to come into farming, should the government get involved? At what level (local? State? National?) or should some other mechanism come into play? For example, someone with wealth who wants a "hunting preserve" could buy land, and rent it at below market value so that someone could "afford" to keep farming it? Or perhaps an older farmer could sell land at lower rates to a new farmer, in order to pass on the land.
2. Entertainment farming, in Terry's words, "was very good to us." However, the "farming" part of the farm was more difficult to show a profit from. Is "agri-tainment" a good way to make a living off the land, or is this not really farming?
3. Terry has criticisms of universities, but also praise for an individual here or there. He is also supportive of universities because "that is where

technological innovations come from." He has also had direct experience with the bias that can be introduced to the university by funders. If the universities are not tax funded, who should fund them? Should private interest groups be allowed to give funds to universities with strings attached?

4. Groups like Rodale and some universities support the idea of on-farm research as a way to involve farmers in the research process, and a way to get realistic, site-specific data. However, research can be time consuming, and as Terry says, some farmers don't want to be involved in "itty bitty plots." What is the best role for on-farm research? Some farmers may be quite good at it, but should they be asked to take this on in addition to farming? What are the best ways for farmers and research entities to team up to get answers?

5. Technology was discussed in this interview as generally a good thing – saves labor, saves time. However, it was also pointed out that with large equipment, most farms are "too small" to use the equipment efficiently, that is to capitalize, or effectively pay for the equipment. Terry also pointed out that if you get in the way of technology, "it will run you over." What is the role of technology and how it might interact with one's value system? Is technology a "given," or can it be selected or rejected?

14

Rich Bennett
Napoleon, Ohio

"And if anyone learned from all the research I did, they did learn that they could reduce inputs and get the same type of yield."

Rich is into a different kind of pork now. He has moved from raising confinement hogs, to pasture hogs, to not raising hogs at all, and is now an elected fourth-term county commissioner in Henry County, Ohio. He continues to farm the cropland of his home farm, 600 acres of sandy soil in the northwest corner of the state. Napoleon, Ohio is on the Mamee River, about an hour west of Toledo, and maybe 60 miles from Lake Erie. Rich smiles a lot, and jokes about how he got started in sustainable agriculture.

I was going broke. I came to that first meeting that Rodale sponsored, registered for the first day but was sure I wouldn't come back for the second. Jim Morgan, at the registration desk, just smiled when he saw me come back.

Sustainable ag worked well for Rich, because he was able to reduce inputs and their costs without reducing yield, thus improving net profits. He was also an active member of Rodale's on-farm research network in the late 1980s, and conducted several on-farm experiments on rye cover crops prior to soybeans, pasture hogs, and other studies related to soil building while using modest amounts of fertilizer. That's how we met originally, at the bi-annual cooperator meetings—one in the winter over a weekend to discuss research results and the other in the summer to tour one of the cooperator's farms. The cooperator network included about 10 farms in nine states, but we all looked forward to the meetings and learned a lot from each other. Rich called himself the "token nozzle head" of the group, as he perceived himself to be the furthest from organic in the group, only a couple of the cooperators were certified organic.

I listed the questions that I was seeking answers to; things we've done right, accomplished in sustainable agriculture, things we could have done better, and where do we go from here? I also mentioned that I was interviewing a couple of the other cooperators on these trips around the Midwest.

Well, there has to be a nozzle-head in there somewhere! What was the first question again?

What were the milestones for sustainable Ag?

For me, or the whole thing? I guess learning how to grow inputs rather than applying inputs to cash crops. That was major for me. If anyone learned from all the research I did, people learned they could reduce inputs, and get the same type of yield. If I accomplished anything over that, my neighbors decided that they could reduce inputs, and they have. Not significantly, but they tried some.

I got a county agent on fire about it. He's over in Wood County now, the next county over from Bowling Green. He's worked hard at trying to promote it. I got a county agent working on it! [Laughter] The new county agent here is very adaptable to new things. If nothing else, we came a long way from Ohio State's stand to begin with to them accepting possibilities of farming in other ways. I never thought it would be possible to be where we are at now at Ohio State. On a scale of one to ten, they are probably still a one. [Laughter] Some of the old timers retire. New people coming in have some fresh ideas, people are making it with organics, that type of thing, maybe we should be helping them, support them.

Do you think sustainable ag information is getting out, making a difference?

Boy, if we try to get the consumers to demand sustainable, as well as trying to encourage farmers to be sustainable, we might just get there in our lifetimes! [Laughter].

Or, it will take at least 20 more years? What do you think is the future of agricultural universities?

They're diversifying too, any opportunity they can. It's not what it used to be here, where there is teaching. It's more learning from what all the big farmers are doing, and spread it out to others. Most farmers have their own people that do testing, crop consultants, that type of thing, agronomists. They're advancing more than what the universities are in many ways. Oh, they'll be around, but they'll be just coming up with other alternatives to remain as a political entity. Many times, you keep yourself there by some form or means. They aren't doing what they used to do; it's not going to be teaching and educating.

Do you see any change in taxpayer support for universities?

I think that is probably going to be a problem here because of the state situation we're in economically. They've been balancing their books by cutting local governments. We have to come up with more revenue locally because of services being cut off. We could be looking at cutting Extension practically in half this year because of that. We aren't going to have the money. We tried to pass a sales tax, and it is back on the ballot, people started a petition against it. It will probably go down. Yeah, there are going to be changes.

In Ohio, does Extension combine counties by districting?

They did that a few years ago, when they had economic problems. They feel like they lost ground, credibility with that. Most farmers identify with the county agent they had, didn't want to go to another county at that time. That's probably going to be the norm, going to come. It's going to make it more difficult for education and teaching, and things like that. If they don't know their county agent, they just aren't going to use him or her. Grain elevators and farm supply co-ops are stepping in, working with farmers, doing what extension used to do, educating and teaching the farmers they deal with.

Chapter 14—Rich Bennett

Of the sustainable ag practices on your farm, how are the rotations working out?

Soybeans planted/drilled into rye residue are still working out, but I'm not working more no-till into cover crops. I gave up on the hairy vetch and just went to red clover. I'm getting real good nitrogen out of that behind wheat. I went to more of a three-crop rotation, wheat-corn-soybeans rather than just corn-soybeans. Like Dick Thompson always said, if you use a three- or four-year rotation, it just averages out—and it does. I've found that to be successful.

Are you still raising the pigs on pasture?

No, I'm not. That turned out to be a wash, a break even. The savings I had with the pasture the mother gilts ate it up in feed. They ate more to keep up with the lactation. It was excellent the way they had their pigs and they were in good shape, but they just burned off the feed.

If I remember the data correctly from your on-farm research, you had fewer pigs per sow, but higher rate of gain, which made up for that, so you got the same pounds of pigs per sow.

Yes, and they got to market sooner, they were healthier, no question. It was healthier for me too, compared to working in those confinements, but with the investment already there with the confinements, it just wasn't justifiable.

So, are you still raising pigs in confinement?

No, I got out of that, I'm into the new pork! *[Much laughter. New pork refers to politics. Rich has a career as a county commissioner.]* It's my fourth term. They love me I guess.

How would you criticize sustainable ag, deficiencies, things we should have done better?

I think sustainable ag tried to mix in with all different kinds of branches, and here in Ohio it ended up mixing in with organic. I never thought that was the route to go for most farmers. They just weren't going to accept organic, but they would accept what I showed them in reducing inputs. You know, they'd have gone to cover crops if they could have gotten a benefit with nitrogen without buying it, but buying nitrogen at the prices that they were is just about a wash. That may change. It also was a wash buying the diesel for the trips over the field to get your nitrogen source worked into the ground to till in the cover crops. They just weren't going to do that with a large acreage. My emphasis was always trying to get something where they would recognize the benefits of added income. Not pushing for the huge yields, but trying to make a profit. That's what sustainable is to me. If anything, we missed it by saying we have to go organic. That's what most of the sustainable people did here in Ohio.

Did you ever go to the OEFFA[1] (Ohio Ecological Food and Farm Association) meetings here? (Note: the people who go to these are fairly enthusiastic about organic practices)

No, those people drove me nuts! [Laughter] They were just so dogmatic. They'd say you have to be in it, and why aren't you in it? I'd just say I don't want to be. Working all that sandy ground with all that phosphorus, I got in hog manure. I'd just be run over with weeds, and I have to protect the moisture. I said I'd do it if I have the right kind of soil conditions. I felt like they just beat me up all the time. Of course, it didn't bother me, I've been around it, but the average person, they aren't just going to put up with that. Nice people, but geez, they do some off the wall things . . . they would do a lot of different things small scale. I just wasn't in a small-scale area, 600 acres isn't very big now but I didn't want a patch of raspberries, or a patch of strawberries, that type of thing, and that is where most of them were. I just didn't fit.

We then talked about how the "snake oil" products for sale at those types of meetings can be a turn-off for some too.

So, did OEFFA have you on any speaking panels? Wasn't it supposed to be a sustainable ag group, not just organic? Then they had another group that started out sustainable, sort of like the Practical Farmers of Iowa?

No, I didn't speak on any panels. [Laughter]. They tried to gear it towards sustainable ag, but then a lot of them just went to OEFFA. They made good money at what they were doing, dairy, that type of thing, selling milk. It all sounded good for them.

It sounds like the niche market stuff is not for you. I know that from personal experience I'm finding that direct marketing is really time consuming.

It is. A lot of them were in an urban area, where they were selling produce where you have a bushel basket, and they pay you once a year, you get vegetables every week. It's neat for many people you get good vegetables that were organically grown.

How else would you criticize sustainable Ag? Where would you go to get questions answered in the early years? Did you just have to figure things out on your own?

You were a main source probably, or anybody at Rodale. The network[2] we had, that was a nice source of information. At Ohio State University, if you go back to the 40s and find those books, you could glean a lot from the cover crop research studies. That's what I usually told people. After I did the on-farm research trials and field days, I thought maybe there might be a larger interest, people calling, but that just never worked out. Just a few people called, mostly about the rye. That was the primary one, people wanted to try that rye, I don't know why they wanted to do it . . .

In Kansas, they won't touch rye with a 10-foot pole. They are afraid of it because of the wheat, the potential for contaminating the seed.

Some people are afraid of it here; it might get away from them. Most of them want to plant too early, and then get such establishment that you can't do anything with it, that is the main problem.

I have to give them credit, [3] there is money available for grants, to do different studies. I don't know how we could have promoted that any better. People just didn't want to do that type of thing. If I used the Extension agent, that was a big help. They could do tissue sampling, that kind of thing; do a lot for you when you are busy. I don't know how we could have gotten Extension involved any more.

Chapter 14—Rich Bennett

Has the recent emphasis on long-term sustainable agriculture rotation trials at Ohio State University been a help to you?

No, not really, I just have that different soil type; it just didn't work out. They had those heavy clay type soils. The researchers there used to do a lot of organic matter tests on my fields; they were trying to build organic matter. I just can't build organic matter in this sand. It's just not going to happen. We spent a lot of time on that. The organic matter did go up, but it was so minute; but it didn't go down. I still need that good manure and I've lost that. I can tell a difference on the sand hills now; they are fading fast. I'm putting chicken manure on now. I get it from a large producer close to Bowling Green. They have it stockpiled and are looking for places to get rid of it. You pay them to bring it. They dump it and I spread it. That way they don't have to deal with EPA. They are restricted if they put it on, but as a farmer I'm not as restricted on getting it covered up as fast. They have to cover it the same day.

It seems like that animal part of it is key for sustainability even with cover crops in the rotation. Cover crops just aren't enough. Manure is still very beneficial; it's more beneficial than the average person thinks. The way Dick (Thompson) has it with a five-crop rotation, and cows and pigs that really helps. But without livestock, I'm really falling backwards on the organic matter. Those hills are just getting lower all the time.

I've had some Kansas growers tell me that they feel that manure is their main way to build up the soil, but that cover crops help to maintain, or fine-tune it. For example, if the soil becomes high in phosphorus from manure, the cover crop can help maintain the nitrogen without adding phosphorus. The two go together.

Cover crops do so much better with manure too; they're spindly otherwise. My phosphorus levels stay high no matter what on these sands for some reason. All I have to put is potash (potassium) on the sand, because it just leaches. Otherwise, we just add the nitrogen we need according to tests. I don't even do those tests anymore, I can see it now [laugh]. I can see by the biomass and everything, you just know what is going on. You know there is plenty there. Part of the experience I guess. When you start out, those (soil) tests are very beneficial.

Do you see any of the federal conservation programs helping you?

I've used CRP (conservation reserve program) along the ditches, since those rows didn't produce anything anyway! [Laughter] There's ground that has been put in CRP that really benefited from it, where it eroded badly. All of the no-tillage we've done in this county, and everything we've done, we have not changed one iota the amount of dirt going down the river into lake Erie. There has not been one benefit from it. This stuff is not coming from farming. I don't know if it is from construction and that type of thing or . . . We've grown in no-till acres and these type of covers along ditches . . . many miles of that, protective measures, but it hasn't changed the amount of dirt going down this river from this watershed one bit. They are still dredging just as much, they've sampled it, tested it. Not one bit.

Farming in the Dark

How about the SARE program? Have you had any of the farmer-research grants?

I've had two or three, worked it through the extension agent. He was able to get some tools, like nitrogen test kits and that type of thing. That was beneficial for him, and he did more tests for other farmers at the time. The results were useful to me too. We did the tours, had a lot of good people come. I think we got quite a bit out of it. I wish more people had applied for them. For what they wanted to do, they were just hesitant on doing it. They asked me, "How much does your county agent do for you?" I said, most of it really, almost all of it. Testing and everything, he had the time when I didn't. They said, "Well our county agent won't do that." But some did. You didn't necessarily have a loss on them if you wrote the grants properly. You learned from them, and that was the purpose.

Have you learned from other peoples' grants, from reading the reports or on the website about other farm trials? Did their information help you out?

No. [Laughter] There again, I guess I'm just different—the way I have to do things on the soil type I have. That's what I always tell people; you have to know your soil type to know what you can do and what you can't do. I can't give you a recipe; you have to have your own recipe.

One of the limitations of sustainable ag seems to be that we can't offer a recipe, and yet they want it.

Yes, that's true. They want it. The best way to work out a recipe is to dig out those old books from the 40s and 50s and say, here is how they suggest it.

We tried to do that with some of the older K-State publications, some back to the late 1800s. One limitation we found is that you can't really use the information directly. You have to convert it to present day situations.

Sustainable ag definitely does take extra management, and extra time to know your farm. I didn't really know my farm until I took that time and did it. Most farmers, lets face it, we're lazy. We don't want to do that. We want to have it fed to us from the university. We don't want to find out on our own. There just isn't that recipe there.

Do you think farmers are more open to that now, because of economics or other reasons?

No, these are huge farms, coming about around here. They are just huge. They are family farms, but three brothers, a father; they are just taking over the ground. They just have big equipment, and they are spraying as hard as they can, as much as they can. That is the trend. I just don't see the family farm being what we think of, a husband, wife, father operation any more. It's huge. Livestock is pretty well gone here from here. Some are vegetable oriented. Big operators will have vegetables for Campbells' and that type of thing. Maybe I'm too pessimistic.

So you don't see the big farms adopting sustainable agriculture?

They are just too big. I was at a point where I was desperate, trying to keep the farm, to keep it going, and sustainable was the route to go, because it generated additional income, by reducing inputs and growing your own inputs. It saved me; as far as I'm concerned, it allowed me to keep farming.

Chapter 14—Rich Bennett

You apparently had the economics of your farm figured out well enough to know that? How many farmers do that?

Not very many, because they are working a job, and trying to farm at the same time. They're just following a recipe somebody gives them, and they're not making anything.

Don't their accountants tell them, or their tax preparers?

Maybe through depreciation and that type of thing. I think they know it, but they just don't want to give it up. They always think that this year there might be $15 beans, and it's not going to happen. We still keep producing huge, huge surpluses. *(Note: Most years the price is about half that or lower for conventional beans.)*

What would happen to those people if the commodity program payments were cut?

It would wash them out. The big guys would take them over. Only the big guys would be left.

What do you see as other challenges for the future?

I think that there will still be small farmers that will continue, like these chicken farmers that can find a niche. If you are doggone determined to farm, the opportunities are there. If you are willing to search and find somebody that can help you, or make suggestions the opportunities are there. I've thought about how a young farmer could get established with me. It would be a significant loss of income if I did help someone get established because I can rent it and make more than what I am now by farming.

You can make more by cash renting?

Yes, I could.

Have you encouraged any of your kids to go into farming?

No, I tried to leave the opportunity there, with the hogs, but then hogs went down to eight cents a pound. Compared to a job in town with all the benefits and everything, they were out of here.

If there were a national health care plan, if people didn't have to buy their own health insurance, would that help more people go in to farming?

That's what's typical around here, to have a job with benefits. Oh boy, I guess I could see that as a possibility maybe, but having that would still make a huge cost some place, whether it is now or in the future. I know what it is like trying to pay for health care for the county employees. We are cutting that thing, cutting it back from what it was. It was a real nice plan, and it's down to practically nothing now, due to increased costs. I don't know how it could ever change down the road with 15 percent increases in medical costs every year automatically now. They are suggesting we plan 15 percent every year without any plateau. I don't know how to answer that one, whether it would be beneficial or not.

Do you have any other thoughts about the future?

I have to come up with something positive! One of the interesting things I have seen is farmers cooperating. Two or three are coming together, trying to reduce equipment costs and trying to farm that way. I've been keeping an eye on it. I always thought that was the best way, the way we used to do it with threshing, four or five guys work together. There is a cooperative venture going on with farmers. These are guys younger than I am. They're going into different crops: popcorn, string beans, other areas, where they aren't just bound to corn and soybeans. They are able to do that in some unique ways. I think that is positive since most farmers are just independent and not willing to work together. This is part of the economics. They are willing to say, ok, I can't afford that combine anymore, but if two or three can afford a combine and work together on it, and they just decide that if we are going to keep farming, we have to do this.

Maybe because of that, there will be an attitude change about better managing. Like me, maybe I would be the one to do the managing of the inputs, purchasing the inputs, do the soil tests, that type of thing. I like doing that. Maybe we can mix and match talents. I can't fix equipment, but another guy could, if he's good at that kind of thing. I'm pretty good at marketing too, so I could do that kind of thing. That spirit of cooperation, that's the sustainable that I see coming back, the cooperative nature in farming. Because the big guys are going to get bigger, so maybe I had better join in, and we'll make a "big." Three or four can make a "big" and have the talents to do some things and make it more sustainable.

I do see people looking for other alternatives. The mindset of corn and beans is starting to dwindle. That is good to see. I see some sustainability in different crops. A young person that works as a mechanic at a place where I do business for equipment has 40 acres and that's his dream. It would be nice to help him get established. He is trying to rent ground in competition with the big guys. You can tell he's determined he is going to farm. How do you help someone like that? He thinks he wants to grow corn and soybeans. He sees the numbers, and he's worked with livestock, he's established himself raising dairy calves. That's another start. You just have to keep looking for other opportunities. They'll open up to you if you just seek them out. I say, "Dan you can make money rotating pasture, take those 40 acres and really grow lots of beef with that." He says there is no way, and I say, "You got to look at the research on that. You are in a climate where we get rains, it will grow most of the year and you can rotate that." That just isn't appealing; he wants to be like everybody else. I tell him not to be like every body else, you have to be unique in some specialty. That is still the mindset. I don't know how to move people off that, to try something different.

How did you personally get started in this?

I was going broke. [Laughter]. When your inputs are more than your returns, something has to give. That's when I started getting the *New Farm*[5] magazine, and seeing some of the things they did. I saw Dick Thompson was going to be in Lima at this meeting. I was there one day, came back the second, Dick Thompson was there, and Donn Klor. I only paid for one day, because I knew this was all snake oil.

Chapter 14—Rich Bennett

This was one of those Rodale-sponsored meetings, and Bill Liebhardt talked that day. He had his act together; it really impressed me. I showed him my soil tests, and he said, "Holy smoke, why are you adding more fertilizer?" That really opened my eyes. From there, I started talking to my dad about cover crops, and how they used to use them, and he said that they were a nuisance because you have to plow them, and still keep the sand from blowing. That's a problem, and I just started generating ideas from there.

About that time, no-till was coming around more. Sprays were available to kill them.

So then, you started experimenting with cover crops? Did you plow them or disk them, or how did you handle the cover crops?

I just worked the tar out of them. I had to work them in so I could get through with the planter. Had some trouble with it bunching up in front of the planter. Then I changed planters so we could work through them. I definitely saw a benefit in the water holding capacity on those sands, it made a difference. Dad could see the difference too. I thought I could get a nitrogen source with hairy vetch, I was determined to make that work, but couldn't get the growth that far north, or enough of it to get a nitrogen source, for all that you had to put into it to get it going.

Did your dad encourage you to go into cover crops, or discourage you?

He was always a believer in them from way back. He said rye was the best thing to grow for the sand, but all he could do was plow it under to get rid of it. There was just no other way, and then the sand would blow, and we didn't want that. Having the right equipment to handle it, and planters to plant through it, to keep it from blowing was important (reduced till equipment).

Anything that you think I should put in this for consumers? How to get them to recognize sustainable as compared to organic?

That's a good question about consumers. They are responsive to quality and the way it is grown, and so forth. Organics, they've grown. There is no question about demand for organic food. That would continue to turn the tide, if the consumer continues to go organic, that would bring them to the idea of sustainable, to reduce chemical use. I believe if I were younger now, I'd make every attempt I could to make organic work. The dollars are definitely there, even with the extra trips over the field.

Notes

[1] Ohio Ecological Food and Farm Association, "A grassroots coalition which supports and promotes a healthful, ecological, accountable and permanent agriculture." They offer educational meetings, apprenticeships and organic certification. See http://www.oeffa.org/

[2] Rodale's on-farm research network, active beginning in 1987 through the early '90s. Included a Midwest network, southern states network, and east coast network, each with a full-time on-farm research coordinator located in one of

the states included in the network, technical backstopping from Rodale Institute and New Farm magazine staff. Goals included research and outreach, the farm data was published, and cooperator farms hosted field days and tours.

[3]Referring to various grating agencies that fund sustainable ag research. The most prominent is the USDA SARE program. See www.sare.org

[4]*New Farm* magazine, published from the 1970s to the mid 1990s and now on the web at www.newfarm.org.

Discussion Questions

1. If sustainable agriculture can't offer a recipe, what would be other ways of communicating information needed by farmers to use some of the practices?
2. Is there a viable way for small and medium sized farmers to compete with the "bigs"?" Rich gave examples of niche markets, and groups of farmers cooperating to share labor and/or equipment. Are there any other examples?
3. On-farm research has been valuable to some, but the results are site-specific. Universities conduct both applied and basic research, and hope that the results will be broadly applicable, but also are limited by working on only one or two soils types, which may or may not be similar to the farmers they serve. Are there other models of research that should be considered? What would they be?
4. Sustainable ag practices, especially soil testing and cutting back on fertilizer and the use of cover crops for soil building and to help with weed control in soybeans, has worked for Rich, and he has realized economic benefits. He has shown this to his neighbors at numerous field days and events, with cooperative extension helping with this outreach effort. However, even today, most Ohio farmers probably don't use these practices. Why not?
5. Rich has pointed out the antagonism between the organic proponents and the "nozzleheads" within the sustainable ag movement. Do you think this debate is helpful, or does it detract from progress that the movement could make overall if it was more inclusive?

15

George and Melanie DeVault
Pheasant Hill Farm, Emmaus, Pennsylvania

"The United States is a food importer now. That's something new in the equation . . . a scary thing."

I've known George since 1986, my first year working at Rodale. George and Melanie are originally from Ohio, but moved from Florida, where they both had newspaper jobs to Pennsylvania in 1981 to work for New Farm[1] magazine. I probably met George before that through his writing in New Farm, as I had started to read that in the early 1980s, along with other graduate students at Cornell interested in ecological agriculture. Other people also probably know George, and his wife Melanie through their writing, as both have had successful careers as journalists, as employees of newspapers, and as free-lance writers. They have also written and edited six books on farming and market gardening. In addition to his work in the United States, in 1989, Bob Rodale asked George to help him get the Russian version of The New Farm, *up and running. George worked on that assignment for more than 15 years, and in 1993-94 he spent a year in Moscow overseeing some of the logistics of the publication, as well as editorial content.*

He and Melanie also have a farm near Emmaus, Pennsylvania. They helped start the farmers' market there. This is actually the first time I've seen their farm, nicely laid out on a gentle slope. I can see the high tunnels and greenhouses, many flower crops, and a beautiful berry patch near the top of the slope on the hill as we drive up to the farmhouse. Fortunately, they both had time on a warm September Sunday evening to sit outside at a picnic table and share their reflections on the past, present, and future of sustainable agriculture from their perspective.

How did you first happen to begin working at Rodale, and writing for The New Farm *magazine?*

I was working for a newspaper in Florida, and was reading "Editor and Publisher" one day, and there was a full-page ad in there with a headline saying "Join Our Conspiracy." I said, "Hmm, I'm going to read this." And here was this guy called Bob Rodale, I'd never heard of, in this place I'd never heard of called Emmaus, talking about disappearing farmland, rising food prices, environmental problems, distribution problems, all of that. He's put together something called the Cornucopia[2] Project to study and analyze this. The ad said to write to him for more free information. This was all the stuff I was writing about in Florida at the time, so I wrote him a little note and sent him a couple clips of what was going on down there. I said I'll give you a call next week, do a quick phone interview, and I'll do a feature story about your Cornucopia Project. That's how I came to *New Farm*. Before I called them, they called me. They said we have this farm magazine, we need an editor, are you interested? That was early 1981 or late 1980. I mean, Bob was way out in front on a lot of this.

What do you see as achievements, and what would you offer as a critique of sustainable agriculture?

There are an awful lot of achievements. One of the big ones, I think is getting across to people the idea that you don't have to make major, drastic changes, and bet the whole farm on abandoning one system and adopting something unfamiliar. There are little management changes, fertilizer rates, herbicide rates, tweaking the rotation, stuff like that can make a tremendous difference. That was one of the big things we encountered at *New Farm*. When we started, people were thinking that this is purely organic farming, and if I use a drop of chemicals, I'm going straight to hell.

We tried to get it away from that very quickly. It's unrealistic to expect people to risk everything on turning the system upside down. That's where Bill Liebhardt helped so much with that fertilizer study[3]. People took a look at that and said, "Hey, I'm going to try cutting back here, see what happens." They figure you can save a few bucks doing this, and still get a good yield. It makes them much more receptive to taking the next step, trying something else. Maybe I can cut back a little on herbicides, maybe I'll try a little cultivation in here; maybe I'll mess around with some cover crops, not on the whole farm, but on a few acres here and there. The more that works, the more confidence they get, the more eager they are to try some new ideas and see how else they can bring their cost down. They started out here, and before they know it, they are way over here, looking around thinking, "How did I get here, this is great!"

Several people that I've interviewed have mentioned New Farm *as a starting point, to get down that path that you talk about, and they consider the magazine,* New Farm, *as an accomplishment. When did it first become independent from* Organic Gardening?

It was 1979. Actually, it broke away first in 1947 or '48, because it was *Organic Gardening and Farming*. The farmers and the gardeners were always an awkward mix, for advertisers especially. Bob's first job at Rodale was to edit a thing called *The Organic Farmer* that they published for five years. They dropped the "and Farming." They just had *The Organic Farmer*. That didn't work so well, chemicals were just starting to be a big thing in the late '40s and early '50s. Then they rolled that back into *Organic Gardening*. That kept up until the late '70s when they broke it off into *New Farm*. *New Farm* did some good stuff, if I do say so myself. I had a lot of fun, and I think the readers had a lot of fun.

You published a lot of case studies, or examples of real farmers. I think people liked that.

Melanie: And everywhere he goes today, around the country, he still runs into people who miss the *New Farm*.

Chapter 15—George and Melanie DeVault

George, quoting something he often hears: "When are you guys going to bring that thing back?" Well, we're not. I guess they still have the website, but it's just not the same. It's OK, but not the same. They stopped the print version in the United States in 1995.

Tell me more about the Russian New Farm.

In the beginning, I was going back and forth eight or nine times a year. It was so much we actually moved there for a year in 1993-94. Hired a new staff, shopped around for printers, paper suppliers and the whole bit. Now I go back about once a year.

Is it still a struggle to keep it on track? I remember in the first few years hearing about the problems with paper supply, distribution, etc.

That's Russia, all over the place. After we got the current staff on board, things settled down. Supply and distribution is always a problem over there. In the middle of all this we had inflation, economic collapse, banking collapse, an attempted coup, a revolution that bordered on civil war, the whole bit—little things.

The little bumps in the road, right?

As far as the future of sustainable agriculture, with what is going on in the aftermath of the hurricane (Katrina), this may be a golden opportunity not to exploit that, but a perfect time to raise a lot of these systems questions that people have been asking behind the scenes for a long time. The entire sustainable agriculture community ought to jump on that with both feet. Get our dander up, and say, "Hey, we are sick and tired of being treated like second class citizens. The whole department of agriculture, from the top down, should be sustainable. If you want homeland security, then you need a sustainable agriculture. There are no two ways about it." Really get strident about it and not let them kick sand in our face anymore.

We have enough established research, other information, case histories and what not. We know it works, we know it makes sense. The more fuel prices go up, the more other problems crop up, the more sense it makes. I think we just ought to beat them over the head with a baseball bat.

The old populists use to say we need to "Raise more hell and less corn."
Absolutely.

Sometimes disasters are sustainable agriculture's best opportunities.

We are seeing that in New Orleans right now, but the ripple effect over the next few months, as the fuel prices translate into higher food prices, possible food shortages. We are a food importer[4] now. That's something new in the equation, and that's really a scary thing.

Look around at how much more we could be doing if we just do a few things differently. One of my pet peeves here in Pennsylvania is we lead the nation in preserving farmland. We have more acres sewed up than any other state in the country, like a quarter of a million acres, the last I heard. However, we aren't doing a damn thing to preserve farming or farmers. It's the same-ol', same-ol'—corn and beans, hogs and beef, commodity programs, and people are going broke left and right. If farmers would start raising high value crops, adding value to what they produce, they could sell it locally, sell it directly. The latest USDA statistic shows that on average the farmers get $0.19 of the food dollar, down from $0.30 in, say 1990. That's down from $0.50 in the 1950s. We're being squeezed and squeezed and squeezed. It's time to say, this picture is not just wrong it is upside down. We need to stop doing what we've been doing, and do things that make sense.

How do we talk to consumers about the food dollar, and that food isn't expensive, but it's too cheap now? They think the federal farm programs are welfare, but people aren't paying the full cost of food at the grocery store right now. If you want people in the United States growing your food, you have to do something. Do they want us to grow and give it away free? Sometimes at the farmers' market, we see people drive up in the most expensive cars, and some of those will question our prices.

We watch cars here too, see which one of our customers come up in a Lexus, or Escalade, or a 20-year old Volvo station wagon, covered with bumper stickers on the back. People are funny. There are many new alliances taking shape. Things like Slow Food[4] are helping get the message out. People are increasingly aware of taste, nutrition, freshness. We see that at the farmers' market. We're not certified organic this year; we dropped our certification because our customers just don't care. They know us and know how we do things. The main thing everybody likes is freshness. Stuff is just hours out of the garden, it tastes like it is supposed to taste, the flowers, they come back a week later, the flowers are still good, but they buy some more. Stuff from the grocery store, it's just jet-lagged and taste-less. We can build on that.

One thing I've been doing the past couple of years with the Kellogg Food and Society Policy Fellowship is speaking and writing op-ed pieces and letters to the editor. A neat thing about that fellowship, I had a chance to go around and study things more, then get the word out, try to educate as many people as you can. I gave a talk to Sierra Club about why farmers' markets make so much more sense. Regional and local food, I got into that. I'm also doing a periodic column for Lancaster Farming Newspaper, which is a weekly for conventional farmers. We didn't title it sustainable agriculture or anything like that, but talked to them about selling at farmers' markets, talked about our experiences, the prices, the cost, how consumers respond, how to do things differently. We tried to open their eyes a little bit, one thing at a time, and the response has been great.

What do you see as the future for collaboration between sustainable agriculture and Sierra Club, Audubon? My understanding of the history is that it has been on again, off again since some of the groups don't see agriculture as part of their mission, since farms aren't wild lands.

I'm a bit disgusted with the sustainable agriculture "movement" right now. It seems that in many ways, it has reached the age and size, where much of it has become so much like the establishment we've been battling. The main things they are interested in sustaining is their job, their health insurance, their other benefits. We've gotten away from much of the grass roots, more activist oriented stuff; the volunteer "lets go get them" kind of thing that we had in the beginning. It's becoming, in many ways, just another dirty business. I have mixed emotions about it.

One of my pet peeves with SARE[7] (Sustainable Agriculture Research and Education), is that they are doing more of the same-ol', same-ol'. The farmers are getting the shaft and the academics are getting the gold mine. That doesn't seem right. That's what we were complaining about in the first place.

Looking ahead, what do you see yourself putting energy into for the next 10 years, or what advice would you have for an organization like PASA?

For any organization like that, keep the "establishment" at an arms length. Maintain your distance, your integrity, because if you climb too far into bed with those people, you become them. You get too many people from the Land Grants industry and what not on your board; suddenly you're almost just like Farm Bureau! It's ok to be different, it's ok to be contrary, and raise a little hell. Question the establishment. The more we question the better. I know the need to get grant money, and matching funds and cooperate and all that, and be friends with the secretary of agriculture, but don't go overboard. There is a very real danger there.

That is a good point. What kinds of new projects should we take on, or new issues? Or, have we completely finished the ones we've started

Some we need to finish up. The cover crops you mentioned, some of that needs to go on. The Farming Systems Trial[8], it would be nice to keep that chugging along a while yet. We don't need to keep repeating a lot of the stuff that has already been researched to death. We've done a lot in the last 20, 30 years. Let's look at the next step. That's what I keep harping about to these people in Pennsylvania. We've preserved all this land; ok, what are we going to do with it now? How are we going to keep farmers on the land? Get them growing, transitioning out of corn and beans, working more on fruits, vegetables and direct marketing! Forget the produce auctions out here.

Are they getting living wages from that?

No, I don't think so. I went to the Kutztown produce auction recently with John Weaver. It was late in the season, he pulled his tractor up there, he had two hay wagons behind it just full of beautiful produce; 25-pound boxes of tomatoes that were wonderful, neatly arranged in lots. It took the auctioneer all of about 13 seconds to sell of this one lot of tomatoes. They sold for $2 per box. John looked at me and he said, "The box cost a dollar, and the auction house gets seven percent, 10 percent if you take it inside under the roof." I said "John, why bother? They're giving this stuff away." We need to work, in the community, somehow establishing links between consumers and producers so the farmers get a better return on what they're doing, and people get better, fresher, tastier, more nutritious food. We can't put too much effort into that area. If it's going to be sustainable, that has to happen. The way it is now, even the vegetable people, who are getting a much better return than are the commodity people, they're just treading water. The closer development comes, the higher land prices and taxes go, the more likelihood they're eventually going to sell to developers.

Here, our little farm, is preserved. We sold the development rights almost 10 years ago. If a sewer goes in here, they cut the wetlands buffer in half like they are talking about, suddenly the economics totally change. If a developer comes in and says, "I don't care what the tax consequences are, I'm going to buy this property, and I'm going to put in seven- or eight-million-dollar homes, and life's going to be good," then your farming is just going to go right down the sewer.

Penn State the last year or two has been holding hearing after hearing after public meeting on what's the future of farming in the Lehigh Valley or in Pennsylvania. I went to some of them. The older, more conventional set says, "There's no future what-so-ever, sell your land, take the money and run." People our age, especially those who are members of PASA, and the younger folks who are doing something a little different, are at least optimistic about the future; they see a future. The more farmers' markets we can set up, the more of a friendly environment we can create for people to do this kind of stuff with on-farm processing, chicken plucking, what-have-you, the more farming is going to have a future, especially in populous areas like this. We can't ignore that. We ignore that at our own peril.

Having sold your development rights, what does that mean exactly? If you retired or moved out of state, would the next person not be able to build anything here?

One other house is allowed as long as it somehow ties in with the operation of the farm, or if other family members lived there, or farm labor lives there, that's ok.

Could somebody buy your house and put all the land into lawn?

It needs to be in some kind of agriculture, but the list of agriculture is as long as my arm. It can be anything from aquaculture to Christmas trees, a horse riding stable. A heart surgeon from Manhattan could buy it, build a fancy horse barn, just pasture his horses here and that's agriculture. It's very loose. Keeps it open, keeps it green.

Will this keep the price in an affordable range for other farmers?

No. If anything, in the short run rather than the long run, it will make it more valuable, more expensive. Because when everything is built up around here, this is going to be the only oasis. In that sense, I think it is going to backfire. We both have mixed emotions, 10 years after the fact, about having sold the development rights.

Melanie: The thing that bothers me is the eminent domain. Because now, with the sewer thing, they were talking about taking one of the properties here for a community on-lot septic system, a giant mega-sand mound. I guess the property behind us and even part of our lot had been considered. If you preserve these big chunks of land and the community has grown up around it, then they can say, "We need a school, or we need a golf course." In a way, it makes us kind of a target, with the recent Supreme Court decision[9] it gets even scarier.

Is the state doing anything to try to protect private property? I had heard that some states were passing new laws to do that.

George: I don't think they are doing anything in Pennsylvania. Selling the development rights gave us the capital we needed to start building real greenhouses, get a modern tractor, other stuff we needed so we could really step into the commercial realm, rather than just an overgrown hobby garden. That helped a bunch. We still think it was the right thing to do, don't we *(turning to Melanie)*?

Melanie: Don't ask me now; I'm kind of sour on it.

George: There is too much other conflict in our lives. Even in 1984 when we bought this, there were people looking at it to turn it into a house farm. They wanted to put 15 or 20 houses back here, but it is a little too rolling, a little too wet.

Melanie: But with the sewer coming, they could surely do that now.

How did you both get interested in farming? On your website, you talk about the gardens that you've had in the various places that you've lived[10].

Melanie: His parents have a summer home in Maine, right down the street from the Nearings[11] and Eliot Coleman. [12] Before we got married, we would go up there, and talk to Helen and Scott Nearing, and to Eliot. We would buy Eliot's produce and we thought, "Wow, this is what we want to do." We've been in love with it that long. How long have we been married, 30 years?

George: Thirty years this past year. We stuck with newspapers, and moved from Ohio to Florida. The funny thing, but that is where I started writing about farming, was in south Florida, writing about the winter vegetable industry, because they were getting hammered by the low priced imports from Mexico. In one year, in Dade County and Homestead, I was writing about these people, they left their tomatoes in the fields to rot. They had a great crop, but they couldn't even afford to harvest it. Mexican tomatoes were crossing the border in Arizona, trucking clear across the country to south Florida, selling in the Winn-Dixie grocery stores for cheaper than they could manage to grow them in Florida. This was the center of a federal anti-dumping suite for years. I wrote hundreds of articles about that. That's what got us sucked into farming.

Melanie: When he went to work at *New Farm,* he would travel all over the country and see all these organic farms and how wonderfully they were working. He would come back excited and say, "We've got to try this; we've got to do this!" We lived in a house in town, because that was all we could afford at the time. On weekends we would put the two children in the car, drive all over the area looking for a farm. We wanted to do this so badly!

What do you see as the future of agriculture? With the World Trade Organization opening up markets with no restrictions, are we going to see more of what you saw in Florida, dumping at low prices, and farmers in the United States unable to make a living wage? Will they have any legal recourse, or are farmers being left to fend for themselves?

George: Yes, especially with the large commodity crops. The sooner we can get out of that, and the farther away from it we can get, the better.

You mean the commodity support payments?

Let's not just raise less corn and more hell, raise no corn! Forget it. It just doesn't make sense anymore. The only thing keeping that afloat in most of the country is the commodity programs. Sooner or later, somebody is going to have to start tightening the screws on that, and eventually, just pull the plug. And there goes the ball game. Brazil and other places are going to take over in corn and beans. Are you familiar with a person, his name is Black, out of the University of California at Davis? He wrote book a few years ago, *The End of Agriculture in the American Portfolio*[13]. I'll dig that stuff out for you. He says the only three things that are going to make sense in agriculture in the future of this country are golf courses, nurseries and turf farms. Forget everything else. Yes, we can grow corn and do all this other stuff, but it is like a Ph.D. playing child's games. Why, what is the point? It is a very interesting read.

It seems that is what the agriculture college in New Jersey has been doing, and the fastest growing program area in the agriculture school at K-State is turf—and we are the "wheat state," in the middle of the grain belt.

Pull the commodity checks out of the equation, and look where you are. That's where studies like the Cornucopia Food Systems become so important again. Look at how much you are importing from where, how much more you could be producing in your local area, and start connecting the dots in a little different pattern and suddenly you have a road map for where you might want to go in the future. And if your farmers can't make money growing corn, beans, wheat, whatever, look at what else they could be growing to sell locally or regionally, or add value to some kind of processing. Cut out foreign oil, cut out some of these food imports, that's going to give us homeland security more than anything.

Melanie: I'm really discouraged with the way the government has not helped the small farmer. (*Turning to George*) What was that Florida heirloom tomato?

George: Oh yeah, the Big Ugly. There was this outfit in Florida; I have a bunch of clips on it around somewhere. A year or two ago there was a tomato famine. These people were growing heirloom tomatoes, they have a huge harvest and the Florida tomato commission wouldn't let them sell them out of state. They couldn't ship them north, because they didn't meet the standards of the marketing order.

I had a student in one of my classes last year tell me about working on a Florida tomato farm. They had a situation like that They would hide certain varieties of tomatoes in the middle of boxes of regular tomatoes in order to ship them out, almost like contraband. There was nothing wrong with the tomatoes.

(Laughter.) I love it. Like bootlegging. You need a perfectly round baseball; a hard, tasteless pink Florida 303, or whatever it is.

Melanie: Another thing for the future of sustainable agriculture, we need more consumer education. It is amazing how many people just don't get it. They don't understand that local food tastes better, it's better for you. If we don't support the small local farmer, everything is going to come from China. They have three-quarters of a million acres of certified organic fields. Now Wal-Mart is buying industrial organic from wherever it's cheapest.

George: A few years ago, we had the worst drought in a hundred years. In the middle of the summer, we have lettuce! We have bags of salad mix!

Melanie: We used shade cloth; we watered three times a day, working like crazy. We're selling it at a fair price.

Chapter 15—George and Melanie DeVault

George: We were at $3.50 a bag, and then went to $4.00. A woman walked up, picks up a bag and said, "Humph that is too much!" She walks away, and Mel not so subtly said, "Well grow your own damn lettuce in a drought!" A couple of our customers saw all this and rushed over and explained to her that this is unreal that they have this salad mix, it's a lot of work and on and on.

Melanie: And it keeps for three weeks in your refrigerator because it didn't come all the way across the country! She came back and bought the bag.

Wow, your customers really stick up for you!

Melanie: I muttered it; I didn't say it loudly.

George: You said it loud enough! Sometimes you have to do it.

The whole farmers' market thing; we've had quite a phenomenon here with the Emmaus farmers market. It was like an overnight success, an overnight sensation. The community, all the pieces fell into place at the right time, it took off and people love it. It's like a weekly community fair family reunion kind of thing. It's only from 10 a.m. until 2 p.m. on Sundays. We have people wander in around 9:30, to start scoping things out. We will set things back for certain people. Then about 10 o'clock there is this mad rush, for a half-hour, hour, sometimes two.

Melanie: Everybody's hurting right now though (referring to the amount of produce that farmers have to sell at this time of the season—late August, early September). It was such a long, hot summer. The next couple of weeks are going to be very hard for us, but then the greens will be coming in and will carry us into fall.

You don't have okra? That's one of our late season favorites.

George (dryly): Yankees don't eat okra.

Melanie: For flowers, we have ornamental okra, burgundy okra, but the deer have been eating it. They are eating the pods off!

George: I need to check that fence charger. Did you turn that back on? I'll pop a new battery in there.

Melanie: Our flowers, we expected them to go longer. This was a rough year. Everything was two or three weeks early. I don't know if we will make it through September this year (referring to how long they will be selling at the farmers' market).

George: You do what you can. But we don't have global warming! No, there is not enough research on that (also with sarcasm). The high tunnels that we were looking at and talking about earlier, they are another low-level technology we can push in the sustainable community that really help. It's not expensive, it's environmentally friendly and it really gives you an economic boost. It's been researched a lot. I don't think we need to research it a whole lot more. What we need are people to see the economics of it. One thing that really helped around here was last year the late blight came early. It hit everything from the Carolinas up to the Canadian Maritimes to Michigan. It wiped out tomatoes and potatoes left and right. The ones that were under cover, in a high tunnel or a Haygrove[14], the environmental conditions inside were just enough different, the blight didn't take off. We had tomatoes for weeks and weeks and weeks. Everybody with tomatoes in the field was wiped out. We know we had blight here, because across the fencerow in the far field, we lost a quarter acre of potatoes over-night. It was here. It was on the tomatoes, but it just didn't catch fire.

As a side question, do you have soil quality concerns in your high tunnels? A student at K-State has been doing a survey of people with high tunnels to see what their concerns are, to see if there are some interesting research questions. Many notice some sort of calcium precipitate on the soil surface, and maybe some salt build up from shallow irrigation.

We have questions about that calcium. I've been meaning to do some soil testing out here inside the high tunnels. I have the kits sitting down in the basement. Just haven't taken the soil samples to see where we really are with that. It (the high tunnel conditions) changes things, and with the compost, you keep piling that on, and watering, and leaching out salts. You could be doing some things you'd rather not have happen.

Do you think regular soil testing might be enough to keep it in balance?

I don't know. I'm disappointed in the Penn State soil tests. They just give you the same old NPK, the pH; they don't even look at organic matter. A&L labs is a lot better. What I might do is take the samples from each, send off the Penn State kits I got, send a bunch off to A&L and see who comes back with what.

We're doing the standard tests but also total N, total P, total organic matter, and labile or particulate organic matter. My suspicion is that we are burning up organic matter in these things faster that we think due to the heat.

I wouldn't doubt it. The heat and the repeated tillage. Whether it is mechanical or by hand, you are still working the soil, mixing it up, and breaking it down.

Melanie: Those two beds with the eggplant, I want to use the tilther and work with it.

George (excitedly): I do too.

What is a tilther?

It's the slickest damn thing in the world. I'll drag it out of the barn for you.

Melanie and I make small talk while George goes to get the tilther to show it to us.

George returns. This summer, I haven't had a chance to use it yet, but we're going to put it to use this fall. This is Eliot Coleman's answer to a tiller to use inside whatever structure you might have. A re-chargeable electric drill fits in here. This does the top two inches of soil. You pull this rope handle to depress the trigger on the drill, and you just go along the top two inches of soil, up one side of a 30-inch bed and down the other. No fossil fuel, no exhaust, probably still beats the hell out of your soil. But, if you are adding enough good stuff to it, it will be ok.

So he does that for the fine seedbed, for the salad mix?

It gives you the fine seedbed, and they have a refined planter, a 6-row precision seeder to go along with this. I have that out in the barn out there too. I just haven't had a chance to use it yet.

My husband Raad then asks; how much does it cost?

Chapter 15—George and Melanie DeVault

I think this was about $300 or $350, and the seeder is about the same. They have developed a couple of other tools, or are in the process; one is a harvesting basket. I did a story on this in *Growing for Market* [15] and the phone started ringing off the wall at Johnny's[16] (the catalog that lists them). All of a sudden, they had too many orders and not enough tools. New stuff like this would be a great area to focus on in a sustainable future, things to reduce labor, because as you know, this doesn't get much more labor intensive. If you can simplify it, speed it up, make each of us more productive, reduce the amount of fossil fuel and other non-renewable resources that go into it; that is barking up the sustainability and the profitability tree.

Melanie: You know where there is a big need for sustainable agriculture education and research information is flowers. There is so little available. Lynn Byczynski's book, *Flower Farmer*[17], is the source. All the big growers, they use a lot of chemicals, Roundup and preservatives. There are organic alternatives. I also had someone from NOFA[18] call me looking for a speaker. They had the names of two big growers in Maryland, but they weren't organic. I tried the Vita-flora[19] products; this year we did a lot of experimenting with them, which is allowable under certified organic standards. We had great success with that. The conventional growers shied away from it, they said it won't work; it's too expensive. We bought a 55 gallon drum, we figured out all the costs, and it was every bit as cost-effective. My flower partner really watches costs and returns. It really worked; it was wonderful. You don't have to go there (meaning the conventional chemicals). We used milk spray on our zinnias for powdery mildew. It worked very well. We used one part whole milk to nine parts water. We used it as a preventative, and it worked fine. We had good, saleable zinnias. You can do so many things. Unfortunately, I didn't do my honeysuckle or mondarda. We didn't spray that, and it . . .

Would it work on that?

I think it would.

Did you use the kaolinite clay on your vegetables?

George: We never have.

Melanie: We were really going to work on that with our apple trees and stuff but we just don't have time. We buy our compost; we don't have time to mess with that. We are busy from the time we get up until after dark. At this point in the summer, we are falling apart. It's been a long, hot summer.

Do you use any pest monitoring traps?

We use sticky traps in the greenhouse. To tell you the truth, we are bad about that stuff. It's not like research, and it's not like . . .

Well, I'm also terrible about monitoring, even though it is what I studied in graduate school, the whole IPM process; know the life cycle. We have a few apples; I put out traps, but I didn't monitor them. I even have a handy pocket pest ID guide, but I didn't go out and monitor, figure out which pests are on which apple.

Melanie: Exactly. I have these guides; I thought this is great. In the winter, when you have time, and you are sitting there looking at all this stuff, and then the season hits, and it is all you can do to cut the flowers, get them to market, and make some money.

George: that is one of the neat things about this kind of farming compared to the commodity operation. You are out there every day, usually on your hands and knees working in this stuff; you are up close and personal with it, and you know what's going on, what's working and what's not. So you are monitoring, maybe not by the book, it's not statistically valid, but you know, you're feeling the pulse all the time.

We get a lot of people stop out here, calls and questions from people who want to get started in farming. There is a good educational opportunity there, because they still have in their head, "I need a tractor, I need 100 acres, I need to make hay, I need to grow corn," when it is probably the last thing in the world they ought to consider. The more alternatives like this that can be explained to them, the economics of doing a few thousand square feet under cover as opposed to trying to do 40 or 50 acres of a grain. "Oh really, you mean I don't need to find 100 acres? I could do 10 or 15 or five and still make some money?" There is a lot we could do in that area.

In my spare time I'm trying to update, do a new version of Whatley's book, *How to make $100,000 farming 25 acres.* I'm working on *"How to make $25,000 farming one acre."*

So have you written up your farm story?

Ummm, bits and pieces here and there, in *New Farm* and various articles, columns off and on for *Successful Farming*, other people. They are receptive to this new stuff; especially not just organic stuff, they are going for it.

Many people still have the dream.

With what is going on in the world and the economy now, I don't think there has ever been a greater need for it. I have a bad feeling about some of this stuff. It will be interesting to see what happens.

I'm seeing that same kind of interest in growing food among people, based on some gut level insecurity or something. Are you seeing more people than during the back to the land movement of the '60s and '70s?

It's not just the old hippies. It is the 20- and 30-somethings. People come up to us at the farmers' market and say, "I am afraid of food from other countries." They like the fact that they know where their food comes from. If it's local, they know they can count on it. We don't have to have a full-blown famine, even a shortage, just a little blip. "Oh my god, we can't get strawberries in January," or something like that, or tomatoes in February, what is the world coming to? Or a disease scare, it doesn't have to be an actual epidemic. There are enough bumps out there to shake up the apple cart. If we could move on some of that stuff, and explain that it doesn't have to be that way folks, here are some viable alternatives, things you should be thinking about anyhow.

Do you think one of our roles in the future through the next 20 years, for those of us that have small farms; our farms will be farm-teaching centers?

I think so. The farmers I've always dealt with, sustainable, organic, whatever you want to call it that has always been a big part of it. Their farm is an open book, and they don't mind answering questions. They are honest and free with information. I think the two have to go together.

Do you want to comment on the future of the land grant universities? Do we still need land grant universities?

Personally, I don't think we do. I think it has reached the point where the public is starting to question that the same way they are questioning the commodity programs. A couple of times in our local counties there have been big budget scares with our local County Extension office. "Oh my God, it's going to get cut; we're going to lose all these people." Most folks say, "Yeah, so what?" We're not agricultural anymore, we're urban, we don't need them, they're just a drain. If the land grants and extension and that whole bunch don't really get their act together and get out in front of some things, I don't see they have much future. Enrollment is down, way down in many places. Nobody wants to be a farmer. They want to sell stuff to farmers.

Melanie: Or they want to be an extension agent and give advice to farmers!

George: I think at this point right now they are probably their own worst enemy, because they are just re-arranging the deck chairs on the titanic.

Are there useful things that extension – either local or state, has helped you with? (George begins laughing)

Melanie: I'm currently the president of the Lehigh County extension board.

So are there any success stories?

They are doing many things with inner city, helping, that type of thing. As far as sustainable agriculture goes, they refer everybody to us. They give us support.

George: They give us a pat on the back every now and then. We were named conservation farmer of the year last year in the county by the conservation district.

Do they pay you for your time, answering the phone, giving tours?

Are you kidding? Farmers work for nothing. Our time is not valuable.

So they (extension) cost you money, they don't make you money!

Yeah. Basically. They really haven't done anything to help us. It's been the other way around.

Melanie: And they are very nice people. We like them all.

George: I've said it many times before, I'll say it again. If they were paid based on the additional amount of money they put in farmers' pockets, most of them would starve. They can't really justify their existence.

I sometimes wonder what would happen if agricultural economics department faculty salaries were indexed to average farm profit in any given year, or their salary would be the same as the farmers they are supposed to be serving.

George: Like physicians in China. As long as I'm well, I keep paying you. If I get sick, you (the doctor) don't get paid.

Everything is backwards here.

George continues: We could do away with a bunch of those titles and a bunch of those departments, throw extension out the window, close USDA, have a U.S. Department of Food, a food commissioner, a food bill instead of a farm bill.

Notes

[1]The *New Farm* magazine. Published by the Rodale Institute from the mid 1970s until 1995. Archived articles and current news at www.newfarm.org

[2]Cornucopia Project, was funded in several states in the mid-1980s by the Rodale Press. The goal was to look at how close each state was or was not, to food self-sufficiency; and to describe what a sustainable food system might look like. These reports are out of print, but some state reports can still be found in local libraries, or in the closets of sustainable agriculture organizations who helped write them.

[3]Bill Liebhardt, while at Rodale in the early 1980s compared 69 different soil test labs' recommendations, including public (University) and private. Results showed that many labs recommended more fertilizer than would be economically or environmentally sound (nitrogen recommendations ranged from 0 to 230 lb/A for the same soil, for example). Results were published in a series of New Farm articles (named one of the 10 best censored major news stories of the year by Project Censored, Sonoma State University), and the book, "The Farmers Fertilizer Handbook," by Rodale Press.

[4]United States as a net food importer. This refers primarily to fruits and vegetables. See current statistics at http://faostat.fao.org/. If grains and other commodity crops are considered, United States is still a net exporter ($). See http://www.ers.usda.gov/Data/FATUS/

[5]Slow Food movement, see http://www.slowfood.com/ and http://www.slowfoodusa.org/

[6]PASA Pennsylvania Association for Sustainable Agriculture. See http://www.pasafarming.org/

[7]USDA/SARE—Sustainable Agriculture Research and Education, funds grants for researchers, farmers, students, and promotes education of agricultural professionals on the topic of sustainable agriculture. See www.sare.org.

[8]Farming Systems Trial, started in 1981 at the Rodale Institute in Pennsylvania to compare three cropping systems, two organic and one conventional. Several papers are now published in refereed agricultural journals about the study, which is still on going. For current information, see http://www.newfarm.org/depts/NFfield_trials/0903/FST.shtml

[9] A September. 2005 Supreme Court decision about the right of eminent domain to take property for the community good, even if it is a good in the business sense.

Chapter 15—George and Melanie DeVault

[10]Their farm, Pheasant Hill Farm, has a great website with photos and articles, etc. www.phforganics.com. Articles published as part of the Kellogg fellowship may be found at www.foodandsocietyfellows.org/fellows.cfm?id=80337.

[11]Nearings—Helen and Scott Nearing have inspired several generations with their books and personal interaction with visitors to their farms, first in Vermont, and later in Maine. Their classic is *Living the Good Life: How to Live Sanely and Simply in a Troubled World*, Galahad Books, 1970.

[12]Eliot Coleman, author of several books including *The New Organic Grower*, Chelsea Green, 1995, and *Four Season Harvest*, with Barbara Damrosch, Chelsea Green, 1999.

[13] *The End of Agriculture in the American Portfolio* by Steven C. Blank, 1997.

[14] Haygrove, one of several designs/brands of high tunnels. For more information, see http://www.hightunnels.org/

[15] *Growing for Market*, monthly newsletter published by Lynn Byczynski for Market Farmers (fruits, vegetables, and flowers primarily). See http://www.growingformarket.com/

[16] Johnny's is a seed company based on Maine. See http://www.johnnyseeds.com/

[17]Lynn Byzcynski's book called *The Flower Farmer, an Organic Grower's Guide to Raising and Selling Cut Flowers*. Chelsea Green, 1997.

[18]NOFA—Natural Organic Farmers Association, the state and regional umbrella organization for several states in the northeast to promote education and organic certification among farmers and growers. http://www.nofa.org/

[19]Vita-Flora products. Hydration and nutrient solution for fresh cut flowers. http://www.vitaproducts.com/

[20]Booker T. Whatley's . *How to make $100,000 farming 25 acres: With special plans for prospering on 10 to 200 acres*. (Paperback) by Booker T. Whatley and George DeVault, Rodale Press, 1987. Most recent edition published by American Botanist, 1996. (Also see http://en.wikipedia.org/wiki/Booker_T._Whatley)

Discussion Questions

1. Do you think consumers realize how much work goes in to growing their vegetables? If so, what percent do you think have an awareness of this?
2. If extension asks farmers to visit with other farmers, either on the phone, or at field days, do you think they should be paid? If so, how much?
3. Do you think ag Extension should have their salaries indexed to local farm profitability? Why or why not?
4. What do you think of the principle that in China people only pay the doctor if they are well, not sick? Should we do that here?
5. Should the next U.S. farm bill be called the food bill? Could/should consumers be involved in drafting the bill, in addition to the usual farm interest lobbies?

16

Carmen Fernholz
A-Frame Farm, Madison, Minnesota

"We have this free enterprise system. It gives you all this freedom to market the way you want, and the reality is that there is no more freedom there, in terms of how you treat the land."

We met for lunch in Willmar, Minn., between the twin cities and his home town of Madison, which is on the western edge of Minnesota. It was a mid-November day, but unseasonably warm. I've known Carmen since he was a part of the Rodale Institute on-farm research network in the late 1980s. I had attended one of his field days and helped analyze some of his data from his randomized, replicated on-farm plots. His interests then (as now) were weed control efficacy in his soybeans, and the nitrogen benefit from some of the "annual" alfalfa varieties that had recently been released. In addition to hog manure from his farm, he adds organic matter to his fields through application of leaves from the city each fall. He is very civic-minded and he contributes to his community through his donation of time as well as taxes and directs plays at the community theatre. In addition to farming, he has served as a staff member in state government in Minneapolis and was chosen to serve in the MISA endowed rotating chair position in1997[1]. He has farmed organically for over 30 years, and last year received the organic farmer of the year award by the Midwest Organic and Sustainable Education Service (MOSES[2]). Carmen has also served in several leadership roles in farming and marketing organizations throughout his years of farming.

I began by asking Carmen to comment on either the successes or failures of sustainable agriculture.

When I look back over the last 25 or 30 years, the frustrations build up and I wonder if we are really getting anything accomplished. If you take a few minutes, you see yes, we have. The most obvious thing is to raise awareness about the fragility of our ecosystem. Because 30 years ago when I would tell people that I was an organic farmer, they would say, "Where is your long hair?" or whatever. Now when I say that I'm certified organic, they are interested—so we have definitely raised the awareness level of organic and sustainable, and why we have to move in a sustainable direction.

We have brought the debate to the table in defining sustainable. We know that sustainable agriculture is a work in progress, but everybody has an image in his or her mind and knows when something isn't sustainable today. It may not be their total image of what it is to be sustainable, but they know if it is the right direction to go.

Failures? My personal mission has always been marketing and not only marketing organic, but marketing conventional food production as well. Even today, I don't think we've resolved marketing.

We've made great strides in organics with the formation of OFARM (Organic Farmers' Agency for Relationship Marketing)[3]. That has really helped a lot. Producers have benefited from the premiums, and we've protected the premiums. I think the educational institutions have not been, and still are not forward enough in talking about the dangers that will come to organic marketing if we don't organize protection. Without protection, the free market system is always looking for the lowest common denominator. That's what got us into trouble with the sustainability of conventional agriculture—cutting back, cutting back—the environment was the thing that took a hit. Now we are seeing the industrial model creeping into organic, especially the dairies[4].

Sustainable, that piece is never defined or emphasized, the rating of stability on the landscape. You can't create stability without adequate income. If you look at the shortcomings for all players in the sustainable game, we have not done our part. Yet it's a hard thing. It's almost like trying to tell a person what religion they should practice. We have this free enterprise system. It gives you all this freedom to market the way you want, and the reality is that there is no more freedom there, in terms of how you treat the land. If you are not marketing properly, you are mistreating your fellow producers. In terms of shortcomings, I'm looking at that piece.

I told Carmen that those were some of my concerns as well, that sustainable agriculture has not done very well at assuring a living wage to farmers. It is one reason I want to write this book.

Exactly—we don't realize that we have to generate income for ourselves, but for the infrastructure within which we live. I have to help support schools, the churches, hospitals, safety, and I have to do that by generating income. All of us should be excited about paying taxes! [Laughter]

It's a contribution to our communities. That is never dealt with. To me, that is the downside.

Do you have any ideas? Any solutions?

To me, the model is OFARM. Not only for organic, but conventional agriculture as well. I just finished reading the book *Tipping Point*[5], by Malcolm Gladwell. He wrote *Blink* as well. He talks about what happens when something goes fast track, spreads like a virus or just takes off. One of the things he talks about is the idea of the Methodist Church; how it was founded. John Wesley (the founder) made connections, not with a lot of people, but with many groups of people. One whole chapter in that book, which solidified this for me, is devoted to the magic of the number 150. Groups are effective, but that effectiveness begins to deteriorate when they get up to 150. All of a sudden I say "yeah," since I've been directly involved with OFARM—it plays out, no question it plays out. It has to do with trust. Effective marketing is based totally on trust. If I trust the local elevator to market my grain, that is one thing. What we have found with OFARM is that each of these groups forms their own trust group, with their own marketer, someone they have anointed. Those anointed people (the marketing representatives), they form a trust level among each other, exchanging information. What that amounts to is information, information, information. That whole concept needs to be developed in conventional agriculture, and it has to be more developed in organics for sure. If we want to make marketing more sustainable, that would be the way to solve it. One can't be successful with one single office and thinking they can market cross-country. It won't work. That's the direction we should be looking.

It is going to boil down to economics anyway. Enough of us are aware of the environmental issues. We know when we are hurting the environment. Not enough of us know how to create economic opportunity. It's a big country, and you have the free enterprise system plugged in there, and competition. We live and die by competition. In fact, I'm just in the process of putting together a new lecture on "The role of the family farm on the rural landscape" and defining that role for the upper Midwest Conference in Lacrosse, Wisc.[6] In previous years I've been doing a presentation called "The Organic University," about transitioning to organic. I've been doing it with Elizabeth Dyck, who was heading up the organic research at the Lamberton Field Station[7] here. For me, that role, obviously, is taking care of the environment, giving to society, but it can only be developed and maintained if you are dealing with a base of economic influence. Every one of us has a base of economic influence. It depends on how big you want to make that influence. As an individual, I can only do so much; I need to work with others.

Over the years, they think I'm a broken record on (the importance of) marketing, because if you are hungry, you aren't going to think about anything else. I have just finished reading *The Working Poor*[8]. It's so obvious, absolutely so obvious. All those things are put together. I'm jumbling it all together for you, but it does fit together.

On Land Grand Universities, Barriers to Entry to Farming

I think enough people have stepped forward, in a few of the land grant and other schools as well, that are willing to tap some of us and ask, "What should we be researching?" We couldn't have said that 15 or 20 years ago. Today they are looking (to us) for research questions. As we look to the future, as we enhance the importance of organic production and sustainable production, the questions will keep surfacing in terms of what we should research. We also have to provide the economic opportunity for people. Just because you believe that sustainable is right, does not mean that you have the means to practice it. You may not have the means to practice it.

Farming in the Dark

What are the barriers?

One of the biggest is peer pressure. We've gotten today so that our farming is judged by how few weeds we have in our soybeans or corn. The (general) fear of weeds is another one. We are also fast losing—how can I define it—the natural means of maintaining soil fertility. There are tons and tons of questions we should be asking about how we should be using our legumes, how to use our cover crops. We are depriving more of our land of animal waste with consolidation of animal production. We are taking that tool away from some people.

Then access to land—that is a big, big one. I have a son. I'll use him as an example. He is an engineer at Ford, but if the opportunity were there tomorrow, he would close the door and walk away, and come to the farm. You know the desire is there. I only use him as an example—there are thousands of people his age, in their late 20s and early 30s, if the barriers weren't there, they would be there instantly (in farming as an occupation).

Are the barriers the access to land or the land prices?

Interesting you should ask. I went into town the other day, walked into a store, and a person said, "There's another one of those rich farmers." He said that some farmland just sold for $2,600 per acre. I said, "Do you want to buy?" [Laughter] For western Minnesota, with only 18 to 24 inches of precipitation per year this is a very high price! And at $1.50 corn?*[note: the cost of production for corn is about $3.50 per bushel, and so farmers selling corn at $1.50 per bushel are clearly loosing money.]*

Do you think the land price is being driven up by investors?

That's all I can see. A young person doesn't have a chance. However, if that young person could look at $7 (per bushel) corn, $7 wheat, $13 soybeans, which would be comparable (in terms of inflation), it's doable. When I talk about the unbalance that we've created, they just look at me askance.

Here is some more background. About six or seven years ago, I was working with Lamberton Field Station, and Elizabeth said, "How would you like to plant golden flax?" I planted 10 acres that year. I mowed down half that got weedy, and harvested the other five; got it cleaned and bagged. It is a classic niche market, selling it out my back door. Doing the math, we arrived at a price of $1 per pound, 56 pounds per bushel, so it's worth $56 per bushel. I have nothing to lose; we pushed it and now it has evolved into a real monetary benefit for me. This past winter I was connected with the Local Harvest[9] website, and now we ship to Alaska, Hawaii, you name it. We are grossing at least $1 per pound plus shipping and handling. If I can harvest 15 cleaned, marketable bushels per acre per year, I can gross between $700 and $1,000 per acre. Obviously, that's a good windfall. You can't expect to do that on everything. Let's look at the value of flax (in terms of marketing potential in the future), the heath benefits, the omega 3 (fatty acids); flax is a naturally occurring nutraceutical—so we look at those kinds of opportunities. That's what it is going to take to say, "Yes, I will walk away from a six figure (income) job and come back to farming."

Is the profit from 15 bushels per acre worth the hassle and cost of cleaning and bagging?

It's a high-risk crop, and it is not competitive with weeds. This year the yield was down. I only ended up with 10-12 bushel cleaned. I'm making that much per acre. Look how much more I can contribute to the infrastructure of the community. Look at what we expect from doctors and lawyers, and higher income people—what we expect them to contribute to the community. Then look at the cutbacks in our schools and the curriculum they offer, and our disastrous health care system in this country that people can't afford. A young person looks at that and says, "I can't walk away from this job...how am I going to take care of my family?"

If we had a national heath care plan, would that make a difference?

The survey that the Center for Rural Affairs a year or so ago showed that the number one reason people can't come back (and farm) is the cost of health care. Health insurance is $10 to $12,000, even $14,000—that's most of their whole income or at least a good share of it, so what are you going to eat? Digressing a second, what I can't figure out with those large corporations is why they didn't endorse that—and industry can't compete internationally because of health insurance cost.

I know that you've worked with your state department of agriculture—are they in a position to help with structural, and policy changes, to encourage local food purchasing for example, or state funded insurance plans?

I think the potential is there, it depends on the bucket we carry the water in. Here is an example of participation at the local level. Look at all the institutions we involve when somebody wants to put up a large livestock facility. We get the county engineers, the county commissioners, the township boards, the environmental boards, and the citizenry involved. That is the approach to take for all those other things, especially when we approach the department of agriculture. Let's get the Chambers (of Commerce), the County Commissioners, our legislative institutions—let's lay out a plan that does in fact define and show economic opportunity. If we can show economic opportunity, then I think we'll start to be taken seriously. This is where we've really failed in the past. I think we really have. We've been out there like this (fist in the air) . . . instead of saying, let's sit down and talk. Not that we aren't getting better at that, but we are still dealing with some of the scars from the past.

We are talking about doing this with the "Green Lands Blue Waters"[10] initiative. One of the things I suggested to them is, let's get these other entities engaged! If we don't, they'll be throwing stones at us, like we've been throwing at them and what do we have to lose? If what we define (as sustainable agriculture), does in fact have all the value we believe it has, it will survive the pressure, and the attempts at compromise, and attempts at dilution. If it is as good as we believe it is, and say it is, it will carry the day. A case in point is the bombardment of organic standards. Some people have really given their life to protect their standards. That's how I would respond there.

What is your opinion about the WTO (World Trade Organization) [11] *negotiations? Is there a silver lining to that cloud?*

My parochial, myopic view is that the WTO, as much as I see of it, is connected only with conventional agriculture. I'm just saying that for the sake of discussion. It might be, depending on how it plays out, the snowflake that breaks the branch of conventional agriculture being dependent on subsidy payments. In other words, we can't get ourselves to limit payments, but this might force us. It might eliminate these million dollar payments. [12] Another thing is that it might enhance the economic opportunities for other crops. (For example) it might become more beneficial to grow field peas instead of soybeans, or oats instead of wheat.

You mean it levels the playing field between program and non-program crops?
Exactly.

What about the difficulty of competition with countries like China, with lower production costs?

Right. That to me is a big unknown. I really don't know because we are a net importer already of so much. What has allowed this whole trade thing to grow the way it has is the fact that the subsidies, like the LDP (loan deficiency payment) [13] and corn for example, allow Cargill and the other exporters to know the amount that they have to pay for corn. Without subsidies, it's not going to be like that any more. Why should they (Cargill) keep the price up? The only money that comes back here (to farmers) is the subsidy. If those aren't going to be there, there is no incentive to export (because the grain won't be cheap anymore). If there isn't incentive to export, who is going to want all the corn? If you don't have a subsidy (to farmers) for buck and half corn, the LDP or counter-cyclical payments, we (farmers) aren't going to make it (make a profit) on a buck and a half. You start taking away those things, then those countries (that now import U.S. grain) will say, "We got to feed our people," and we aren't going to export over there. I may not understand some things, and I may be wrong, but those things that cross my mind when I'm sitting on the tractor. You bet. I'll be the last one to say I understand the whole trade thing.

One thing, one question that I haven't answered for myself yet, was raised in the book, *The World is Flat* by Thomas Friedman[14]. He talks about everything except agriculture, as globalizable. To me, it is conspicuously missing there. I'm not sure why. I have a couple of opinions or speculations. One has to do with energy costs, the processing cost, (the ability) to keep it fresh and ability to buy it. A lot, so many people can produce it. I'm not sure. You take computers, televisions that stuff—a few years from now, even our taxes will be done in

Chapter 16—Carmen Fernholz

India, not locally. Agriculture is not easy to globalize. In the end, food will be the great leveler of civilization. I really think so.

Another big unknown—and this is where research has always been so interesting and intriguing—our research tools get more sophisticated every year, parts per million, parts per billion, parts per quadrillion, etc. The more we do that (research), the more we begin to understand human anatomy, the closer we make the ties between what we eat and how healthy we are. The more we make those ties, and make those ties more and more obvious, we will see that is the main reason we can't have globalization of food. My own humble opinion is that the food we eat should come from where we live, primarily. There is a reason why bananas don't grow in Minnesota or corn on the equator or whatever. As we continue, (the research), our level of awareness keeps rising.

I mention that on our farm, we eat something that we grew almost every day.

And seasonal, we haven't learned how to eat seasonal yet, really.

We both notice that on our plates at lunch, we were served melons, out of season, instead of apples and cranberries, which could have been local and in season.

You have to think our diets are out of whack, with the whole obesity issue.

What do you advise people to do, for the next 25 years in sustainable agriculture?

I have to confess, I haven't thought ahead 25 years. I'm always trying to look in the opposite direction of where you see things going. Personally, in my operation, I'm looking for different cropping systems. For example, for the first time this year, I planted field peas. I'm looking for how that enhances my rotation. It adds another crop to the five or six I already have. If you don't have the right attitude, you aren't going to get anywhere anyway. And self image. Believe in yourself. Once you've figured those out, then go where you want to go. I don't think I could tell anybody to go farming, or do this, or do that, but do the other things first, and then whatever is supposed to happen, will.

I had to have felt sure of myself when I started farming, because I threw away a security blanket in the form of a steady job as a teacher. I jumped into farming, and I didn't know a year ahead what was going to happen. If you like food production, if you like to get down and dirty, you will find a way to do it, even if the odds may not look good. The only downside is that we keep making this opportunity less and less available. Even if I have all the desire in the world, the opportunity just isn't there to make a living for my family and myself.

Even with organic prices?

Yes, because you can't generate enough! Look what you need to generate to pay $12,000 just for health insurance, and that's over and above what you need to eat and raise your family. What we've done, is we've looked to size to solve our money problems, instead of looking at money as a way to solve our size

243

problems. I always use the example, what if a doctor had to do what a farmer does? How many patients would he have to see? How many kids would a teacher have to have in a classroom? They say, "Well, in agriculture you can do it," but then you aren't really farming sustainably. How attentive can you be if you are farming 2,000 acres instead of 200?

I am going through that now with my oldest son. How do you advise him? One thing I would definitely look at for them to get started, is beef. Not only is it an enterprise that can generate income, but you have multiple ways to market— organic, as a finished product, or a product to be finished. There are many variables in the market. More importantly, it gives you the safety net for the farm, which is the buffer for your cropping system. With alfalfa in your pasture, you get nutrient build-up, weed management and maintenance of the soil. If you use the livestock to take care of those expenses, it saves you output dollars. In fact, my son and I are talking about that. I'd love to see policies developed for these CRP acres...for farmers to keep land in CRP, have a partial payment but allow grazing. Those lands shouldn't be worked anyway. I know there are programs out there I haven't studied. The native grasses—talk about quality grazing—if you learn how to graze them so you got your winter grazing, your summer grazing. It will also keep the animal wastes spread out and avoid a lot of pollution. Even to finish, beef are a lot less grain dependent than swine. In fact, if I were 20 to 25 years younger I'd be doing the beef thing.

Are you still raising the pigs?

Yes, though it is strictly a conventional operation, it does provide an income. It's an economic thing. I can buy $1.50 conventional corn, and sell mine for $5.00. Part of the reason we sell pigs conventionally is that the market for organic pork is still not reliable. With beef, even though the market is not reliable yet, the other benefits that you gain from beef include the omega-3s, and LLCs or whatever you call them. You will see many beneficial connections of grass-fed (beef) to health. That is the same reason for the flax. In fact, I personally lowered my cholesterol 40 points. We know our diets in this country are atrocious. We just know it. Once you get in a habit, it's hard to break it.

How did you first get in to sustainable agriculture?

I think any time anybody asks me they get a different answer. I think one thing is my Dad. It's so ironic, because when he started farming he was organic by default, then he went into conventional. When I started, I was conventional the first year, and from then on moved in the opposite direction. He raised nine children. We didn't live top of the line, but we had a nice living, plenty to eat, each got a college education. He did it on 240 acres, and did it for most of my time at home using pretty close to organic practices; we had crop rotations, we had livestock; all that. I thought to myself, he did that, and raised nine children. I have four children, I have modern technology, more current agronomic information and it's a challenge to do that. The challenge, where the organic and the conventional part ways came to be, can I compete on yield? If I can, what is it going to take?

Chapter 16—Carmen Fernholz

I think just having a college education in liberal arts, a broad perspective, taught me to teach myself, to look to sources myself. That kept evolving, and it's amazing, it's still there today. In fact, three or four weeks ago, I picked up Marianne Sarrantonio's book on legumes[15], and just started paging through it. No matter how many times you see those things, you see something new. To start with, my Dad was always someone who was pushing the envelope. Your parents give you things you don't know they gave you. I have the teacher instinct, and by taking it out to other people, you take back more than you took out. All those things flowed together. I was humbled, but appreciated winning the organic farmer of the year award at LaCrosse. That meant an awful lot to me; maybe I might be going in the right direction. It's like teaching; you never know for sure—only feedback gives you that reassurance. I guess that's really how it started."

How did you learn the techniques to farm organically?

I read books. In fact, I have the first issue of Acres USA[16] When I first came back to Madison, and we started farming I went to NFO (National Farmers Organization) [17] meetings. There were farmers here and there that were looking at organics. My Dad did get Organic Gardening and Farming[18]. As a kid, you read the non-verbal. He was always conservation-minded, sensitive to the environment. He was the first person (in our area) to use the chisel plow. We're talking 1943 or '44. Those are the things he was telling me that I didn't know he was telling me. He was always out there, pushing the envelope. The organic thing resonated. The first or second year I got into farming, I was connected with a group in Southwest Minnesota, the Minnesota Soil Association, around the town of Marshall. We'd get together three or four times a year, a lot like the Rodale farmers did. We'd exchange information.

Is that group still going?

It dissipated about five or six years after I had been with it, each of us got our families going, but I needed that support at the time. It's funny how you connect with someone, who connects to someone else. I'll never forget when Martin Culik and I were on the same panel at the University of Minnesota, in 1981 I think. Someone asked me where I get my information, and I said that where I don't go is the University. Martin turned to me and said, "You know, we are starting these trials at Rodale…," and he told me about the Farming Systems trials that they were just starting (comparing conventional to organic). To go back to your question, how do you advise people? The most important thing is that you have to keep an open mind, keep looking. Just because you listen doesn't mean you have to buy, but you have to listen. It's important to listen, to be non-judgmental and to ask questions.

Because you never know where your next idea will come from?

I'll give you a quick case in point. I had 150 acres of small grain crops this year that I under seeded. I had oats, barley, flax and peas. We had an excellent fall, good rains in August and September, and the legumes were coming back lush. I pulled out Mariannne's book. I have to study this green manure thing. How should I get the most benefit—should I leave it grow, not mow it off, till in the fall, mow it off, both? I talked to Craig Schafer on campus two weeks ago (forage specialist at the University of Minnesota). He said, "Let's do something, lets take biomass samples, get a nitrogen reading this fall. Next spring, do the same thing and see what we got. See how much went in the ground." We did it on a new seeding and one where I had taken off this years' alfalfa crop. That's the inquisitiveness you need.

I asked Carmen if he was still planting the annual alfalfa, since he had a nice variety trial going with it when I was at his field day many years ago. He still grows it, and stated that a principle that he has had for at least 10 years is to under seed every small grain on the farm with some sort of legume cover crop. He likes alfalfa and red clover, which has a bigger leaf and suppresses grassy weeds better. We also talked about global warming/climate change, and whether that is changing the timing of his farming operations any, since one of the keys to successful cover crop establishment is timing.

We didn't have a killing frost this year until October 20, and that's nearly a month late. When I started farming, after the first 15 years, I was beginning to know the general trends of the weather—I can expect rain here, this two-week period. You can't bet on it any more, absolutely not. In fact, I looked at soil temperatures; we are getting 50-degree soil temperatures in April. I never would have guessed that (which is considered warm for Minnesota in April).

I then asked Carmen what his current views on the Land Grant system are, since he has come from a point of not going to them for advice in 1981, to having the Endowed Chair for Sustainable Agriculture in 1997.

I'm so glad you asked that question . . . I didn't want to leave it that like that. Today I have the opposite answer. Now I go to key people in the University, like the example with Craig Schaeffer. It was a natural for me to go talk to Craig about this legume—he knows a whole lot. On the other hand, I've been doing randomized research plots on my farm in cooperation with the University for the last 10 years. When you look at the University research overall, we may get the impression they don't understand organic, but now they are attuned to sustainable. Thirty years ago, they wouldn't say the "s" word, and today they are even saying the "o" word. It's like a big ship that turns slowly. As the constituency, it is our responsibility—we help them create the research agendas. It is my responsibility to connect with them, and say hey, let's look in this direction. Not that it will take all the research, but we can start with little pieces of it. As Mark Twain said, when I was 17, my parents were really stupid, but now that I'm 21, they've learned a lot. What I've learned about the university is that I've seen that there are people who have the same ideas at heart that I have—its just a matter of connecting with them."

If you were to advise other farmers in other states about how to get their voices (and research agendas) heard at their universities, do you advise them to go in as individuals or groups? What is the most effective strategy?

Chapter 16—Carmen Fernholz

To be honest about it, I'd say it depends on the group. Some groups automatically raise flags with the universities; you might not want to go in with them. In that case, it would be safest to go in as an individual. Go in seeking information. Ask a research question, or ask about research ideas. Go in seeking out a person, and keep in mind that the person you are seeing is a teacher, is serving, and wanting to serve. I have an example of that working for me back in the late '80s during a grasshopper infestation. The entomologist had recommended spraying. I happened to know the guy, I saw him on campus, asked him, are there any natural pesticides? Then next year, he had research plots out to evaluate Noseema (a biological control for grasshopper that is approved for organic use). That's the approach you have to take—come at it as a student to an instructor, saying let's talk, help me, that is always the safest way. If you have a good organization that has that same attitude, you can do the same thing. Like it or not, university people are under pressure, publish or perish, whatever that is. A lot of them at heart want to do what they feel is the right thing, but job descriptions might not allow it. One person I know is a weed ecologist. He is one of the ones we used to call nozzle heads, but at the same time, he and I can sit down and talk about weed ecology. I can learn so much from him about weeds. I don't bad-mouth him for doing research for Monsanto or whatever, because that is part of his job too, and I go and tap his information on weeds. In other cases, I know of some graduate students who are sustainable agriculture oriented, but they had to go where the money was to make a living. I think we can't be blinded by prejudice. We have to look beyond that. When I think back to the '60s, '70s and '80s, we thought we were a minority that was being discriminated against, which we were, but we didn't know how to deal with it. We came out fighting, and kept throwing rocks at the university instead saying, hey, let's sit down and talk.

That's what I learned. Another example is that most farmers, when you talk about agricultural economists, they just roll their eyes. That's why, to me, when I connected with Dick Levins (agricultural economist at the University of Minnesotta) that was a first. When I connected with Dick, and he could teach me the economics of this whole thing, and we could throw questions at each other. I enjoy when I can sit down with a prof, and throw questions at him or her, that give and take, that student-teacher relationship. It's there if you let it happen.

Background information about the farm[19]: about 400 acres—350 tillable, 30 acres restored prairie, 10 or 12 wetlands. They (Carmen, his brother, and nephew) usually end up finishing out about 1,000 hogs per year. His brother does the breeding and farrowing, and his nephew has a hot nursery (for recently weaned pigs), and Carmen has facilities for fattening the older pigs. They worked out a pricing formula for each stage of the pig's life, and then divide the gross returns according to those percentages. Carmen grows several different crops: corn, soybeans, oats, but not all of these every year. The soybeans are grown as seed beans for Blue River Organics. The oats are sold as oat hay to a local organic beef farmer. In some years, he raises organic wheat; this year had 100 acres of barley and 60 to 80 acres of alfalfa, as a cash crop. Though the alfalfa hay is technically organic, there is not an organic market out there for it yet, but Carmen gets a good price for it based on its feed value, and also appreciates the non-monetary benefits (soil quality, nitrogen). He is thinking of moving more and more away from corn, and growing more of crops like field peas. He notices that corn is a heavy nutrient user, and after two years of alfalfa on his better soils, he gets a good crop of corn, but on his lighter soils, it is difficult no matter what soil amendments he uses. He also notices that on corn ground, the structure of the soil seems to change, gets hard, and clods up. Other crops leave the ground mellower. He mentions that he sometimes uses the oats as part of his weed control strategy, in addition to growing it as a hay crop for his neighbor. If a lot of sunflowers come up with the oats, he'll work the ground in July, to encourage a second flush of sunflowers. He'll till that, and then plant a fall cover crop in September. Last year his soybean yield was 40 bushels per acre, even with a food grade variety (which often don't yield as well as a feed variety). His weed control was excellent. [19]

You know you are where you should be, that if I had one, I had 20 compliments on that field . . . when they start complementing organic fields . . .

Notes

[1]MISA endowed chair—from website www. Misa.umn.edu: "The College of Agricultural, Food, and Environmental Sciences at the University of Minnesota recognizes that individuals both within and outside of the University environment can make meaningful contributions to the activities and scholarship of the institution. The Chair is intended to serve as a catalyst for innovation and progress on agricultural and rural issues within the College and Minnesota, and can be filled by individuals or project teams, which rotate through the position, serving flexible, varying-length terms as appropriate for their proposed activities.

[2]MOSES Organic Farmer of the Year. The Midwest Organic and Sustainable Education Service presents an annual award to an outstanding organic farmer at the Upper Midwest Organic Farming Conference each year in LaCrosse, Wisconsin.

[3]OFARM (Organic Farmers' Agency for Relationship Marketing) see website http://www.ofarm.org. Mission statement – To coordinate the efforts of producer marketing groups to benefit and sustain organic producers. Aims and objectives – Strengthen the marketing programs of member organizations. Inventory production and manage organic marketing in a responsible manner. Exchange pricing and marketing information among member organizations. Develop and support communications among organic producers. Research, support, and enhance market development. Assist producers and consumers in broadening their knowledge of organic marketing concepts. Promote public policy, research and education in support of sustainable agriculture.

[4]For an example of contrasting strategies in organic dairy farms see descriptions of Horizon and Organic Valley marketing networks, in "What is Organic? Powerful Players Want a Say," by Melanie Warner, Nov. 1 2005, New York Times.

[5]The Tipping Point: How Little Things Can make a Big Difference. By Malcolm Gladwell. 2002. Back Bay Books.

[6]LaCross, Wisconsin Upper Midwest organic conference. http://www.moses-organic.org/umofc/workshops.htm

[7]Lamberton Field Station, University of Minnesota—Southwest Research and Outreach Center http://www.maes.umn.edu/components/ROCSW.asp or http://swroc.coafes.umn.edu/ Research at the Center is directed toward improving environmental quality by focusing on nitrogen cycling and nitrogen management, organic management systems, water quality, and environmentally sound pest management research. Research also includes diversified agricultural production systems, sustaining the value added contributions of livestock production to the region while addressing community and environmental concerns.

[8]The Working Poor: Invisible in America. By David KI. Shipler. 2005. Vintage (publisher).

[9]Local harvest - Maintains a public nationwide directory of small farms, farmers markets, and other local food sources in the US. http://www.localharvest.org/

[10]Green Lands Blue Waters Initiative—http://www.greenlandsbluewaters.org/ Mission and Vision: Green Lands, Blue Waters is a long-term effort in the Mississippi River Basin to integrate more perennial plants and other continuous living cover into the agricultural landscape. Participants include five Land-Grant Institutions in the upper Midwest and over fifteen non-governmental organization participants

[11]WTO (World Trade Organization) rules are under discussion that would limit the amount of direct support for agriculture that any country could provide, which would include our current commodity support payments. Instead, countries could opt for more "green payments" in the form of subsidies for conservation practices.

[12]Reference to "million dollar payments" – the Environmental Working Group made public the amount of public dollars supporting farmers from 2003 to 2005 in the form of a searchable data base. To look up farmers in your county see http://farm.ewg.org/sites/farmbill2007

[13]LDP and counter cyclical payment explanations – for more explanations, see the publication "The Non-Wonk Guide to Understanding Federal Commodity Payments – 2005." http://www.rafiusa.org/pubs/nonwonkguide

[14]The World is Flat: A Brief History of the Twenty-first Century, by Thomas L. Friedman. 200x Farrar, Straus and Giroux, publisher.

[15]Methodologies for Screening Soil-Improving Legumes, 1991, by Marianne Sarrantonio, Rodale Institute, Kutztown, PA.

[16]Acres USA, monthly magazine on Eco-Agriculture. http://www.acresusa.com/magazines/magazine.htm

[17]National Farmers Organization – a non-profit organization begun in 1955 as a way for producers to communicate agricultural and rural concerns to congressional and local leaders. Their work towards a better marketing system began in the 1970's. Currently, they negotiate prices and sales terms for millions of dollars of commodities for their farmer-members. .http://www.nfo.org

[18]Organic Farming and Gardening, published by Rodale Press since the 1940's.Current Rodale publications include *Organic Gardening*, and several books on the topic of organic gardening and self-sufficiency. From 1979 until the mid 1990s they also published *New Farm* magazine, about organic and sustainable farming practices, and this publication continues today in an online version at www.newfarm.org.

[19]Note: can also find additional information and a summary of the Fernholz farm at the SARE website, www.sare.org/publications/explore/profile1.htm.

Discussion Questions:

1. Carmen states that one accomplishment is that "we've raised awareness about the fragility of our ecosystem," but that we haven't figured out how to create economic opportunity. What would be your ideas for addressing Carmen's question?
2. In collaboration with researchers at the University of Minnesota, Carmen planted ten acres of golden flax. He got a good return the first year, and a more marginal returns the second year. He also had to do all his own seed cleaning and marketing (web-based) in order to achieve good prices. This is a classic "niche marketing" example, with some trade-offs in terms of profit vs. time spent. It is also serves a market with some limits. Given these constraints, what is the future role of niche markets in sustainable agriculture? Do you think they should be promoted as route to farm profitability? Why or why not? Could niche crops and/or marketing be improved if there were a more organized effort some-how?

3. Carmen states that "without protection, the free market system is always looking for the lowest common denominator." Do we now have a free market system in the United States? In agriculture? Why or why not? Should it be "free" market? What are some other marketing models or examples (note: one includes our neighbor to the north, with wheat marketing boards).

4. Another quote from Carmen is "We've looked to size to solve our money problems, instead of looking at money as a way to solve our size problems." What did he mean? How could this conundrum be resolved in the current market system? Could it be solved if we had a different marketing system? Is bigger always better in agriculture, or is there a place for the small and medium sized farms?

5. The importance of nutrition, and eating close to home was also mentioned. "The more we begin to understand human anatomy, the closer we make the ties between what we eat and how healthy we are." In Chapter 19 of this book, you'll find almost exactly the sentiment echoed by Chef Nora Poullion. Do you think what you eat affects your health? Do you believe that eating locally is a part of that? Are there scientific studies to back that up? Should more research be done in this area?

17
Bob and Elaine Mohr
Slow Food Pioneers, Manhattan, Kansas

"Slow food gives. Fast food only takes. Slow food gives back to the community, and helps foster the connections between people in the larger system."

I've known Bob and Elaine for several years, and have marketed my produce cooperatively with Elaine, first at the Saturday market with the "Blue Earth Organic Growers," a small group of us that decided to market together. Later we created a Wednesday "Green Market" setting at our local food co-op. We are gathered on Bob and Elaine's new limestone patio for an "oven party." My parents were confused when I told them about the event. "Yeah, I said, they just built an oven, and they want to have their friends over to celebrate." More confused looks. Built an oven? Celebrate? What kind of people has their daughter taken up with now? In their circle of friends, Ovens come from the store, and use either gas or electricity, not wood. I asked Bob why he built the oven.

Well, Elaine had this book about how to build cob or clay ovens, and we thought since we just finished this patio, why not build an oven from stone, bigger, longer lasting.

Why did you have that book in your house? What gave you the idea in the first place? I pressed to get to the real reason.

In a tone of voice usually reserved for revealing the deepest, darkest, sometimes most embarrassing secrets, Bob confessed, "Well, my wife and I are into slow food." *Ah hah, the truth at last. He continues,* "It is a movement started in Europe to support local growers, artisan cheeses and so forth. The food tastes better, is usually organic, and we wanted to do artisan bread baking with wood from our place, in an oven that makes an ideal loaf of bread." *He shares the three books[2] with me, and explains some of the things he learned from each. They used a combination of concepts in creating their own, wood-fired beauty.*

Bob and Elaine's backyard is adjacent to a nicely wooded area on a flood protection levee, now Linear Trail Park that goes around the outside of Manhattan. It is used by hikers, bicyclist and dog-walkers. Their home is within the city limits, they are only 10 minutes by car from any location in Manhattan and biking distance is reasonable too. This isn't where you would normally expect to find a farm.

Mohr's stone patio is lined with native limestone walls, comfortable sitting height, and many of Manhattan's "foodies" have gathered to "celebrate the oven." Two of the guests, Bruce and Kirk, have at various times been bakers and chefs at local coffee shops, and a third, Linda, is now creating the deli at People's grocery, coming to Manhattan with experience in establishing a microbrewery and restaurant in Lincoln, Nebraska. The food at the potluck is excellent, mostly homegrown, as you might expect from a crowd like this. I even get to taste two products from my own farm, as one of the guests, an employee at People's, has brought sliced apples and pears she purchased from our farm last Wednesday, and Bruce brought a cas-

Farming in the Dark

serole including okra from our farm. This is a first for me, to get to taste my food as prepared by others at a local potluck.

Then the pizza begins to come out of the oven. It is fabulous, topped with black olives, onions, peppers and a sprinkling of cheese. The olives, obviously, aren't locally grown, but Elaine grew the onions and peppers, and probably the wheat comes from Kansas. Then the loaves of bread go into the oven. Bob explained how the Wheat Fields[3] bakery in Lawrence played an important role in helping him learn how to use the oven. Wheat Fields' brick oven came from Spain, and they employed a master mason to put it together. Wheat Fields also learned their baking techniques from the masters. During a recent visit, they showed Bob the tools of the trade, including the linen cloth, or couche, used to shape the baguettes, and the wooden bowels used to hold the dough during the rising phase. Bob said that they even shared their sourdough starter with him, and showed him some of their other techniques. They just thought it was cool that Bob was doing this. Bob won't be competition for them, at least in the sense of marketing his bread in their sphere. He and Elaine are offering their oven for our use, and are teaching us how to build an oven in our own backyard.

This whole thing only cost me $650 in materials. The books said you could build an oven in three weekends. It ended up taking me three months, but it is ok, because it was labor of love. *The base of the oven rests on concrete blocks, and the floor of the firing area is lined with kiln brick. The walls and ceiling are a dome of firebrick held together with a special mortar that can withstand temperatures of 1000° F or higher. The walls are insulated with vermiculite between the brick and the outer layers of limestone and stucco. On the top, they've placed a sheet of cement board, which can be removed for inspecting the condition of the bricks from time to time. He offers the leftover vermiculite to anyone in the group that is ready to build the next oven. The whole thing rests at the same height as the stonewall, and looks like it belongs there, like the native stone in the walls, and even the guests at the gathering.*

He reports that it takes two wheelbarrows of wood, cut from their acreage, to fire the oven. They start it on a Sunday morning, and in a few hours, the oven temperature is more than 800° F. They take the coals out, use them in their grill for grilled meats and vegetables and begin baking breads. The first few don't take long. At then end of six to eight hours, they still have a useable 250° F temperature, for fruit pies, baking tomatoes or other garden produce they want to cook before freezing.

Bob demonstrates how he chose the height (about three feet) of the oven floor to be comfortable for taking loaves in and out of the oven using the flat paddled tool known as a "peel." Even though he has to lean over to look in the oven to see if it is baked, it is at a comfortable viewing height from a chair he places a few feet away from the oven. He mentions that on his baking days, he enjoys watching the oven, like some people enjoy watching TV. Now I know he is crazy and why my parents wonder sometimes about my crazy friends, since not many people would sit and watch their oven just for fun. Bob goes on to tell me more about the books he is reading.

One of the books says never to invite people over while you are baking the bread. It destroys your concentration. Pizza is ok though, except that it is hard to figure out how to get the dough from the kitchen where it is rolled out, onto the peel

254

to place it in the oven. The author even says that you will have to figure out how to do that on your own, so I don't feel so bad that it took me a while to get it right. The consistency of the dough is key, as is the liberal use of corn meal to keep it from sticking.

Bob and Elaine have bravely invited us over for both bread and pizza, and we are invited to bring over our own creations to make this a community oven. Elaine continues; Here is a small tool called the "lame." *It looks like a sharp razor or roller with a yellow plastic handle. It's used in areas of France where a whole village of people might bake in the same oven. Each person marks, or scores their bread with a unique signature. That way, when the bread comes out, each person knows which is theirs.*

Her specialty is the 100 percent rye bread, which takes more than four days to prepare from scratch with the sourdough-based starter. One guest comments, "Now *that* is slow food!"

They continue to share their baking experiences with us, as the meal progresses. I brought my friend, Ahlam, with me, since she and I have been discussing the idea of a clay oven in my backyard almost since the first year we met and butchered chickens together at my house. She explains that in Iraq, her home country, the ovens are cone-shaped, and the bread is made in flat sheets that are "thrown" on the sides to stick there, while the fire and coals are still in the base of the oven chamber. She is making fresh "baba ghanouj" from three eggplants, roasted in the coals from the oven firing this day.

We talk about the idea of a restaurant based on slow food ideals. Bob expands on the restaurant idea;

There is this one guy in Australia that feeds people in his backyard, based on local foods cooked in his oven. They don't know the menu ahead of time, and the price is based on the honor system, the money is just put in a jar. Once a week the community would come together, bake and cook the meals together. They knew they were headed towards an excellent meal, that's how he operates his business."

He mentions a similar group that comes together to bake in New York City

They don't just let anyone in. This one guy, who was very wealthy wanted to join, and they told him no, "you're a jerk. You can buy us out, but you aren't coming because you're a jerk." They also didn't want to invite the bad karma this person would bring with him. This whole thing can only really work if everyone trusts everyone, if they are nice people.

Linda comments that U.S. health codes would never allow something like this to exist as a commercial business, but then we decided it could happen, if either it was a sort of "underground" operation, and/or it was run as a non-business, that is, based on donations only.[4]

Bob mentions other political frontiers that the slow food people are getting into, for example trying to protect artisan cheese makers and truffle hunters.

I heard a story about a goat cheese maker in Italy, he was going to be put out of business by the laws they were trying to change. They want people to buy only corporate. They managed to change it into a business so that legally he could still exist, but it would be something else, a recycling business for products off the farm. The thing was that he had a whole following of people who wanted to buy his product, and he just said, "I worked so hard for it, and the people love it so much."

Farming in the Dark

Bob credits the slow food movement with inspiring him to do something like this, and we talk about how maybe it is time for a slow food chapter to materialize in Manhattan, since many of us in the group are aware of the movement independently, but just haven't gotten together to discuss an organized effort. He points out some of the photos of incredible ovens, for example a beautiful large brick community oven in France, centuries old, with outdoor communal washbasins near by, mountains in the background. Other community ovens from around the world are made of clay or other more simple materials, and are beautifully illustrated in these books. We talk about the idea of community ovens here, and more "oven parties" to share the heat from a firing. Linda wonders how anyone can ever be bored, with so many interesting projects like this to participate in, and reads a quote from one of the books; "only 4% of the flour made in the U.S. is used for home baking." *Bob comments that* "the US and Australia are notorious for the worst bread in the world," *since most of it comes from a factory.*

I see an abandoned, non-functional stopwatch laying on one of the limestone walls near the oven, apparently used to time the bread-baking during the learning phase over the past month or so they have been using the oven.

Oh yeah, I guess the battery ran out, *Bob comments. The find is ironic, since Bob is a self-employed repairing watches and clocks. His wife Elaine is also self-employed, in home remodeling and tile setting, primarily using an all-woman crew when she can. The garden and fruit crops are also her realm, though Bob seems to be the baker.*

Do you sell any of your bread?

I usually give it away, or maybe freeze some, but it is so good when it is fresh, it often just is eaten. My neighbor always knows when bread is coming his way, because he can smell it baking.

The meal winds down to a close. The parents of some of the younger children have left the party, and a few guests have retreated to the screened in porch and are eating apple pie from the oven, with melting ice cream. Some of the kids are in the raspberry patch, at Elaine's invitation, to eat their dessert fresh from the bush. A few of us are taking the garden tour and peek under the polyester row cover tunnels, and see glorious stands of at least a dozen different varieties and colors of lettuce, spinach, and oriental greens. I can't wait to buy some from Elaine next Wednesday, as I've only just planted my fall greens. They won't be ready for at least four or five weeks. She and I often trade or barter one vegetable for another. One advantage of being a market gardener is the ability to barter, or participate in the non-cash economy, while enjoying excellent food. Bob mentions that sometimes he doesn't want Elaine to sell her vegetables, until he has eaten his fill. He tells her that he will buy them…"please don't sell them."

Elaine is my marketing colleague at our newly created "Green Market" site, located under the awning of our local food cooperative, "People's Grocery." We set up our tables every Wednesday, from four o'clock to six, to meet, greet and feed our customers. This is immensely more satisfying to us than our previous arrangement at the Manhattan Downtown Farmers' Market. The community market has existed for more 20 years, and Elaine was, in fact, a founding member. It has grown steadily, and has a loyal following of customers, but recently the "local politics" of the situation have become burdensome, and both of us have tired of dealing with the authoritarian nature of some of the leaders of the organization that decide

Chapter 17—Bob and Elaine Mohr

the stall fees, stall locations, etc. I asked Elaine if we needed any rules or guide-lines for our new, renegade market. "Only one" she said, based on the assumption that bringing only organic produce was a given. "You have to be nice." That sounded good to me, and we have proceeded under those guidelines ever since.

I asked Elaine my "big three" questions about what we've accomplished regarding sustainable agriculture, what we haven't achieved and where we go from here.

I think that we've influenced people's health, there's been a definite increase in the number of healthy people, or at least more people demanding healthy food. I know there's also been an increase in obesity too, but maybe they are eating too much healthy food?" (*Elaine has a knack for sarcasm.*) Or, really not enough of it.

Do you notice that your customers are healthier? There is a downtrend in national health statistics?

But not for my customers . . . I have some of the best-looking customers you could ever ask for! Another accomplishment is that organic has made major inroads into produce departments. When you see major grocery stores having an organic section, that's a big step."

Do you see deficiencies?

There is still a perception that organic/sustainable costs more in price. But then everything is costing more…Sometimes its true, that you have to do extra hand work instead of going out and spraying therefore you have to charge more because of the extra work. *We discuss briefly a recent article that found that at least half of the items at an organic farmers' market were cheaper than even the Super Wal-Mart in Tulsa, Oklahoma.[5] Elaine follows up with a comment on the difference in quality between her produce and other sources.* I hear over and over again how long the produce lasts that they (my customers) buy from me. It is picked today, which is a big difference compared to what you buy at a store. Like you say, its perception, the higher prices from organic farmers, but my produce is fresher, lasts longer and tastes better.

Do you consider it a deficiency that farmers aren't making a living? Are the prices high enough?

Obviously not high enough to keep farmers in business. They have to find their niche market, and that's what it is. They can't make a living with their regular wheat, corn and soybeans type of thing. *Would you call sustainable agriculture a movement?* I don't know. If people are doing it, whether it is a movement or not, we are doing it because it needs to be done. And the time is right.

What other gaps do you see?

We need to start a CSA (here in Manhattan). The one we had was so popular here.[6] Our problem is just finding enough growers that can produce reliably. That's always been the problem. I think we'd find more restaurants that would be willing to buy the food if we could be more reliable in our production. I've sold to a couple of chefs here in town. I tried, but it was hard. They are very open to it too, if we could do it reliably. No one wants to do it full time, or can do it full time as a sole source of income. That's one of the deficiencies, making a living at it. It's a real good supplemental income, as you know.

I commented that for me, I benefited from the healthy food and healthy exercise more than the economic return, and asked her if it had truly been good for her eco-nomically.

Farming in the Dark

I probably don't even get minimum wage if you count the time picking it, packing it, coming here to sell, in spite of the 'high' prices, but I'd be doing it anyway. Its just kind of extra money to pay for the seeds, fertilizer, the tools, everything. It works out well. If we can get more people into the Wednesday habit. Next year we'll probably end up changing the day we have our market. Maybe by next year we'll have a CSA.

We continued to discuss ways that a KSU student farm CSA could coordinate production and marketing with local organic growers like Elaine and me, and whether that should involve a group of growers, where each does more specialization, vs. planting diverse crops, as we do now.

I followed up by asking Elaine how she got into this in the first place.

I've always been interested in food. I've gardened everywhere I've lived, even in southern California in a hard-as- concrete dirt backyard. It was so rock hard. I tried putting down newspapers to soften up the hardpan, everybody just laughed at me but that was back in the 1970's.

Did you learn to garden from anyone in your family?

No, I taught myself, figured it out myself. It was in the '70s, everyone was getting into it, back to the land. I gardened in California, Idaho, New Zealand . . . even transplanted carrots from one place to another . . .

Did it work? "

No, they forked.

What was your favorite place to garden?

New Zealand was good, the climate, (and except for the hardpan) California was good; everything goes nuts out there. Kansas, my favorite place now is Kansas.

Back at the party, Bob shares his definition of "good food" with me.

Local, fresh ingredients; fresh is probably the first step, and local goes with it, grown by yourself hopefully, but if not, you go to a farmers market. It's an attitude about food where you take your time making the food, and you have your friends over preparing the food together, and it's a whole approach to enjoyment of food, good flavors, healthy food. The slow food movement just says that you are slowing down; it's just the opposite of fast food. Fast food is generally not good for you, and doesn't have the flavor.

Then he pauses to remember the local welder, the suppliers of the materials for the oven, the helpful staff at the library when he was looking for books and photos of ovens, and how it became possible to build this oven with help from others in the community. His closing comments were:

"Slow food gives. Fast food only takes. Slow food gives back to the community, and helps foster the connections between people in the larger system," *including those of us that are there to eat.*

Notes

[1] Slow Food Movement websites: slowfoodmovement.com, slowfood.com and slowfoodusa.org. A general definition of slow food is a cultural shift towards keeping life in balance by deciding to cook more meals at home or with friends, purchase locally grown and processed foods, savor the flavors. The slow food move-

ment began in the 1980's in Italy and emphasized local foods and wines, but has expanded to include environmental concerns related to food, biological diversity, and to promote school food education programs.

[2]The 3 books are: *Your Brick Oven* by Russell Jeavons (Australian), Marlowe & Company, 2005, *Build your own Earth Oven* by Kiko Denzer, 1998, Earth Ovens, and *The Breadbuilders* by Daniel Wing and Alan Scott, 1999, Chelsea Green.

[3]Wheatfields Bakery is an artisan bakery and restaurant located in Lawrence, KS. See www.wheatfieldsbakery.com. It has been rated as one of the best bakeries in the United States.

[4]Health Department rules are generally thought to be created for protecting consumer health. For small farmers and food handlers, some of the rules don't make sense, can be quite restrictive, and some seem as though they exist to be economic barriers to entry for small, less capitalized farms and businesses. For an interesting read (rant) see Joel Salatin's latest book, *Everything I want to do is Illegal: War Stories from the Local Front*, Polyface, 2007.

[5] "Are supermarkets cheaper than farmers' markets?" by Emily Oakley and Mike Appel, published in *Growing for Market*, Sept. 2005, Vol 14 No. 9. (They compared organic prices at their own stand and the Tulsa Oklahoma Wal-Mart, Albertsons, and Wild Oats on the same day in May and again in July 2005).

[5]"Wind River Organic Growers" was a CSA (community supported agriculture) marketing partnership among three growers, and 20 client families in Manhattan, Kansas during the summer of 2003. It was successful in the sense that most of the client families wanted to continue to participate in 2004. There was a waiting list of at least 20 families willing to join when the operation was ready to expand. We were unable to continue because the primary grower (70 percent of the total proceeds) and organizer, had a baby in 2004, and needed to take a summer off for parenting. She later moved out of state due to her husband's job. The other two growers, (I was one of them) were unable to expand our own operations or to find another primary grower.

Discussion Questions:

1. Slow Food started as an international movement (in Italy), and now has a foothold in the U.S. with several local chapters or conviviums. Do you think it will really catch on in a country and culture so geared to fast food and instant gratification? Why or why not?
2. Bob and Elaine are urban farmers, and combine farming with other careers. Is this a viable model for other people, or are they an exception?
3. Research local health codes in your area, especially those that regulate egg handling for local sales, milk (and cheese) handling and the legality of selling locally, and also meat butchering, especially small animals such as chickens and rabbits. Could a small grower, with a net profit of say $500 to

$5000 afford to implement procedures necessary to do these things "legally?" Should there be a small farm exemption of some sort?

4. Research similar health codes for starting a small restaurant, catering business, or for selling processed or baked goods at a farmers market. Do you consider these to be restrictive, or would they allow for small "incubator businesses" to survive and thrive? If they are restrictive, are there ways to exempt small businesses and still protect public health? What types of laws or policies would need to change? (OK, now go change them!).

5. Do a little research on outdoor ovens in different cultures and food traditions. Do you think an "oven revival" would improve food quality (or at least bread quality) in the United States?

18

Laurie Pieper
Morningstar Bed and Breakfast[1]

"The experience with Bill—with the milk and the eggs—made me realize that even someone who didn't think he was interested in organic foods could tell the difference."

I first met Laurie when she was organizing a tour for the Riley County Master Gardeners at the Olathe, Kansas research site, where I have my herb research plots. The second time we met, she was helping to choose the sites for the Riley County Master Gardener home and garden tours, which is a spring fundraiser for the group. They pick about a half a dozen gardens to tour each year, sometimes with a theme. This year's theme was "historic gardens," since the city of Manhattan was celebrating its 150th anniversary. Since I live in a stone house, built in 1870, this qualified me to be on the list, even though my garden is not really in the same category of "neatness" as most that are on the tour. Laurie thought it would be a good stop since we also have chickens, vegetables, sheep, etc.; it was the only garden on the tour was also a "farm." At about this point, Laurie began buying our eggs, which worked out great for us, especially in the winter, when we don't have our usual marketing outlet, the farmers' market. Her need for eggs almost exactly matched our supply (about five dozen a week) and we could drop them off on our way to work. She is also buying vegetables from us during the growing season, but grows most of her own herbs.

Laurie's academic background includes philosophy and she taught for a number of years before changing careers. I didn't realize this for several weeks after I met her, because when I dropped off the eggs we talked about food, recipes, and various life experiences. She is originally from upstate New York, but did her academic training (Ph.D. at UCLA) on the west coast, and lived in Oregon before moving to Kansas. She also has traveled extensively, most recently to New Zealand in the spring of 2005. I felt that her views on food, gardening and consumers would be relevant for this book, and the interview captures several angles. We met for lunch at the new deli at the Peoples' Grocery Co-op[2]. The special of the day was piquant split pea soup, so we had that, along with a tofu salad (like egg salad) sandwich, and settled into conversation at the lunch counter nestled in a sunny window.

How did you get interested in local/organic food, and buying eggs from me?

This is an odd route, not one most people have, but after I hurt my back I started to have other health problems. I thought I have enough to deal with, with my back. I need to do everything I can to make sure I don't have other health problems that could be avoided. I became vegetarian, and I wanted to do it the right way, to get enough nutrition. It was through reading about nutritional requirements and how to go about meeting them as a vegetarian that I learned about organic . . . . So, the back injury got me started.

Farming in the Dark

Something I noticed when I started to eat more healthful food was my taste buds changed. Those were the foods I wanted, not the surgery or fatty foods. Even the simple things, like ice cream made with whole milk tasted too rich to me. I had gotten used to eating frozen yogurt and low-fat ice cream. After not eating meat for a while, I lost interest in it, it no longer appealed to me. I've read that there are studies on how what children eat affect their food preferences for their whole lifetime, but it seems to me from in my experience, if you make a change, and stick with it after a little while, your taste buds will change. I guess it is hard to get yourself to make the change; it takes something dramatic.

As she became more conscious of food, she also wanted her husband Bill to eat a healthful diet. She began purchasing non-fat milk, since he drinks a lot of milk. She was concerned about the hormones and chemicals that might be in the milk-fat, more than simply the calories. He didn't like it at first, and complained that it tasted "like water." When organic milk became available in Manhattan, Laurie began buying the non-fat organic brands, instead of the conventional. Bill immediately noticed a difference, and asked if she had switched back to the higher-fat milk, since the new milk had so much more flavor.

Then we started getting your eggs. On occasion, when I couldn't get them from you and I bought grocery store eggs, even the expensive "cage free" eggs, I noticed they were much paler in color, not flavorful. Bill even accused me one day of watering down the eggs. The experience with Bill—with the milk and the eggs, made me realize that even someone who didn't think he was interested in organic foods could tell the difference. If I had told him before hand that the milk was organic, he would have decided that it wasn't different tasting, because he believed that milk can't taste good without the fat, but not knowing that I was getting non-fat organic milk, he could taste a difference. It was the same thing with the eggs. I didn't tell him, but he could tell they tasted better.

How do you make decisions about whether to buy local or not, organic or not?

The first consideration is always what is available. Because we live in Kansas, we don't always have fresh produce. So in season, I try to get as much as I can locally, like tomatoes. With some things, the difference in quality is much more noticeable, the tomatoes are a good example. I think everybody recognizes that homegrown tomatoes are just an entirely different food. Strawberries are the same way, and I think that with the potatoes we notice it too. I don't know if it is because we are buying different varieties than ones grown for the commercial usage, or if there something about having them local, but even with the potatoes also there is a big difference. When it comes to things like organic cereals, or sodas, or juices, I try to go as much as possible with organic varieties when they are available, but once in a while, they are just too expensive.

Do you have a cut-off point when you decide if it is too expensive?

With cereal, if it is a dollar more a box, I don't mind. I've been paying attention and I often pay up to 25 percent more for the organic items. Pasta, is about 200 percent more for organic than non-organic, so I only buy organic once in a while.

Sometimes with the organic produce, when it is not local, the quality isn't there, because it has been on the shelf longer. In those cases, I end up getting the non-organic. I also get your eggs, and whatever type of produce you have, and greens. From local farmers, you can get varieties that you can't get from the grocery store, like a certain variety of strawberries, or different varieties of tomatoes, so that is a treat.

Are there other things we should try to grow locally?

That's a little bit harder. I did tell you about our trip to New Zealand, and about how the foods were very different down there. Since they don't have interstate highways, everything down there was local; the eggs, the foods sold at a restaurant . Your plate of food comes with vegetables that I don't see here, or see in restaurants here. I thought that everything there tasted really fresh, and it was very colorful, not like the food here, except for maybe in California. I think we could do some of that here. For example, in restaurants, the foods were almost always served with beets that had been swirl cut; a big pile of fresh beets on your plate, and sweet potatoes, were on all of the menus. The types of breads that were served were much more whole grain. We were there at the end of summer, so I don't know what their foods were like in the middle of winter, but I think that we could grow more variety of everything, varieties of corn, native fruits and nuts. I'd suggest more variety in general, which would be good not just for the consumer, but also for preserving genetic diversity in crops.

I did think of something we don't grow here, mushrooms. I have a friend that just got back from a tour of a mushroom farm in California. They import wheat straw and manure; compost it and grow mushrooms. I thought, why don't they do this in Kansas[4]?

I mentioned my experiment in growing shitake mushrooms in my backyard, which I found a bit difficult during our hot dry summers, in terms of keeping them a little wet and shaded.

Every year we have a couple of morel mushrooms that just pop up in our yard if I have shade in just the right area.

I was curious if anyone else that she serves at the Bed & Breakfast besides Bill noticed a difference in the food she serves, and asked if she tells them that it is local.

I tell them that the eggs are from a local organic farm as well as the greens and potatoes. I also tell them that the milk is organic. Most people have a positive response. Every now and then we'll have someone who is in agriculture, but not sustainable agriculture, and it starts them thinking, "What is her political bent," is she trying to proselytize? That usually comes from cattle people. We've had some ranchers stay with us. They will ask what sustainable agriculture is. We start talking about your research, and they look at me as if they are testing me, "Does she know what she is talking about?" I think we tend not to get many small family farmer types staying at the B&B because it is hard for them to get away. We did have one couple from the east coast who was retiring and just couldn't wait to get away from the farm; it had tied them down for so many years. I was surprised. They were coming to Junction City to get a custom RV made. They were selling the farm to travel the country for the rest of their lives. They loved the farm; but were tied down by it. I would say it is from the faculty on campus that I get the most positive response, when they come over to meet with job candidates. Part of it is that they are surprised there is some place in Manhattan that makes the effort to purchase organic. They are also surprised that local organic is available.

Now I move on to my big picture questions, and ask what she considers the successes or milestones in sustainable agriculture?

Farming in the Dark

I have noticed more in other parts of the country—not the Midwest—but in California and Oregon, that you can go into restaurants, and get organic food, and free-range meats. It's nice to see you can get that in restaurants. At least now, in the Midwest there is more awareness of organic food. There is a big section at Dillons (a major local grocery store), which means there is enough of a local population looking for that at the big grocery stores. So while the food co-op has many more varieties of things than Dillons has, and is a nice place to shop, the fact that the big stores have organic shows that there is increasing general consciousness about organic. Probably going along with that is concern from people in public health, about nutrition, about peoples' weight, and about how obese Americans have become. There is more attention being paid to what children are eating; not just to what foods, but the quality of the food. There is certainly a long way to go, but there is more awareness of it.

What are the shortcomings? What should we have done better?

Somehow, it seems like there needs to be a different public relations effort, because it doesn't make sense that more people aren't concerned about having organic food. Certainly, the information is out there about preservatives and food colorings, and all kinds of things we are eating that aren't good for us. But the majority of people are still eating these foods. Somehow, even though the information is available, it isn't motivating people. I don't know what the explanation is for why it isn't. I don't know if the country as a whole has just gotten more conservative, or if it's the cost of organic foods, but somehow it seems like this information should be more of a motivational factor for the general public, at least the portion that can afford it. Not that only the wealthy should have organic food and poor people shouldn't . . . but we have become a Wal-Mart society, and want to get things as cheaply as possible . . . something has gone wrong.

Do you have any ideas why the connection hasn't been made in the public mind about food and health?

People are aware of it, but not motivated by it. I haven't researched it, but that is my guess. Maybe they just don't believe it. They think, "If is really bad for me, it wouldn't be in the grocery store." Equally important is probably the fact that it takes an effort to change, to look for local and to look for organic rather than picking up what is handy. It's like I was before becoming vegetarian. I knew the moral facts, but it took me a while to make the effort to go against the dominant culture and not eat meat even though I believed that was what I should do.

We then talked about a communications class on campus that is going to take on the task of trying to educate the public about food prior to a conference coming up next February on organic foods. I asked if she had any ideas about how to motivate people, especially on the non-health, environmental issues.

This is difficult. Especially, as fewer people have connections to farmers, they could care less about the environmental issues, like soil quality, animal cruelty. I grew up in the suburbs of Albany, New York, and my neighborhood had only recently become a suburb. Prior that it had been farmland, and it bordered on a local farm; our street ended at their produce stand. Every autumn, we would walk down the street, and my mom would give me a couple of bucks, and say, "Go buy some corn and tomatoes." I don't know how many people have memories of being able to do that kind of thing. I hadn't thought of that . . . we just always got zucchini, corn, tomatoes, pumpkins, homemade preserves, strawberry rhubarb pie made that day. There was something about being able to just walk down the street, something nice about it. We would get a little bit of exercise, it was not very far, it made the whole thing seem that much more fresh because we had that little bit of outdoor air.

Do you remember it tasting better?

The tomatoes, corn, and strawberries would be the ones I have the strongest memories . . . and there was just something appealing about buying it outside. If she didn't have corn, she would say, "Wait a minute, I'll go pick some." Many people don't have those types of memories. So finding a way to make a personal connection to the value of the environment will be important.

We talked a little about college-age kids, and their reputation as the entitlement generation, and wondered whether the food quality message might be one that would appeal to them.

It is definitely a message that would appeal, but persuading them that it is better quality . . . I think that is something that has to start younger than college-age. There have been a couple of chefs in London and in New York City providing quality food for public schools, cooked for the kids. It is an intriguing idea, if you can get someone who knows how to prepare fresh food working in school cafeterias. Instead of wanting the greasy pizza or a bowl of French fries for lunch, the kids get to taste fresh food.

We used to spend our summers in the mountains, and there were wild berry patches, blueberries, blackberries and raspberries. I loved to eat them, much better than at the grocery store. I've been surprised at the B&B that many people don't like fruit. I really think it is because they haven't had fresh and/or real fruit. I had one person who didn't know what fresh cherries were when I served them—they couldn't identify it at all! Then there are people, who have kids and don't have fruit at home, and they have canned fruit at school, they think peaches are slimy, canned cocktail fruit salad, and then they don't want to eat fruit. I think that you have to get kids used to good foods earlier. So many of them grow up with things like French fries, and then they crave the fatty foods.

For me, something that we haven't talked about, but I see as related to this is how animals are treated on commercial farms. I know that the farms probably wouldn't have to use so much in the way of antibiotics if the animals weren't crowded. If they were kept in conditions that are more humane, they would probably have healthier immune systems. I also care about the cruelty, because I don't want the animals to suffer. In terms of agriculture itself, it seems like creating more humane conditions would ease up on the need for antibiotics.

Before I started buying your eggs, I had never had farm eggs before. I didn't know there was going to be a difference in flavor, but the reason I started getting them from you was that I thought that you took good care of your chickens, and that is important to me. It bothers me when I read about the de-beaking, the cramped quarters; they are not allowed roost, and all of those things. I'm very happy to have eggs from your chickens. Then with your lambs, I'm a vegetarian, which is why I don't buy meat from you, but it is important that the animals be treated well. I don't know why more people aren't concerned with that particular issue. If they were motivated by the connection between animal health and their

Farming in the Dark

health—there has to be a linkage—then maybe they would start caring about conditions on commercial farms.

To me, we should be concerned about not just more healthful fruits and vegetables, but for the animals too, meats and fish. I just really hadn't thought about the food I ate as a child for a long time. Another thing I miss is that we would have fresh caught fish. My father was a fly fisherman in the Adirondacks. We had brook trout and lake trout, and perch, and fresh water salmon and pike . . . he would troll for pike, but fly fish for the trout. He would go out and catch fish, and that's what we would have for dinner. When I was young I would go with him, I could never clean it. If I caught it, he would make me clean it. I love fresh fish, and it is hard for me to buy fish in the store. This is a regional bias on my part, but it is hard to think about eating the fish that come out of the waterways here, because the water is murky; there is a lot of agricultural runoff. These are not the clear trout streams of the east coast, but some of those streams have been ruined too due to development. I have more of a connection to fresh food than I had realized . . . the fish, the orchards . . . then my mom having her small vegetable border . . . I hadn't thought about all that.

Has your background in philosophy played a role in your thinking about food systems? Has it helped or hindered?

As a philosopher, I thought about many issues that weren't fun to think about, such as abortion and euthanasia. When I started thinking about food systems, I didn't have the reaction that many have, which is dismissive. That's how many people are when it comes to the environment. They don't want to know because they don't want to change. As philosophers, we deal with uncomfortable issues. When confronted with such things, many people react not by wishing it wasn't true, but they wish you hadn't told them, "But now that you told me . . ." I guess many people miss that step. They are stuck in "I don't want to know," rather than wishing it could be otherwise, and that makes people stop thinking about the issues. Philosophy taught me to keep thinking about them.

What are some of your ideas for the future?

What should we do next? This is not an original idea, but it has to start with changing children's tastes, with fresher foods. I haven't worked with this, but one of the aims of the Junior Master Gardener[5] program, is working with gardens. If the schools would become more interested in food that would help. I heard on the news this morning, that school lunches do not come under department of education, but under the department of agriculture, and I thought that was interesting. I know one of the schools in town, has an OWLS[6] grant. The classrooms each have a garden, and each one has its own theme; one is a pizza garden, one is an Asian food garden, and the students go out and work in the garden, learn about horticulture and about foods. That's one approach, and of course, it depends on funding and on making room in the school curricula.

When my mother was a teacher—she taught grade school for more than 30 years—she always wanted to introduce her kids to different foods. She thought a lot of them would have a large bottle of soda and potato chips before going to bed . . . so she would do things like bring fresh baked bread into the classroom . . . fresh coconut, crack it . . . just with different foods. One time I talked with her about this. She had theme food days, a tropical food day with fresh coconut, mangos, kiwis, etc. and invite the parents in. She'd try to tie the food in with something from social studies or one of the other subjects she taught. Probably now they would need printed consent forms!

We talked about how very few families have home gardens anymore, so children aren't exposed to that at home.

Growing up, we always had carrots and lettuce and radishes. Most of the property was ornamental, but my mother always had a vegetable border, and a grape arbor. She had grapes because her parents had a grape; that was her connection with her parents. We had a guest here at the B&B, from San Diego Institute for Community Health, who talked about community gardens. She thinks they are very important, but if we are going to have community gardens, they have to be in a very safe place, where the neighborhood bullies aren't going to come in.

At the master gardener conference one year, I took some classes on edible landscaping. The instructor, Rosalind Creasy, put it in such blunt terms, "if you can grow two plants, both of which are very attractive, and one is edible, and one isn't, why not grow the one that is edible!" (Laughter.) I grow many edible flowers. I can serve them at the B&B, and guests enjoy the ornamental garden. I put some eggplants, peppers and tomatoes in the middle of the ornamental patches; mix in herbs, and when people are looking at the flowers, they'll say, "Oh, look, here are tomatoes." My other suggestion is if there is some way not to marginalize, but to make food gardens a focal point and show that they can be attractive. For practical purposes on a farm, you might have to have rows, but the all around aesthetic appeal could be brought out. That could get people more interested, instead of, "You have to mow the lawn."

I think it is great the UFM[7] has cooking classes, where people come in and do Japanese, Chinese and Russian cooking. In many other cultures, fresh food is more important. Cooking classes are a way to get people interested in food, and an interest in sustainable agriculture would come after they get interested in food. I don't know how successful the eco-tourism movement has been, but I hear about it from the Chamber of Commerce. I suppose if there were some way to get people to go out and see what depleted soil is like and what healthy soil is like. Obviously, that isn't enough to attract public interest, but you could incorporate that kind of thing in a tour where they get to do some food tasting, pet the animals, pick berries along the way . . . something. What I thought was great last spring was when you took your class out to your farm and they actually saw your chickens in the orchard. For some reason, people don't think about chickens being pretty birds that like to be around trees. This is a real thing. They don't think of that when they buy the eggs. It's the same thing with cows.

People are somehow largely unaware of the dangers, of the commercial big farms. When you think about the amount of waste that they produce that isn't composted, that leaches out into the water, or accumulates in the cow manure/feedlot. I guess they have just done a good job of keeping that out of view. It is hard to get people to read books like *Fast Food Nation*.[8] They say, "I don't want to know, then I might need to change my habits, and I don't want to change my habits." There is that kind of inertia to overcome. If somehow the restaurants could get more of the local food out . . . I did have one restaurant tell me they aren't supposed to use locally grown tomatoes. He sneaks them in. I have never heard that was true; I have never heard that from the health department. He at least thought he wasn't supposed to get them because they aren't inspected. Even the restaurants have some odd misinformation. To educate the people who work in school kitchens, and in restaurants, that would be something.

I haven't thought about many of these things since I was a kid, but they probably do influence my views now. In upstate New York, where I grew up, there were a lot of apple orchards. A popular activity with children was apple picking. We also had strawberry farms. In the autumn, we would go apple picking, and in spring, in June, the strawberry farms also had peas. I remember doing field trips to the farms. Do they have anything like that round here?

I mentioned a few that I knew about; especially some set up with pumpkin patches and corn mazes.

They should actually pick and cook the pumpkins, so they know they are food.[9] Another thing that hurts this area—at the farmers' market here, the food they sell doesn't have to be local. People buy Colorado peaches. It seems to me that it would undermine some of the purpose of the market if the food were shipped in; if it's not local, it doesn't have that just picked flavor. Then people are going to think that this is the best they can get. It would also be nice if more fruits that are native were grown locally. I took a class on native fruits, nuts and berries, like the Sandhill plums, persimmons, black walnuts, gooseberries that can be used in cooking. It would be nice to have a combination of local, native fruits grown again. One problem is that we haven't had them in so long; we need to be taught how to use them again. You could teach the cooking classes, but if you don't have access to the ingredients, it won't do much good.

I asked her to tell me more about a book she had been reading, that discussed some of the history of food.

It's actually a big fat book about making sauces; the title is, *Sauces*[10]. The author talks about how difficult it is actually to tell much about ancient and medieval foods; we have very little in the way of recipes. They didn't have recipes written down for the people who were actually preparing the food, since they were common people, and were illiterate. There is enough information in various written texts to know about the ingredients that were used. Going back in history, there was this tension between local food and imported food, because local food was considered peasant food. They didn't have the money for imported food, so the imported foods, dried foods, preserved lemons, things that could be shipped on ships or later on trains, were part of a wealthier person's diet; not what the peasants could get. Then it would go back and forth, with some interest in local, and then something would happen, turn people back to imported foods. It wasn't until we had hotels and restaurants where food needed to be standardized, if people ordered the same sauce at the Hilton, it would be the same thing no matter where they were in the world.

That changed again with more preservatives, artificial flavorings, and the flavors they manufacture in this place in New Jersey—like the French fry flavor. It's interesting, how the two movements went back and forth. The author tries to produce an account of how the flavors have changed in foods. It sounds like ancient Roman foods, as close as we can tell, were a bit sourer. One of the ingredients they used was something called "verjuice," and was made from un-ripened grapes.

Chapter 18—Laurie Pieper

We talked a little bit about family histories, which also included more food memories.

My father's father was from Canada; was a local grocer, butcher: he ground coffee beans, sold dry goods, stuff like that. My father's mother was always in the kitchen. When they built their house in the Adirondacks, she wanted the kitchen at the front of the house. She didn't want to be way in the back. In the front, she could look out at the lake, and you couldn't walk in the house without walking through the kitchen. She was always cooking, and had this gorgeous view of the lake. She made absolutely everything from scratch, nothing that she didn't make.

My mother's parents had simpler taste in food. Every morning of their lives, they had eggs fried in bacon fat, and they lived to be in their nineties! I figure that when they started eating bacon, it probably didn't have the nitrates. I was talking with one of my helpers at the B&B about food that my grandparents ate. They grew up in a rural, coastal area of Scotland. Fish was of course a major part of their diet, and they also always ate cheese and jelly sandwiches. She thought that was strange. They didn't have a supermarket where they could go and get fruit in the winter, so their way of having cheese and fruit was to have cheese and jelly. Where we live, here in Kansas, we have to recognize the limitations of what's available at what time of year. Having homemade preserved food, or preserved food that doesn't have the chemical preservatives is something people don't think about a lot any more. They also came from a more active generation; my grandfather worked until he was 84. He was very proud that he never missed a day of work due to illness.

We talked about other food customs in other countries, and then Laurie's impressions of food in New Zealand.

When we were in New Zealand, we tried to make conversation whenever we were. We were talking with a server, and she talked about her experience in the United States. She hated the food here because she couldn't get vegetables. She was here touring with an athletic team, maybe soccer, staying with host families, and none of the families gave her vegetables. She found herself pleading with them, "I'm an athlete, I need vegetables!" People wouldn't say that here. She also mentioned that everyplace she went before the events they would have a food table and all they would have is bagels. A good New York bagel is a good thing, but . . .

In New Zealand, did you find there was a lot of food consciousness?

Not everyplace, but often in the small towns, the local bars, where there were no restaurants, the food was just incredible. One place, I would not have gone for dinner, except that it was the only place in town, I ordered the veggie burger plate. I expected something like what you would get here on a bun, with fries. Instead, it was on a plate with brie cheese, fruit, veggie burgers, fresh vegetables, topped with chili hot sauce, and no bun. Any place you went if you ordered a sandwich, it would have a mixed salad that tall (about 6 inches) on top of it. That wasn't just at the fancy restaurants, but also at the coffee shops, there was a lot more vegetarian fare. They also have a lot of meat there, venison, chicken, seafood, but they had more vegetarian offerings too. They had something called a veggie pie at a lot of places, which would be layers of eggplant, pumpkin, tomatoes, kind of like a quiche, but not nearly that much binder. In a fancy restaurant, you would get a big plate with an arrangement of fruits and flowers, and a couple of vegetables sides with your main dish. A popular dish was coconut with mixed vegetables. There is a lot of farmland, a lot of ocean, and then they have a different ethnic mix, many

Farming in the Dark

British, in certain places, many Greek, Asian and Indian foods. Their fusion was a different fusion. The kind of Mediterranean foods and Indian foods come from cultures with more vegetables. Everything was incredibly fresh and beautiful—passion fruit, melons I'd never heard of, kiwi orchards, avocado orchards, and just a vast amount of farmland.

There seemed to be more diversity of what was grown. In Kansas, it is wheat and milo, corn. There you would go from kiwi orchard to avocado orchard to a dairy farm, all within the same region.

Did they have any special programs to support the local farms and this diversity?

You would need to verify this, but my belief is that when land is sold, it is designated for a certain type of usage. Here we just have residential and commercial, with varying intensities of commercial. There it specifies whether it is crops, or whether it is going to be animals. I think they are making sure that a certain percentage of the land stays farmland. I remember seeing property listed as "agricultural or lifestyle usage."

Another thing that was cool was in the "quicky marts," right inside the door would be a refrigerated case with fresh produce. It would have apples, oranges, pears, avocados, mushrooms, sweet potatoes. That was right inside the door. I saw that everywhere.

I asked if she thought the food was expensive.

I would say that it was a little bit more expensive. To us it seemed a little more expensive because the U.S. dollar was low when we were down there. If you could find a way to factor that out, it probably wasn't much more expensive, but it was much better quality. We went into a couple of grocery stores when we were down there. I couldn't believe how flavorful the produce was, and what a variety they had. You could tell the produce was picked ripe, and the quality, the fruit was actually ripe."

She went on to describe a typical supermarket, where a much larger portion of the store was dedicated to fruits and vegetables than the ones we have here. They also include artisanal breads and cheeses, and a large fish counter. A much smaller percentage of the store was for packaged foods.

They did have some weird junk food items. The people we visited wanted to have s'mores (a desert made with roasted marshmallows, chocolate, and graham crackers). The only marshmallows we found were cherry flavored pink ones. It's not as if they didn't have junk food down there. You could buy bags of chips, and they like ice cream. They do have indulgences, but I think their indulgences are more balanced out. The kids have much more in the way of athletic activities than our kids do; they all run around in wet suits, they all surf, and they play rugby, and cricket. Whenever we went by the schools, afternoons, weekends, they were outside.

We didn't see very many elderly people. I don't know whether it is they have worse medical care system, and they didn't live as long, or if they just look younger, because they are healthier?

(Note: statistics show that New Zealand has an age distribution very similar to the United States, longer life expectancy for both men and women, and lower infant mortality, as compared to the United States.).

Chapter 18—Laurie Pieper

Maybe there were more on the south island and not in the tourist areas. We did have a friend down there, a visiting physician from the United States, working there for six months. She said that when she did health consultations with the Maori, the extended family, including second and third cousins would come. Not just parents or siblings, but more general members of the tribe, the whole extended family would make decisions about health care. That's not the way Europeans would do it. They also have socialized medicine there.

At the end of our conversation, I realized we had covered more topics than I had expected. I originally thought we'd discuss local foods, philosophy, recipes and ideas for outreach to consumers. In addition, we had touched on cultural attitudes towards health and food, how they can be connected at the societal level, and within a single individual, and how important those connections are.

Notes

[1] Morningstar Bed & Breakfast—located in the heart of Manhattan's historic downtown neighborhood, in a beautiful, three-story, Queen Anne-style home, extensively renovated for guests' comfort and convenience. Five bedrooms with private baths are featured in this historic home with high ceilings, tall windows, old woodwork, tile fireplace, leather and oak furniture, and stained glass lamps. The garden in the front features flowers, herbs and vegetables around a fountain. http://www.morningstaronthepark.com

[2] K-State Master Gardeners are associated with Agriculture Extension. In return for receiving training in various gardening topics, they serve as volunteers on community garden projects, answer garden help phone lines, and other projects in the community. http://www.oznet.ksu.edu/horticulture/MG/Welcome.asp

[3] People's Grocery Cooperative Exchange—the local health food store in Manhattan, Kansas, founded more than 25 years ago. Currently located at 17th and Yuma Streets, http://www.peoplesgrocery.biz/

[4] More about mushrooms in Kansas: There are more mushroom growers in Kansas all the time. "Wakarusa Valley Farm," south of Lawrence, now has a wide selection (http://wakarusavalleyfarm.com/mushrooms.html), as does the "ShroomHeads Organic Farm" in Freeman, Missouri near Kansas City (http://www.shroomheads.com/). Subsequent to this interview, Laurie received an inoculated shitake mushroom log from a "mushroom starting party" at our farm, where a friend from forestry brought the fresh oak logs, we provided the spawn and wax (ordered from a catalog), and friends provided power drills and labor. Everyone went home with some logs, which are now producing. Our climate is not ideal for most mushrooms, but with some slight modification (shade cloth, watering, etc.) can be suitable for some.

[5] Junior Master Gardeners: An international youth gardening program of the University Cooperative Extension network. Curricula includes: a) independent and group learning experiences b) life/skill and career exploration, c) service learning oppor-

Farming in the Dark

tunities for youth, and d) correlation to state teaching standards. A national network of 28 universities implements the program at the state level. http://www.jmgkids.us/

[6]Outdoor Wildlife Learning Sites (OWLS)—(from their website) Provides grants to any public or private school for the creation of outdoor learning sites at or near school grounds. http://www.wvdnr.gov/Wildlife/OWLS.shtm

[7]UFM Community Learning Center—a "free" university established in Manhattan (Kansas) in the 1970's. Offers courses taught by community members for community members. www.tryufm.org

[8]*Fast Food Nation*—The dark side of the all-American Meal, by Eric Schlosser. Houghton Mifflin Company, 2001.

[9]Pick-Your-Own Farms in Kansas. See list at:http://www.pickyourown.org/KS.htm At the time of writing, thirteen orchards/pumpkin patches in twelve counties in Kansas are listed with hot links to websites, and/or contact information for finding the farm. Additional farm directories for direct-marketed produce can be found at www.localharvest.org.

[10]*Sauces* by James Peterson, John Wiley & Sons, 2nd Ed. 1998.

Discussion Questions:

1. In contrasting the United States and New Zealand health statistics, there are several differences between the two countries that could account for longer life expectancy in New Zealand. List as many as you think might apply in this situation. Do you think there is room for the United States to improve its statistics? Would diet play a role?

2. Are organic prices higher in the stores where you shop? Compare 10 typical items, both organic and non-organic, in two or more different types of stores (grocery store chain, box store and/or locally owned food co-op). If one wanted to shop for organic items on a limited budget, what would they purchase? What would it be best to avoid?

3. Conduct your own taste test of two or more organic products vs. non-organic, or local vs. non-local. Include eggs, milk, meat, and/or vegetables and fruit on your list. Try to control for "freshness" if possible. What did you learn from this exercise?

4. Do you think children's exposure to gardens and new foods determines their eating habits and taste preferences? Do you agree with Laurie that a person can consciously change their eating habits/preferences if they want to? What are some of your own eating patterns or preferences that you can trace to your early childhood memories?

5. Laurie states that here background in philosophy helps her to think about some of the "hard issues" in agriculture and food. What are some hard issues that you'd rather not think about, or even know about?

19

Nora Pouillon

Chef, Washington, D.C.

*"It's just mind-boggling to me that people don't make the connection
between what they eat and how they feel."*

*I first became aware of Nora's work a few years ago, when serving on a USDA
review panel for the newly created organic transitions funding program. About a
dozen of us had flown to DC after reading and reviewing the stacks of proposals
that had been submitted, which needed to be ranked, in order to decide which ones
to fund. Our meeting lasted for 2½ days, to give each proposal a fair hearing, and
to make final decisions based on the quality of the work in the proposal. At the end
of the first day, we were trying to think of an interesting place to eat dinner. There
were so many choices in Washington DC it was ridiculous, especially when compared to the small mid-western college towns that most of us had come from.
Should we eat ethnic food? Seafood?*

*Someone suggested Nora's, since it was a well-known organic restaurant. We
all decided that it was both appropriate, and sounded like a lot of fun. We called to
make reservations, and unfortunately, there weren't any openings that night. Bum-
mer. The hostess suggested we try Asia Nora, the new restaurant with a more Asian
flavor, but still organic, and still including seasonal produce. It turned out that
Asia Nora's was within walking distance of our hotel, and we were in for a totally
delightful experience. Who was this culinary genius with a conscience? She is also
a founding board member of the Chefs collaborative,[1] an organization that promotes buying directly from farmers, artisanal products, and seasonal cuisine.*

*I was able to interview Nora on the phone, so we started at the beginning. My
first questions were; how did you first get involved with organic food, and with the
organic movement?*

I'm Austrian, and I came to the U.S. in the late 1960s, as the young bride of a
French man who put a lot of importance on the flavor and taste of food. I was as-
tonished how, in the richest country in the world, food was treated as though it was
just second hand. The large supermarkets carried mostly packaged and frozen food,
seasons were not taken into consideration, and the fresh food section was the smallest in the store.

I also noticed that people had many more diseases than I remember seeing in
Europe; heart problems, obesity, and a lot of cancer. I made the connection between
the food and the environment we lived in, and how people felt. Everyone had to
breathe, drink, and eat, so the quality of the air that we breathe, the food we eat, and
the water we drink has an impact on our well-being.

First I looked into beef cattle, and became aware of the use of all the chemical
additives used in raising the beef, hormones and antibiotics that were used to feed
the cattle. When I researched the agriculture practices a little further about vegeta-

ble production and chicken farming, I started to realize that they too used an enormous amount of chemical additives.

This is when I decided to try and find clean, wholesome products. During this time in the '70s with the hippie era, it was a big movement—back to nature with the food co-ops and direct farmer connections. It was here I found cleaner products and contacts to the farmers. I was already eating "natural" at home, so when I started a cooking school and catering business, I decided to carry the same principles into my business.

While I was teaching, one of my students asked me to start a restaurant, the Tabard Inn, and that is where I first met my partners, Stephan and Thomas Damato. Three years later we opened Restaurant Nora—in 1979. When it opened, at first it was difficult to find organic food. There were some local farmers who tried to do things more sustainably and wanted to go back to the land, but had no idea how to do it. It was a struggle. But I preferred to purchase from them because it is important for our health, and also for the environment as compared to the wholesalers who got their vegetables from far away, unripe and out-of-season.

We continued to talk about what it was like before the 2002 USDA organic standards were in place, before the 1990 farm bill defined organic, before the 1980 first USDA report on organic, and before there were even independent organic certifiers. I can remember those days, as I was an impressionable graduate student at that time, and I remember seeing the birth and early growing pains of NOFA-NY in the mid-1980s, one of the first certifiers on the east coast. Nora recalls:

At first I was this lonely person. I called the food "additive free," since no one knew what organic meant. The biggest turning point here [in Washington DC] was when Fresh Fields opened a store, which has now been bought out by Whole Foods. They had an enormous budget, and could do what I could not do. They had the means to educate the public about the differences and benefits between organic and conventional food, and why it is worth spending the extra dollar to have healthier, organic products. As Fresh Fields/Whole Foods became bigger and more mainstream more farmers were needed to supply these stores.

Twenty five years ago I had to change my menu every day, because there were not enough organic products out there for a weekly or even a seasonal menu. Now in the last ten years, it has become so much easier. *Were your early customers appreciative?* I think they liked it from day one, because the food really tasted good, but they didn't make the connection between taste and the fact that it was local, seasonal and organic. I was also popular because I cooked Mediterranean style food, which is lighter and healthier, French food was in fashion then, which is much heavier. I like to think that my menu is an educational tool. I explain to my customers what organic means, and list all my sources and suppliers on the back of the menu. Many people come to the restaurant because the food is delicious, but they have no idea what organic is.

When I became the first certified organic restaurant in the country in 1999, nine years ago, the papers didn't pick it up; the newspapers didn't know what it meant. The food journalists are not agricultural people; their knowledge is mostly geared toward recipes and cooking techniques. They are good writers, but perhaps they too don't realize the connections between what we put in our mouths, how it affects our bodies, our health, and the health of our planet.

Chapter 19—Nora Poullion

Have you noticed any difference since the USDA national organic standards became law in 2002?
Yes it had a very positive impact. It was very educational and informative for the consumer. People never understood what organic meant, and now with the USDA certified organic label on every product, food has more credibility because the government stands behind it. Now my customers give me credit for being certified organic because they trust the USDA certified organic seal. From now on, I cannot buy from non-certified growers.

Nobody had ever certified a restaurant before. I had to find a certifying agency who would agree to create organic standards for a restaurant. It took two years to put everything together. I have to prove with an enormous amount of paperwork that at least 95 percent of the food that comes in to my restaurant is certified organic. This means that I have to show the certificates, which is like a 400-page book. All the certificates expire every year, so I have a person on staff that takes care of the book. He is the buyer, and makes sure I have all the certificates and that they are up to date. If the grower doesn't have a certificate, I have to find another supplier. For specialty items like goat cheese and asparagus, it's still a struggle, but nothing like it was 20 years ago.

What are some of the things the organic movement should address? Education, it's really what needs to be addressed. It's so sad, but the only time people wake up is when a disaster is happening. If you have a couple of cases of mad cow disease, suddenly organic beef is in great demand. People don't know about the quality of water that they drink, about genetically modified organisms (GMOs), and they don't complain. But I predict that perhaps in 10 years the general public will sue the government for contaminating their food supply and their water and their air. People are not aware of the health risks, and nobody informs them. An example of what is to come; a woman sued McDonalds for making her kids obese and sick.

My kids go to an expensive private school, and they don't even have a lunch program; they go to Safeway to buy their lunch. In schools that have cafeterias, the food is government surplus, mostly fatty processed foods, from companies that got the contract. It's a billion dollar money-maker and business. Kids do better when fed a diet that is low in sugar and fats, and when food is shared in a quiet environment. It makes so much sense. Look how you feel after you eat a fatty meal, when you go to sleep that night you have only nightmares. All your energy is needed to just digest the food; your liver and kidneys and heart can't handle it all. I would love to have more education on nutrition for the general public, for kids and for me too. Alice Waters is doing a wonderful job getting the message across with her school lunch program.[2] Another great example of an organization that gets the message across is the conference, 'Food as Medicine,[3] where educators speak about the unbelievable results they receive by providing healthy choices, especially for children.

In addition to improving the school lunch program, because it would affect so many kids, the other thing I wanted to say is that the health care costs are enormous due to poor nutrition. People are sick. It's just mind boggling to me that people don't make the connection between what they eat and how they feel.

Within the schools, educators could combine school lunches with a curriculum and with gardens, which is a great idea. If kids say they hate broccoli, having the children grow it, cook it, will change their opinion. Inner city kids have no chance

Farming in the Dark

to see a farm, to see what it looks like; some think milk comes from a carton. This can all be combined with community service, let the students work on a farm, and work in the kitchen. Just like they have sex education, they should have nutritional education, it's so important. Teachers can make the nutrition program interesting; tell them how much better they will feel, how much their learning capabilities will improve, performance in sports will be greater, whatever they are interested in, connect it to what they put in their bodies.

I hope that in the next 20 years, conventional food will be in the minority, because we are destroying our planet, our health, and the future. Hopefully all of this will change, possibly through a grassroots movement. There are environmental groups—the Environmental Defense Fund, the Environmental Working group and others—whose mission it is to save the oceans, the coral reefs, the white lions.

Do you think it is valid to criticize the sustainability movement as too broad, or is that its strength?

I don't think you can divide it up. You have to see things all together. What do you do if your food is good, but your air is bad, or if you cook your food well but the water is contaminated. You can't compartmentalize. Everything is connected, even to the smallest atom.

Do you consider sustainability a movement?

Sure, it is a movement. People perhaps only see some sides of it that interest them. I could never have been a certified organic restaurant 20 years ago, or even 15 years, this only happened nine years ago, and then it was still very difficult. Even now, I say I am organic, and they say 'yes, my daughter is vegetarian.' When I go to Austria and tell them I'm organic, they say 'you have to be organic in the U.S., because everything there is so messed up.' But now organic is big in Europe too. I recently went to the BIOFACH trade show in Germany; everything is organic; sausage, beef, clothes, which is another big thing now.

How would you compare what is going on in Europe and the United States in terms of progress towards becoming organic?

It was started earlier in the states, but we took a long time to get the movement going. Europeans picked up the organic movement and developed it faster than the US. The awareness of antibiotics and growth hormones came to Americans when Europeans refused to buy their beef. The same thing is happening now with GMOs. I think Europe is ahead with the awareness and the commitment to organic. In America there are not as many people that are committed, it's a bigger country.

In Europe, it would be more difficult to start an organic restaurant than in the US. It's not set up there like it is in America; there are no Whole Food Chains and it is more difficult to find strictly organic wholesalers. There are only some organic items sold in the stores. A big organic supermarket in Vienna only opened a couple of years ago.

To summarize, would you say that you are hopeful for the future?

Yes, I'm very hopeful for the future. It will become better and better, but I'm afraid it will be for the wrong reasons. It won't be because of the beliefs of people or the will to change, that won't be the biggest draw. It will be because we will have one disaster after the next, and people will realize that they have to change. I can't believe that we still go into wars and fight like in the middle-ages, or that people ac-

Chapter 19—Nora Poullion

cept that they may get cancer. It's like we are brain dead, and we forget to look at the bigger picture. Looking at the big picture will happen one day; unfortunately it will only be because of the major disasters that continue to occur. It's sad to think that a disaster must happen before we see the big picture; the disaster needs to be prevented. We will need to go back to the natural, agricultural practices of farming like we once did, except now it will be much better due to all the knowledge we have gained.

Students need to demand healthier food, especially in the more progressive schools. My son went to Cal Poly because he wanted to study soil science and organic farming. Can you believe that an agricultural college doesn't have an organic section? Finally, an organic curriculum was started, but the school wasn't awarded any money for the program because it is financed by drug companies.

We ended the conversation by talking about the upcoming Eco-Ag conference[5] at Asilomar, near Monterey, California. She was a speaker last year, and praised some of the other talks, the great location, and the people she met there.

The whole thing is so beautiful . . . all these houses in the pine trees, five minutes from the ocean . . . .and it blends into nature . . . connected by paths . . . the longer you can stay there the better, and surrounded by people who think like you do, so empowering . . . I loved it.

Notes

Nora Pouillon; Restaurant Nora, 2132 Florida Ave, NW, Washington DC, in the DuPont Circle area. www.noras.com
    In addition to opening the first certified organic restaurant in the nation, Nora has consulted with Fresh Fields, Wholefoods Market and Walnut Acres, where she has developed a number of organic products. She is a founding board member, of Chefs collaborative 2000 and is a leading spokesperson for the NRDC/SeaWeb Give North Atlantic Swordfish a Break campaign.
Nora is a member of the Organic Trade Association, serves on the advisory board of the Greener Business Guide and is a board member of the Amazon Conservation Team, Wholesome Wave, the Environmental Film Festival and the Earth Day Network.[6] Nora received the Chef of the Year Award of Excellence by the International Association of Culinary Professionals. Nora is an active member of Washington chapter of Les Dames d'Escoffier, and served on the board of Women Chefs & Restaurateurs[7]. She is author of 'Cooking with Nora,' a seasonal menu cookbook with nutritional information and personal notes.

[1]Chefs Collaborative—www.chefscollaborative.org. The Chefs Collaborative is a national network of more than 1,000 members of the food community who promote sustainable cuisine by celebrating the joys of local, seasonal and artisanal cooking. The Collaborative is dedicated to local growers, artisanal producers, sustainable agriculture and aquaculture, humane animal husbandry, and well-managed fisheries, and conservation practices that lessen our impact on the environment.

Farming in the Dark

[2]Alice Waters—Founder and owner of Chez Panisse since 1971. It is based on a philosophy of serving only the highest quality products, only when they are in season. Chez Panisse is a network of mostly local farmers and ranchers whose dedication to sustainable agriculture assures Chez Panisse a steady supply of pure and fresh ingredients. Alice is a strong advocate for farmers' markets and for sound and sustainable agriculture, and in 1996, created the Chez Panisse Foundation to help underwrite cultural and educational programs such as the Edible Schoolyard.

[3]Food as Medicine—www.cmbm.org/trainings/Food-AsMedicine/index.htm Working to create a world in which children go to pediatricians in private practices and clinics knowledgeable about the benefits of good nutrition, and can help teach families how to eat for optimum health.

[4]Environmental Defense Fund—http://www.environmentaldefense.org Founded in 1967, they combine strong science, innovative markets, corporate partnership and effective laws and policies to address environmental problems. Environmental Working Group—http://www.ewg.org/ The mission of the Environmental Working Group is to use the power of public information to protect public health and the environment.

[5]Asiolomar Eco-Ag conference. Held annually. For more information, see http://www.eco-farm.org/

[6]The Amazon Conservation Team mission is to work in partnership with indigenous people in conserving biodiversity, health and culture in tropical America. The mission of Wholesome Wave Foundation Charitable Ventures, Inc. is to generate and manage programs that support and promote local and regional, sustainable, healthful and culturally significant food production and delivery systems. The Environmental Film Festival opens for its sixteenth year in Washington, D.C. to play a role in addressing the challenges facing our planet through the artistry of film. Founded by the organizers of the first Earth Day in 1970, the Earth Day Network promotes environmental citizenship and year- round progressive action worldwide.

[7]The mission of Women Chefs and Restaurateurs is to promote and enhance the education, advancement and connection of women in the culinary industry.

Chapter 19—Nora Poullion

Discussion Questions:

1. How does contrasting one country's or continent's food system with another's (United States vs. Europe, for example) teach us about our own food system? What did Nora Pouillon learn from her background and experiences in this area? What are your own experiences in contrasting food cultures?
2. How hard would it be in your area to supply a restaurant, or home, with 100 percent organic ingredients? How hard would it be to find local/seasonal foods? Is it possible to find local/seasonal and organic foods? Enough for a restaurant?
3. Do you believe that the taste of food, its quality and its health benefits are linked? Why or why not? If you have time, do a little background reading about total (plant) phenolics, anti-oxidants and food quality to help you answer this question.
4. Should nutritional education be a required subject in public schools? How could this topic best be presented to engage student's interest?
5. Do you think citizens should be allowed to sue their government for lack of clean water, clean air or healthy food? Why or why not?

20

Starting a Discussion about the Future of Sustainable Agriculture

"Can we create sustainable agriculture within the context of a society that is itself, probably unsustainable?"

I've known myself for almost as long as I can remember. As I sat down one day to summarize all of the chapters and interviews I had collected, I decided to have a conversation with myself.

What do you think people are really saying here?

Well, coming at it from my own research background, it seems like they are saying that research isn't enough. They are saying that universities have not only been less than helpful, they've been part of the problem! They are saying that "I'm not making it with just farm income, and neither are my neighbors." They are also saying that "get big or get out" is not a satisfactory choice. They are choosing to farm differently.

They've also said that sustainable agriculture has been helpful to them. They've found a "new thought process" as Dan Howell would say it. Some saved money, and thus made more money using sustainable agriculture (Bennett). Some were able to survive some difficult years in farming because of things they learned and tried (Holsapple). Others have helped create new marketing channels and opportunities for themselves and others with Kansas Organic Producers[1] (KOP) (Reznicek) and Organic Farmers' Agency for Relationship Marketing[2] (OFARM) (Fernholz). Some found that direct marketing suits them best (Scharplaz, Vogelsberg-Busch, Cusick, Johnson and Jon Cherniss), though two of the three of these vegetable growers also found that the community supported agriculture (CSA) marketing model did not fit for them over the long haul.

What do you think sustainable agriculture should do about all this?

The main thing is I think we need to take our blinders off. It is convenient to look through our little window on the world, and focus only on that. Many of these people were saying, "Hey, we all need to work together here." That takes people out of their comfort zone. In academia we used to think it was a big deal if a soil chemist collaborated with a plant physiologist. Later we found that soil and plant scientists benefited from collaborating with sociologists, economists *and* farmers! Now we need to bring nutritionists into the picture, and more-over, consumers, end users. We've learned to move from talking about "farming systems" to "food systems," especially as we begin to work on developing local food networks. Now we need to push the envelope a little further. What's next? What's needed? Perhaps some global policy/political analysis, which organizations like Oxfam America[3] and IATP (Institute for Agricultural Trade Policy[4]) are doing.

What do you think the message to students from all this should be?

We used to have bumper stickers that said "Question Authority." I'd say that is still relevant, but now we need a new one, one that says "Question Reality." What reality are you being fed by the mainstream media? How can you get to the real truth behind things? In sustainable agriculture in the early years, we were accused of having a bias. That implied that conventional agriculture research was somehow more pure, unbiased. If you get to the fundamentals of this question, *all* research has a bias, or point of view. This could also be called a certain set of assumptions.

For example, a conventional research experiment may be based on the assumption that all farmers could and should use fertilizer, thus the experiment will consist of five treatments, all including fertilizer. The "control" treatment with no fertilizer, for many years, was jokingly called the "organic" treatment.

Someone coming in with another set of assumptions might ask the question, "What sort of fertility program is best for my soil?" That person might have five treatments including chemical fertilizers, animal based manures, plant-based green manures, rock powders and a no-treatment control plot. This person might find that animal and plant-based manures improve the soil more than chemical fertilizers. And this research is biased? And the other isn't?

Now I think we need to also examine some of the larger, societal assumptions. These would be to question the status quo on issues like how goods are traded, who makes money and who doesn't, who controls our ports, and who sets the prices.

What about those larger, looming, political realities out there?

Yes, those are annoying, aren't they? [laughter] Several were pointed out in these interviews, including the contradictions embodied in the current farm bill program, e.g., commodity payments that reward growing conventional crops with conventional inputs while the conservation programs promote some sustainable practices. Things that we *say* we value, such as clean air, clean water and pure food are apparently not valued, at least in monetary terms, at least not in the farm bill. The worst food possible is being donated or sold to schools and fed to our children. Fruits and vegetables are now seen as one of the keys to reducing the obesity epidemic in our country, and studies have also shown that current prices, though they aren't enough to keep many farmers in business, are too high for many low income families in the United States. Are fruits and vegetables subsidized? No! Most schools, at least at this point, say they "cannot afford" to purchase locally, or it is inconvenient. That needs to change.

Chapter 20 – The Future of Sustainable Agriculture

Reasons for the inconsistencies in what we as a society say we want, and what our government does on our behalf are innumerable, and if one follows the thread of logic, it takes us back in time, and also to many root causes. One might be the initial legislation that allowed corporations to exist in the first place, and to have the same rights as a person (Teske); a physical non-entity, or economic entity "without a soul." Another might be legislation that was originally created to keep some of the corporations under control, for example anti-trust legislation, which has been weakened to the point that in many of the meat industries—poultry, swine, beef—only a few companies control most of the market[5].

Some people have pointed to cracks in the façade of what we call democracy. Special interest groups apparently control politics at both the state and national levels. Serious campaign finance reform could help limit this control, as well as other rules that would restrict special interest donations. Most interviewed did not think that this will ever happen. Even with campaign finance reform, one may not be able to control good old-fashioned greed and corruption, but at least let's try not to institutionalize it!

Did these interviews make you more hopeful or less hopeful about the future?

Yes and no. In some ways, it was gratifying to see the resilience of these people, their dedication, and their ability to survive and thrive under somewhat tough odds. Many brought up new factors that are in play now, that weren't present 25 years ago, so now we have a whole new list of things to worry about. We used to worry about chemical drift on organic farms. Now we also need to worry about pollen drift from GMO crops (genetically modified organisms), and also being sued by Monsanto if they come and sample our crops in the middle of the night and find their genes in our crops. We also know now that in Kansas, and probably several other states, hundreds of test plots of pharmaceutical GMO crops are being grown, so that pollen drift onto our sweet corn plot might contain someone else's heart medication!

Twenty-five years ago researchers were starting to take notice of more carbon dioxide in their monitoring stations, and some were suggesting we could experience global warming. Well, now we clearly have evidence in terms of average global temperatures, ice caps and glaciers melting, accompanied by an extreme lack of political will (denial) currently in the U.S. administration. This could change in a matter of months, but we are years, decades behind if we want to implement any policies that could slow or reverse current trends. What this means to farmers is even more unpredictability in the weather and climate—the possibility of long term droughts, water shortages, and bizarre growing seasons.

And then there is the energy question. The concern about peak oil and the need for energy alternatives has been in the public mindset since the 1970s, and a few have been tinkering away, but apparently no serious funding has been put into research and development of these alternatives. The only thing receiving federal subsidy right now is corn-based ethanol, and this is not sustainable environmentally, much less if one looks at the energy budget. In general, U.S. agriculture is *less* efficient energy-wise than it ever has been. Most of the material in these interviews is not too dated, even after two years. However one statement is—one interview refers to $60 barrel oil. This week (March 2008) it oil prices hit $103 per barrel. One person also mentioned that if farm diesel gets over $2 per gallon, things would be tough. Guess what?

And food appears to be less safe. I don't remember hearing about regular recalls of beef, spinach, and etc. 25 years ago. Now instead of just worrying about *E. coli*, we have to worry about a lethal strain, the H:0157. Instead of worrying about a local outbreak of food poisoning, we might have more than a dozen states affected by a recall. We talk about local food, but at the national level, we are moving fast in the other direction, which reminds me of Mary Fund's dream about being in a car going backwards without brakes. When I asked some of my husband's 10-year old soccer team members about their food experiences, most could recall a case of illness related to food they had eaten. A school district employee that I recently talked to on the phone about more local food in schools said he thought it was a good idea, and exclaimed: "The lunches here are so bad! When three o'clock rolls around watch out!" That apparently is when both the students and the teachers begin to feel the digestive effects of the overly greasy fare.

So, the often-cited slogan, "The United States has the most abundant, safest food supply in the world," is blatantly false. We have a food deficit now in terms of fruits and vegetables. We are producing food with less energy efficiency than ever before, and at the same time have moved many people and jobs off the land, by increasing the number of acres a person can cover per hour, with the larger equipment. Some call this a good thing, others don't. These are all forces larger than just what we usually think of within sustainable agriculture, and as one interviewee stated, referring to 25 years ago (in another set of interviews), "We didn't know what we were up against."

OK, how about a more thorough and thoughtful list of what these interviews teach us about the successes, failures, and next steps in sustainable agriculture?

Here are the lists I've come up with so far to highlight key issues within the responses. This isn't a comprehensive list, and many of the readers of this book may find others they think should be on this list, so this list reveals perhaps some of my personal biases, but this is the best I can do at this time. One of my biases is that I'm in a job now that emphasizes research and education, so I've highlighted those issues first, and then move into the other, in some ways larger issues. The structure of this list was originally going to be in the same format as the questions that I asked; what we did right, what we didn't do so well, and things we still need to do in the next 25 years. As I examined the responses, they don't fit neatly into those lists, as many of the pluses are also minuses, or the net effect is mixed. The structure of this list then is:

1. What things were like about 25 years ago, in 1980 (or the 1978 Census of Agriculture, for purposes of data comparison),
2. What things are like in approximately 2005, the year I did the interviews (2002 Census of Agriculture data), and

3. Responses from the interviews, or what I learned, or thought I heard. Following these sections, I have a series of things on my "to do" list for the next 25 years, under the same 10 headings used for the responses.

****

Training

In 1980, there were few classes (19 identified in the USDA report[6]), and virtually no training programs in universities in sustainable or organic agriculture.

By 2005, there are courses in 157 colleges[7] (not all of them agricultural colleges). Several offer degree programs, some undergraduate, and some graduate level.

–Several of the farmers interviewed felt that their backgrounds in philosophy (Bender, Reznicek) or liberal arts (Fernholz) gave them a better preparation for sustainable agriculture than did an agricultural degree. Two others received their formal training in international agriculture/relations (Cherniss and Keller).

–Others felt that it was important to reach everyone, not just future farmers with information about sustainable farming practices. They also feel it is important to start young (Cusick and Poullion). *"I want to train the future doctors, lawyers and county commissioners."*

–Consumer education is as important as anything else. Though consumer awareness has increased, there are still many misunderstandings. Many who do direct marketing felt that was an important part of what they do ("I'm not sure people make the connection between organic and sustainable," —Jim Scharplaz).

Research

In 1980, there were many research gaps identified by the USDA Report on Organic Farming. Virtually no funding at the national or state level was going into specifically organic or sustainable programs. Much research was counted as scale/farm type neutral, such as improved varieties, nitrogen fixation, but that is debatable.

Since then, much progress has been made (see Kirschenman and others), and though we continue to identify more research needs[8], the Organic Farming Research Foundation (OFRF)[9] has also funded, and documented successes in this area. At the national level, we have Sustainable Agriculture Research and Education[10] (SARE) funding for more than 20 years, and new USDA/National Research Initiative is now funding research programs in ecosystems, organic transition and alternatives to methyl bromide, as well as Integrated Pest Management, etc. though these are still a very small slice of the total USDA pie.

–Many interviewed had personal experiences in asking their universities for information, or expressing research needs, and being ignored.

–Some found the old literature on cover crops (pre-1950s) the most helpful.

–Many also knew one or more individuals within the university system who was helpful to them, or related to them well, but they also saw the system constraints on that person, in terms of requiring more grant money, and the need to do research where the money is.

–There were mixed reviews about marketing research—some felt like there needs to be more research in this area, a whole new system (Howell) while others felt like the university was only borrowing ideas already generated by farmers, and extending it to other farmers (Cherniss and Wander).

–Some of the new research from the past 25 years has been useful (University of Missouri grazing studies, for example—Howell) while others found it frustrating. (Cherniss—"I don't have time to use all the information I have now to monitor and control pests—how am I supposed to use even more information?") Others felt like it is important to continue to do more research on topics like biodiversity and how that influences the ecology of the farm (Cusick).

–Some felt like it was helpful for the university to "document what we already know," (Holsapple), simply as a validation of the practices they had been using all along.

Support for/from Universities in General

In the 1980s, the relationship between universities and sustainable/organic agriculture advocates was often contentious. Several of the Kansas farmers remembered a specific meeting in Frankfort, where Kansas State University administrators were confronted with their lack of responsiveness at that time. Other states probably have had similar experiences.

Since then, relationships have been built, first by the administrators who were confronted ("I remember after that meeting in Salina, things got a lot better" Jim Scharplz). Later, faculty were hired, sustainable agriculture centers were created, and partnerships have been built around various projects, field days, and other joint efforts. At the national level, USDA/SARE grants specify NGO/university partnerships as a priority, and provide funding for "professional development training" for state and county level extension staff. The SARE farmer grants program also helped build those partnerships. However, the reviews of the land grant universities continue to be mixed.

–Rich Bennett, in Ohio, was a SARE grant beneficiary for some of his cover crop work with rye and soybeans. He found the project satisfying, and an opportunity to talk to his neighbors about his practices, but didn't seem to gain any new knowledge that was valuable to him, and found that most of his neighbors' farms were so large that they weren't able or not willing, to adopt the cover cropping practices.

–Rich also had a close working relationship with a sustainable agriculture research project in Wooster, Ohio, but found their data didn't fit his farm too well because of the differing soil types.

–Terry Holsapple's on-farm research experience helped him "prove what we already know" about hairy vetch, but after attending a university in-service training with a speaker from the Potash and Phosphate institute, he realized that the university system is more influenced by the companies that give them the money, than by his data.

–Carmen Fernholz in Minnesota has had a positive working relationship with the university's Lamberton Field station in recent years, but described his earlier interactions with the university as not so positive. He also has good things to say about their forage agronomist, and an ag-economist with which he has also worked/consulted. He suggested that going to the university with a certain attitude for learning was better than confrontation.

–Kansas farmers often referred to the new Kansas Center for Sustainable Agriculture and Alternative Crops as a positive, but didn't discuss any particular collaboration with K-State, with the exception of one of the forage agronomists, and those that had participated in the River Friendly Farm program.

–Kansas Rural Center staff (Fund) had found K-State less than helpful in the 70s and 80s when they were trying to determine if conventional agriculture practices were having a detrimental effect on water quality in Kansas. [Note: K-State now has an active research program on water quality issues due to agriculture, largely because of an influx of EPA program funds targeted to this problem].

–Several of the Kansas farmers had stories to relate regarding K-State's advice on farm expansion and/or adoption of organic. Basically, they did the opposite of what K-State advised, and did well (Kirk Cusick was advised not to go into organic, and Jim Scharplaz was advised to expand prior to the 1980s farm crisis.)

–Dan Howell suggests that if universities want to survive, they need farmers' support and tax dollars. If they don't start listening to farmers soon, there won't be any farmers left to support them.

–Fred Kirschenmann had similar advice, that if universities continue to only work on the grant funded biotech and not listen to farmers, they will be "complicit in their own demise."

Marketing—Organic and Sustainable

The 1980 Report on Organic Farming noted that there were few marketing outlets at that time specifically for organic, and many organic farmers were selling their products into conventional channels.

Since that time, many organic market opportunities have been created. The Kansas Organic Producers became a marketing co-op, and several similar organizations have joined together to form OFARM. Many smaller organic companies that started in the '70s and '80s have now expanded, some have been bought out by larger corporations, and many conventional food companies now have an organic label of some sort. Consumer demand continues to grow at about 20 percent per year, and doesn't appear to be leveling off yet. Major retail chains now often carry organic, and whole new chains (Wild Oats, Whole Foods) have been created and are thriving. This new era of larger companies and larger volume has led to accusations of the system of promoting "industrial organic" rather than the small/medium sized farm and food co-op scale that began the movement in the '70s.

—The farmers interviewed generally felt like the expansion of the organic markets and increased consumer demand is a positive thing.

—Many felt the new marketing structures that were created, such as KOP and OFARM, as well as web-based marketing information such as localharvest.org are very helpful.

—Most had mixed reviews on the new USDA organic standards (see comments below on certified organic), and wished that sustainable practices were recognized by consumers as well as organic. [Note: there is movement in that direction for "sustainable certification" [11] and also for "fair trade" and "local" certification standards[12].]

—The three vegetable growers interviewed had all been involved in CSA marketing at one point or another, but only one still markets through a CSA today as a member of a group of growers, so the burden is not all on one person. All three had positive things to say about the marketing model, but it is a very intense relationship with a potentially large group of customers, with a lot of pressure to have a certain amount of product of very high quality every week. Other marketing venues (farmers' markets and restaurants) can be just as profitable (Cherniss) and other relationships to consumers, such as teaching (Cusick) can be just as satisfying.

Organic Certification

Organic certification was not widely available in 1980, but by the mid-1980s had become more common as groups like California Certified Organic Farmers (CCOF) on the west coast, Natural Organic Farmers Association (NOFA) in the Northeast, and Organic Crop Improvement Association (OCIA) nationally began certification programs.

The National Organic Standards Board was created in the 1990 farm bill to unify the various groups' certification standard, and come up with one national standard. Private organizations that wished to certify would apply to USDA to become accredited. These rules and standards came into effect in 2002. Now there are close to 100 certifying organizations nationally and internationally. The new rules also restrict the use of the word organic. As a response, a new "eco-label industry" seems to have been spawned.

—There were definitely mixed reactions to the national standards. Some appreciated the value of having one national standard to simplify things for consumers. At the other end of the spectrum, some felt that it "was the worst mistake we ever made, to get the government involved." (Holsapple).

–Some acknowledged that the cost of certification has gone up with the additional costs to the organizations doing the certifying, but were also appreciative of the opportunity for the cost-share for organic certification now available, though there wasn't always enough for all those that applied (Keller).

–Two of the beef producers felt like going organic was not possible for them, in part because of the limited supply of organic grain and availability (Howell and Scharplaz). However, Scharplaz didn't feel like organic was necessary for most of their customers, as they had direct contact with all of them and could explain the more humane and environmentally friendly practices that they used.

–Some felt that the additional cost wasn't worth it for their local market, and had certified in the past, but didn't plan to continue (DeVault).

–There is some concern that there will always be pressure to "water down" the national standards to allow more of the "industrial organic" type of operations in the door. This fear is well-founded, as this has happened several times since the standards were adopted.

–Others find the national standards unnecessarily rigid (including myself), for example the strict compost guidelines that make it nearly impossible to produce at a small scale, and the requirement to source all seed "when possible" from organic sources (Keller and others).

Farm Profitability

In 1980, (using 1978 Census of Agriculture estimates[14]) there were 2.3 million farms in the United States (one percent of the population), the average age of the principle operator was 50.3, and 53 percent worked at least part-time off the farm. The "middle-sized" farms (between $10,000 and $100,000 gross sales) made up 42 percent of farms.

In 2005 (2002 Census of Agriculture) there were 2.1 million farms (0.7 percent of the population), the average age of the principal operator was 55.3, and 55 percent worked off-farm at least part time. The mid-sized farms dropped to 26 percent of farms, while the percentage of larger farms grew from 10 to 15 percent, and the small farm segment (less than $10,000 gross sales) grew from 48 to 59 percent.

USDA Census Year	Total Farms U.S.	Gross Income <$10,000	Gross Income $10,000 - $100,000	Gross Income > $100,000	Average Age of Operator (years)
1978	2,257,775	48%	42%	10%	50.3
2002	2,128,982	59%	26%	15%	55.3

Farming in the Dark

By way of contrast, the peak number of farms in the United States seems to have been recorded in the 1920 census, with 6.4 million farms, which was 6.1 percent of the U.S. population at that time. There were 29 percent of principal farm operators working off-farm an average of 137 days per year.

Farm profitability would be difficult to compare between 1980 and 2005, as it would depend on the region of the country, the product sold, and the particular price cycle of that product. Other generalizations about this 25 year period though include that high (inflated?) land prices in the 1970s allowed farmers to borrow money using the land as collateral for expansion. When prices of crops (and land) declined in the 1980s, the farm crisis was triggered.

Commodity grain prices have been fairly constant and at low levels, usually below the cost of production, and the only profitability in growing crops like corn and soybeans is from the farm bill program payment benefits. Meat prices have fluctuated during this time period, but pork prices plummeted in the late 90s, driving many independent small hog farmers out of business. Vegetable prices have not risen significantly during this time period, and if anything have become more competitive (lower) due to increasing imports from non-U.S. sources. Prices for organic grains have fluctuated considerably, but in general bring a premium price as compared to conventional grain, if one is able to identify a market, and produce up to quality standards. Comments about profitability from those interviewed include:

–Many of the farmers mentioned the farm crisis of the 1980s in their interviews, even though that was 20 years ago.

–Some survived this period in part by cutting costs, using practices they learned about through sustainable agriculture (Holsapple, Bennett, Reznicek and others).

–Most did ok during this period because they had some or all of their land paid for, were buying land from a family member, and/or had off-farm employment. None were in a period of expansion at this time, except that one (Howell) mentioned being able to purchase land at rock bottom prices after the crash.

–Some had particular empathy towards their fellow farmers, as they saw what it did to relatives (Johnson) or were involved in providing credit and financial counseling services (Teske and Reznicek).

–When asked about their own farms' financial success, the majority (with only a couple of exceptions) mentioned that they needed off-farm employment to either help pay for the farm, or to obtain health insurance benefits. When asked if new farmers could do what they are doing now, and do it full time, the response was often, "Well, it depends on what kind of standard of living you want." (Cherniss and Cusick for example).

–Many saw the high price of land, often driven up by non-farm investors, as a barrier to entry to new farmers. Some suggested that alternatives to land ownership should be considered, such as the model used now in Cuba (Reznicek).

Chapter 20 – The Future of Sustainable Agriculture

U.S. Farm Policy

In 1980, there were several conservation programs in place (cost share for terraces, waterways, etc.) and Natural Resource Conservation Service (NRCS) offices have been in every county since the 1930s, providing technical assistance, and recognition to good stewards in the form of annual awards for "conservation farmer of the year." The farm bill payments were supporting primarily commodity crops—wheat, corn, soybeans, rice and cotton—and in fact, created a financial disincentive to crop rotation at that time because payments were based on historical acreage of these crops, called "base acres."

The "base acres" requirement had been removed in 1996, and by 2005, crop rotation was possible, but still often limited to "program crops" due to low prices overall, and the need for the farm bill supplementation to the price. Conservation compliance became mandatory with the 1985 Farm Bill, and a plethora of new conservation programs to replant perennial grasses, preserve wetlands and endangered species, and eventually the Conservation Security Program (CSP) came into play.

  –Even with removal of the base acres requirement and addition of new conservation programs, farmers still see the majority of the commodity payments going to support unsustainable agriculture.
  –The commodity payments don't really benefit farmers as much as use farmers as a pass-through to benefit the agricultural input suppliers and livestock feeders with cheap grain prices.
  –Some have suggested that the only way to loosen the grip of the commodity organizations is to get urban consumers involved in the Farm Bill debate ("Lets have a food bill, not a farm bill," Paul Johnson).
  –The CSP program was a good step in the right direction, but was under-funded and poorly administrated. Perhaps more "green payment" models could be considered in the next farm bill and/or CSP expanded.

International Agriculture Policy

Since the 1960s, grain donations have been used as foreign aid and also as a way to achieve a positive balance of trade for the United States. Some food items were imported, but the balance of trade for agriculture was positive (more exports than imports).

Grain donations to some countries have been rejected in recent years due to GMO content. The 1990s brought the development of NAFTA, which lowered tariffs and import barriers among Canada, the United States and Mexico. Later, CAFTA did the same thing between the United States and several Central American countries. The impact of these agreements on agriculture has been largely negative, as imports from countries with lower labor costs and lower environmental standards compete with U.S. products, often selling in U.S. markets at below the cost of production here. Current balance of trade for agriculture overall is still positive, but declining, and for fruits and vegetables the United States imports more than it exports.

  –The specific example of tomatoes coming into Florida was mentioned by DeVault. He and others are concerned about the negative effect of imports on U.S. production, and the increasingly negative balance of trade for food in the U.S.

  –Most also saw the weakening of tariffs as negative for other countries' farmers, not just U.S. farmers. Some even went so far as to see WTO and similar organizations as weakening the power of nation-states as compared to multi-national corporations.

  –Many felt that these trends are all going in the wrong direction, especially if one considers national security, and the possible future shortage of fuels for transportation of food. Not only the U.S., but many regions within the U.S. would not be able to feed themselves, and the number of farmers continues to go down, not up.

Widespread Adoption of Sustainable Agriculture

In the 1980s, the primary practitioners of sustainable and organic agriculture were the "back-to-the-land" proponents from the 1970s, and also a certain group of farmers who just had never farmed any other way (Nancy Vogelsberg-Busch's father, for example). The OFRF has estimated approximately 2500 to 3000 organic farmers in the United States in 1992.

By 2005, there are more than 12,000 certified organic farmers in the United States according to the census of agriculture, and probably just as many non-certified organic. There is also a certain percent of farmers using sustainable practices, but this number won't show up in any national survey, so we have no numbers by which to measure the success of something like "cover crop adoption" or "diverse crop rotations." The popularity and growth of organizations like the Practical Farmers of Iowa show that there is interest, but my own anecdotal evidence has been that I seem to see the same farmers and speakers at sustainable agriculture conferences, and not a huge influx of new blood. Certainly, they report that their neighbors have not adopted these practices, even though they are in close proximity, and can see the work-ability of the ideas. It would be hard to say that we are even a blip on the radar at this point, with still less than 1 percent of farms showing up in any kind of census data. By contrast, no-till, a heavy user of pesticides and genetically modified crops, and generally considered un-sustainable despite the potential for soil savings, claims something like 10 percent adoption rate, depending on the crop and region of the country.

  –Farmers comment that sustainable agriculture practices are not a cookie-cutter recipe that one can hand-off to someone who wants a simple approach to farming (Bennett). A person also needs a desire to learn, and a desire to change.

  –No-till, on the other hand, has made a relatively simple system (corn-soybean rotation) even simpler with the addition of the herbicide Roundup, and GMO Round-up tolerant soybeans. ("It has changed the face of agriculture more than any other thing, even the large equipment." Terry Holsapple)

–The scale at which many farmers are operating limits their ability and/or willingness to try new approaches, especially if it is complex, and not well known. Most mentioned that their neighbors are farming five to six times the land they would have farmed 25 years ago, due to increasing equipment size and cost, and low profit margins per acre.

–Up until now, the cost savings of many sustainable agriculture practices were small compared to the time spent learning the new practice. For example, legume cover crops or legumes in a crop rotation—the cost of the seed, the planting, tilling, and lost profit from having a cash crop was just barely offset by the savings in nitrogen fertilizer.[15] One also had the additional risk of the cover crop fixing enough nitrogen, given weather and other conditions, compared to the fact that nitrogen fertilizer is a sure thing. Subsequent to these interviews, the cost of nitrogen fertilizer has been escalating due to world oil prices, so this equation could be changing soon.

–When asked how they first got into sustainable agriculture, not one said "because my county agent talked to me about it," and similarly, none said it was because their chemical dealer proposed it. Most came to it through personal experiences, through gardening, and often through books and magazines they were reading (such as *Organic Gardening, New Farm, Acres*, and others).

–A few said that the organic zealots they met at meetings (Bennett) or the eccentric practitioners were a turn-off at first, but Teske also admitted that they were his friends too! Teske also suggested that some organic farmers may not want everyone going into it, because they want to protect their markets. Others, like Reznicek said that wouldn't be a problem due to the complexity of organic, and the three-year transition period, and others (such as Keller and Howell) have benefited from the mentoring from experienced organic farmers.

The Future of Agriculture

Average *net* farm income was only $19,032 in 2002, averaging all farm sizes. Of the farms in the middle and large size range, state-wide statistics in the grain-belt states generally show that these farms are only profitable when government payments are considered. In 2002, 707,596 farms received government payments, at an average rate of $9,251 per farm.

–One person interviewed from Nebraska (not in this book, see Vol. II), said that his university's board of reagents had declared agriculture a "sunset industry."

–Several have in their lifetimes seen poultry farming, then pork turn from agricultural enterprises to something more akin to serf farming where the company provides everything except the barn and labor, and the producer must have a contract to sell. There seems to be little in the way of bargaining power for the farmers involved in these operations. Some see the situation for beef moving in the same direction.

–Most were concerned with the demographics of agriculture, with the average age of farmers continuing to go up. They also see significant barriers to entry for new farmers, including inflated land prices, beyond what farming could support. Few were encouraging their offspring to come back to the farm, though some were open to it if there was a value-added component, or they could see a clear opportunity for income as compared to off-farm employment.

–Many see a role for mixed crop-livestock farms, and see the sustainability benefits of a diverse operation. Some found it difficult to achieve that balance, and tended to specialize in one or the other. A few felt that new farmers "don't want to work that hard" as it combines a year-round occupation (livestock) with intense seasonal work (crops).

–The person who was able to enhance income by adding "farm entertainment" in the form of a pumpkin patch, hay bale maze, archery, etc. seemed to enjoy the experience and the added income, but also added that he was ready to discontinue the operation once the barn had burned down. He also said that, combined with his other farming enterprises, it all occupied his mind so much that he "doesn't even remember his sons' early years."

–In general, the lack of clear profitability, combined with the significant risk and cost of living expenses, such as health insurance, sounds discouraging, and should be addressed by sustainable agriculture if we want agriculture to have a prosperous future.

****

The next 25 years—ideas generated by these interviews

Training

–Universities and colleges should re-think the design of their sustainable agriculture curricula, and consider combination degree programs—liberal arts plus agriculture. This is already happening at the non-agriculture schools, for example, see the New Farm website listing of training programs in sustainable agriculture; two-thirds of them are *not* at agricultural colleges. Same thing with student farms, one-third are at agricultural universities, and two-thirds are at small colleges and also at larger non-ag universities like Stanford. Perhaps this needs to happen at land grant universities too. [Note: as someone trained in agriculture but having an interest in music, art, etc., I can see the value of the liberal arts too, but also think it would be nice for farmers and others to have some soils knowledge and fundamentals.]

–Create more farm-to-school training programs for all age groups, especially younger kids. This is also already happening, and some of these are connected to sustainable agriculture groups, but many are created by parents, gardeners, and teachers. As a group of farmers, we could be helpful in this effort, using farms like Kirk Cusick's as models.

–Outreach to consumers—one farmer has suggested to me (Paul Johnson) that we ask our state Department of Agriculture or Department of Commerce to sponsor television or radio info-ads that feature a local farm, and a product that is in-season. Many people don't even have a sense of the seasonality of their local fruits and vegetables, as they are available year-round in the grocery store. There are many other forms of consumer education that could be considered. The Organic Trade Association website has a plethora of background information and articles that could help. I've also seen calendars that feature local farms, local chefs, and delicious in-season recipes for each month.

Research

–Perhaps it is time to put together a serious chunk of change that isn't tied to the university system or to the government that should be specifically allocated for sustainable agriculture research. Yes, there are some models out there, with Kellogg funding some food systems projects, and OFRF funding organic research. Perhaps SARE's administrative structure could be used as a model for grass-roots involvement in decision-making about how funds are allocated, and there are more applications for SARE funds than there is funding to go around. Ask the Ted Turners and Willie Nelsons of the world to help create this pool of funding? As Dan Howell said, "I wish I had a million dollars to take to the university, or whatever amount of money it would take to get them to listen to me . . ."

-Taxpayers should insist that their state-supported universities start playing a leadership role in all issues related to sustainability, rather than playing catch-up, or just following the money. Otherwise they will continue to be seen as intellectual prostitutes, only following the dollar, and not the key issues that affect their citizens.

Support for/from Universities in General

–A good next step would be to continue to monitor at both the national and state levels the amount of funding going to sustainable and to "non-sustainable" agriculture research. This includes research that props up cropping or farming systems that have now been shown to use more calories of energy than they produce, or to be inefficient at nutrient absorption, allowing excess nitrogen and phosphorus to pollute U.S. streams, rivers, and now the Gulf of Mexico. This has been done several times in the past, especially in looking at whether organic farming was being served by the research establishment, but targeting "unsustainable research" might be a new twist. University administrators don't like coming up with these figures, but if the public doesn't ask, they will continue providing office space and faculty position to the highest bidder, not the best projects, or even projects that serve their constituents well.

–I'm at a loss as to how to lesson the influence of "big money" grants, especially in the era of diminished public support. One of Kansas' Board of Regents members in a recent letter to the editor (Manhattan Mercury, February 19, 2008) stated that "20 years ago, Kansas State University and the University of Kansas received 52 percent of their budgets from the Kansas Legislature, but today that figure is only 26 percent." It seems as though the universities are already in a downward spiral of diminishing public support and public service, and perhaps this trend is irreversible?

Marketing – Organic and Sustainable

–In addition to the cooperative marketing examples from Minnesota and Kansas, the "Ag of the Middle" program initiated out of the Leopold Center in Iowa seems to hold promise. It encourages one to look a the whole food chain in marketing, and promotes transparency and living wages at all levels, and cooperation rather than competition. [13]

–As an example of this, Nancy Vogelsberg-Busch points out the importance of including her local butcher as an integral part of her marketing and farming plan, and also the need to plan for his profitability and survival as a business. Her appreciation was expressed in an interesting way; "He transforms that life into an entirely different life with a knife."

–Several people mentioned the need to sell more local foods to schools, and one even suggested that a quota system could be created, where 50 percent of the food needed to come from within the state and perhaps 10 percent from the local county.

–Some pointed out the futility of trying to survive in today's markets (especially conventional) without government regulation and enforcement of the anti-trust laws on the books. This would require political action, and perhaps lawsuits similar to those that have been needed to force the government to enforce existing water quality laws in many states.

–As an example of "eating in the dark," consumers in the United States usually don't know exactly what they are eating or where it came from, because of the pathetic labeling requirements we have now. Even the "country of origin" labeling for meat seems to be controversial, and isn't enforced, GMO products are not labeled, and now there appears to be an orchestrated effort by Monsanto in multiple states to pass legislation making it illegal for milk processors to label milk that comes from cows not treated with the BST hormone. Perhaps something the sustainable agriculture community could do is to orchestrate their own campaign for more truth in labeling, beyond having just "organic" and "everything else."

-Some local marketing and value-added programs are discouraged or prevented because of state or local health-code and/or liability insurance requirements. These are always presented as there to "protect the public" but they should also be re-examined for their potential to be a barrier to small businesses (farms, bakeries, restaurants) competing with larger businesses, who aren't necessarily serving safer or more nutritious food than the small, local alternatives.

Organic Certification

–Due to the small number of organic farmers, it almost goes without saying that there will need to be consumer involvement (in massive numbers at times) and constant vigilance to keep the standards from being watered down at the national level. There probably is also a need for farmer involvement at the grass-roots level to keep the system workable and not too burdensome.

–Consumer awareness of organic seems to be largely limited to the effect the food has on them (safer, fewer pesticides, possibly more nutritious) while the farmers interviewed suggested that we have a long way to go in consumer education about the environmental and other benefits of organic.

–Those who weren't certified organic but using sustainable practices would like to see some effort put into "certified sustainable" or some other recognition of their product, or product differentiation. However, this also opens the door to even more consumer confusion, and the possibility of being co-opted by imitators, or green-washing.

–Some felt that getting the government involved is misguided, and perhaps other methods should be used for creating and protecting labels in the future?

Farm Profitability

–One specific suggestion (Cherniss) was to create a website or other ways for vegetable growers to share information about costs and returns for portions of their operations, not just the whole thing, so one could track how their farm compares to others, and also get ideas for improvement.

–Some have suggested looking at totally new marketing models as a path to profitability (discussed above under marketing).

–Niche markets were occasionally mentioned (golden flax for Fernholz, for example), but not seen as the silver bullet for their farm, or for others to necessarily emulate. Some farm profitability enterprises, such as the pumpkin patch and entertainment on the Holsapple farm can also lead to burn-out.

Farming in the Dark

–Simply being smart about knowing one's costs of production, not taking on un-necessary expenses (fertilizers and pesticides) and not getting bogged down in interest payments due to expansion was the advise of the farmers who also provided counseling to farmers during the farm crisis. One individual commented that he hadn't calculated his cost of production for some farm products because he hadn't been in a position to adjust his asking price—he just had to take what the market price offered. Perhaps a goal of sustainable agriculture should be to encourage everyone to calculate and share with others their cost of production figures. Consumers too need to know what it takes to bring a product to them, as many don't realize how much effort, water, etc, goes into each pound of grain, meat or produce.

U.S. Farm Policy

–The sustainable agriculture movement continues to lobby on behalf of portions of the farm bill they think will support sustainable agriculture on farms, but it is under-funded and under-staffed compared to the larger commodity groups. Unless the commodity support payments are either not allowed by World Trade Organization (WTO) rules and enforcement, or the commodity groups become weaker and/or are not allowed check-off funds, the only recourse in terms of balance of strength within sustainable agriculture is for farmers to ask consumers, citizens to get involved on their behalf. Some within sustainable agriculture have suggested trying to get Senators and Representatives from urban districts more involved in the agriculture committees, as they are currently heavily weighted with farm state delegates that serve the interests of their commodity groups (but not necessarily their farmers).

International Agriculture Policy

–This is an area just crying out for more consumer education. If "buy local" is to ever be more than just a slogan or buzz-word, not just people as consumers, but as citizens as tax-payers need to consider the long-term consequences of importing food, and loosing the expertise and infrastructure needed to grow it in their own communities.

Widespread Adoption of Sustainable Agriculture

–To document whether sustainable agriculture practices are being adopted, not just organic, surveys could be conducted and included in the U.S. Census of Agriculture asking how many are using nitrogen-fixing legumes in their rotation, following diverse crop rotations, or doing rotational grazing.

Chapter 20 – The Future of Sustainable Agriculture

–As for wide-spread adoption by larger farms, there needs to be financial incentive, and perhaps one-on-one coaching provided, since many practices are site-specific, farm specific. University extension systems are not prepared to play this coaching role, as their staff time is limited (and being cut further), and they are encouraged to work with farmers as groups, not to serve as consultants to individuals.

–Current incentives towards "un-sustainable agriculture" need to be removed for there to be any serious adoption of sustainable agriculture.

The Future of Agriculture

–It seems that at some point, someone somewhere is going to have to address the health insurance issue. This may happen at the national level with a health care plan. If it doesn't, perhaps a role for non-profits could be to help start some sort of group health insurance plan for sustainable agriculture farmers? As individuals, farmers are fairly powerless to deal with this, with the exception of choosing to have health insurance, or choosing not to, and literally risking the farm in that gamble.

–The profitability question is equally as difficult. Just because grain farmers receive subsidies does not necessarily mean that fruit and vegetable growers should too, but clearly the "market" is not encouraging young people to go into the profession of farming. And as skilled managers of assets in the thousands or millions of dollars, they are not being compensated for their skill or risk. As Nancy Vogelsberg-Busch said, "Curse me like your doctor or lawyer, but pay me the same or at least a living wage so I can get out of this factory job." Kirk Cusick suggested that so many doctors and lawyers are getting interested in local and healthy food, that perhaps a bartering system could be established? Maybe food has a value to some beyond monetary value? Some communities have started local currency systems to facilitate this bartering among a network of participants. In Ithaca NY, this currency is called "Ithaca Hours." What if my hour of time as a farmer was worth the same as an hour of time at the dentist or doctor?

In summary

This "to do" list for the next 25 years is not comprehensive, and doesn't even begin to capture the wealth of information generated by the 18 interviews contained in this book. I encourage the reader to dig further, both into these interviews, and to ask some of these same questions of your colleagues, neighbors and food providers. We also have several things that I think "we" (as a community of people involved in sustainable agriculture) don't agree on. I call these the dilemmas still to resolve:

–The tension between sustainable and organic agriculture (broad vs. narrow definition)

Farming in the Dark

−How the universities, agencies, non-profits and farmers can work together most productively.

−Do we need to be a more "organized movement," or are we doing pretty well as a loose-knit group of like-minded people? Are we a movement or an ideology? Should we be a movement? Do we need to create a think-tank, or get together on this some-how?

−The cost of organic food? People need to pay the full cost of food production, including paying farmers a living wage, but people would also like to see it serve more than just the high-end market. For example, "Whole Foods Market" in some cities is jokingly referred to as "Whole Paycheck Market." How do we make it affordable as a society to everyone who needs it, without asking farmers to make all the sacrifice?

−Will sustainable agriculture be limited by our unsustainable culture, land values and marketing system? Or, as Michelle Wander stated, will sustainable agriculture *lead* the way to a more sustainable culture?

****

Farming in the Dark—Conclusion

As I'm writing this conclusion, I'm remembering the recent ice-storm here in Kansas (December 2007) where large numbers of people were without electricity between three and ten days. That week, I was literally "writing in the dark," as I would read by candle-light, and write on my battery-powered lap-top, using the light from the screen to see the keys. Reading by candle-light is harder than I thought, as I had to strategically place several candles at different heights, and then try to not catch my book on fire.

What else don't we know? What will the "energy future," bring life without electricity? Books like Peak Oil and others have warned us of this upcoming bottleneck. Wind, solar, and now (unfortunately) ethanol are finally receiving some attention and funding, though nothing like they will need in the future to replace the fossil fuels we now use, at their current rate of per capita use.

Students especially can lend a hand here, helping us see possibilities we haven't seen before. Ironically, there seems to be a new category of literature out there about the "end of life as we know it," including the book *End Game*[2], which a student in my sustainable agriculture class chose to review for her class project. Hopefully these students feel they have a bright, not a dark future. Either way, a realistic look at how things are is better than pretending we can go on forever without making a change. Another book, *Blessed Unrest*[17], can provide an antidote for books like *End Game*, and encourages networking among the million or so of us involved in sustainability issues, peace and justice issues, not only in the United States but also world-wide.

I'd like to encourage a similar networking activity, but at a smaller scale, and focused more on what we can do within the sustainable agriculture movement. At a minimum, we need to look at several points of view, and I'd like to see how people respond to what I've presented here. These chapters are by farmers, farmer/academics, farmer/activists, and three consumers/chefs. A website will be created to continue this dialog, www.farminginthedark.net. Please feel free to contribute comments, your views of the future, or interviews that you might conduct with colleagues in the farming, academic, or other realms.

Caveats: People are not stuck in time, and all of these people are different people than when the interviews were conducted about two years ago. However, when all of them were asked to review their manuscript, and given a chance to change or delete anything they wanted, nearly all had only minor changes, mostly grammatical, and none asked to remove phrases they thought would be controversial, or because they had changed their mind 180 degrees on something. However, the world continues to change, and as I write this, the price of oil, and also some of the commodity crops discussed in this book have just reached levels that I've never seen in my lifetime. The next 25 years will be "interesting," as we say in Kansas, when we don't know what else to say, and don't want to come off as negative.

Another joke (from Ed Reznicek): A man was visiting a relative in a nursing home. She had her good days and her bad days, and was becoming forgetful. When the man went up to her, she didn't seem to recognize him, so he asked her, "Do you know who I am?" She said, "No, but if you go up to the front desk, they can tell you." Sometimes you just need to know who to consult!

Notes

[1]Kansas Organic Producers (KOP)—A marketing/bargaining cooperative for about 60 organic grain and livestock farmers located primarily in Kansas. http://www.kansasruralcenter.org/kop.htm

[2]OFARM—Organic Farmers Agency for Relationship Marketing see website http://www.ofarm.org. Mission statement—To coordinate the efforts of producer marketing groups to benefit and sustain organic producers

[3]Oxfam is committed to creating lasting solutions to global poverty, hunger, and social injustice. www.oxfamamerica.org/

[4]IATP—Institute for Agricultural Trade Policy, Minneapolis, Minnesota; www.iatp.org.

[5]Many examples and reports exist. See for example, "Consolidation in the Food and Agriculture System," W. Heffernan, M. Hendrickson and R. Gronski, University of Missouri- Columbia, Feb. 1999. http://www.foodcircles.missouri.edu/consol.htm

[6]Report and Recommendations on Organic Farming. 1980. USDA Study Team on Organic Farming, USDA.

[7]National Agriculture Library Publication—*Educational and Training Opportunities in Sustainable Agriculture* [reference from 2005] www.nal.usda.gov/afsic

[8]2007 National Organic Research Agenda http://ofrf.org/publications/pubs/nora2007.pdf Scientific Congress on Organic Agriculture Research

[9]OFRF (Organic Farming Research Foundation) report on research State of the States, 2nd Edition: Organic Systems Research at Land Grant Institutions, 2001—2003, Organic Farming Research Foundation, 2003. http://ofrf.org

[10]SARE (Sustainable Agriculture Research and Extension) Funds research, extension, and training programs throughout the United States See www.usda.sare.org.

[11]Sustainable Agriculture Practice Standard for Food, Fiber and Biofuel Crop Producers and Agricultural Product Handlers and Processors. Draft Standard for Trial Use, April 2007. Scientific Certification Systems, Emeryville, CA.

[12]The Agricultural Justice Project. http://www.cata-farmworkers.org.

[13]Ag of the Middle—a program created to provide research-based information to support the business development and public policy change components to "value-chain" based food systems. See www.agofthemiddle.org.

[14]USDA Census of Agriculture. Note that the census is conducted only about every five years, so the closest estimates for 1980 and 2005 were the 1978 and 2002 census data. http://www.agcensus.usda.gov/

[15] Janke, R.R., M.M. Claassen, W.F. Heer, J. Jost, S. Freyenberger and D. Norman. 2002.

"The Use of Winter Annual Legume Cover Crops in a Wheat-Grain Sorghum Rotation in South Central Kansas." Journal of Sustainable Agriculture. Vol 20:69-88.

[16] *End Game—The Problem of Civilization,* Vol. 1 by Derrick Jensen. Seven Stories Press. 2006.

[17]Hawken, Paul, 2007. Blessed Unrest: How the Largest Movement in the World Came into Being and Why No One Saw It Coming. Viking. To link to current network of organizations, see www.wiserearth.org.

Appendix: Recommended Reading

All of these categories contain partial lists of books that have been published on these topics. In addition, see the Notes sections of the chapters, and the National Agriculture Library lists of sustainable books in print. Exclusion of a particular book or author is not intentional, but simply due to lack of space.

Early writers and influence (pre1980):

Farmers of Forty Centuries, F.H. King, 1911, Harcourt, Brace

Agriculture. A Course of Eight Lectures, Rudolf Steiner, 1924

An Agricultural Testament, Sir Albert Howard, 1940, Oxford University Press

The Living Soil, Lady Evelyn Balfour, 1943, Faber and Faber

Plowman's Folly, E.H. Faulkner, 1943, Grossett and Dunlap

Pay Dirt: Farming and Gardening with Composts, J.I. Rodale, 1945, Devin Adair Co.

Malabar Farm, Louis Bromfield, 1947, Harper

A Sand County Almanac and Sketches Here and There, Aldo Leopold, 1949, Oxford University Press

Silent Spring, Rachel Carson, 1962.

Living the Good Life, Helen (and Scott) Nearing, 1970.

The Unsettling of America: Culture and Agriculture, by Wendell Berry. 1977. San Francisco: Sierra Club Books. (also author of numerous books since then, this is his classic).

International Federation of Organic Agriculture Movements (IFOAM), 1st Conference Proceedings: Toward a Sustainable Agriculture." 1978. J.M. Besson and H. Vogtmann, eds.

Hard Tomatoes, Hard Times, Jim Hightower, 1978, Transaction Publishers

How to Grow More Vegetables Than You ever Thought Possible on Less Land Than You Can Imagine, J. Jeavons, 1979

New Roots for Agriculture, Wes Jackson, 1980, Univ. Nebraska Press.

Farming in the Dark

Books written by farmers or from their point of view:

Animal, Vegetable, Miracle – A year of food life. By Barbara Kingsolver. Harper Collins, 2007

Women of the Harvest – Inspiring storied of contemporary Farmers, by Holly L. Bollinger and Cathy Phillips. Voyageur Press, 2007

This Common Ground: Seasons on an Organic Farm, by Scott Chaskey. 2005. Viking Press.

Fields of Plenty: A Farmer's Journey in Search of Real Food and the People Who Grow It, byMichael Ableman, 2005. Chronicle Books. (previous books by Ableman also include On Good Land: The Autobiography of an Urban Farm, 1998, and From the Good Earth: A Celebration of Growing Food Around the World, 1993.

Harvest: A Year in the Life of an Organic Farm, by Nicole Smith and Goeff Hansen, 2004. The Lyons Press. (about their neighbors in Vermont)

Four Seasons in Five Senses: Things Worth Savoring, by David Masumoto. 2004. W.W. Norton and Company. (previous books by Masumoto also include: Letter to the Valley: A Harvest of Memories, 2004, Harvest Son: Planting Roots in American Soil, 1999, and Epitaph for a Peach: Four Seasons on My Family Farm, 1996.)

Clearing Land – Legacies of the American Farm, by Jane Brox. 2004. North Point Press.

Micro EcoFarming – Prospering from Backyard to Small Acreage in Partnership with the Earth. By Barbara Adams, 2004. New World Publishing.

Fields Without Dreams: Defending the Agrarian Ideal, by Victor Davis Hanson. 1997. Free Press. (California farmer)

The Contrary Farmer, by Gene Logsdon. 1995. Chelsea Green.

The Orchard – a Memoir. 1995. by Adele Crockett Robertson. Metropolitan Books, Henry Holt & Company, Inc. NY. (about a woman trying to save the family orchard from 1932-1934, near Ipswich, MA).

Future Harvest: Pesticide-free Farming, by Jim Bender. 1994. University of Nebraska Press. (organic grain farmer in Nebraska)

The Farming Game, by Bryan Jones. 1987. Ballantine Books.

Appendix: Recommended Reading

The Natural Way of Farming: the Theory and Practice of Green Philosophy. Masanobu Fukuoka, 1985, Japan Publications

Books about dangers/politics in the food system (a small sample):

The Omnivores' Dilemma – a natural history of four meals, by Michael Pollen, Penguin 2007 (paperback version)

Raising Less Corn, More Hell, by George Pyle, 2005. Public Affairs.

Diet for a Dead Planet – How the Food Industry is Killing Us, by Christopher Cook. 2004. The New Press.

Fatal Harvest, by Andrew Kimbrell, 2002, Island Press

Food Politics, by Marion Nestle. 2002. University of California Press

Fast Food Nation – the Dark Side of the All American Meal, by Eric Schlosser,2001. Houghton Mifflin Company.

Hungry for Profit – The Agribusiness Threat to Farmers, Food, and the Environment, Ed. By Fred Magdoff, John Bellamy Foster, and Frederick H. Buttel. 2000. Monthly Review Press.

Mad Cowboy – Plain Truth from the Cattle Rancher who Won't Eat Meat, by Howard Lyman, with Glen Merzer. 1998. Scribner.

Books that are hopeful, or present a plan:

Blessed Unrest – How the largest movement in the world came into being and why no one saw it coming, by Paul Hawken, Viking, 2007

Good Growing – Why Organic Farming Works. By Leslie Duram. 2005. University of Nebraska Press. (includes interviews with 5 farmers)

Agrarian Dreams – the Paradox of Organic Farming in California, by Julie Guthman,2004, University of California Press.

Women and Sustainable Agriculture—Interviews with 14 Agents of Change. 2004. Anna Anderson. McFarland and Company.

Rural Renaissance: Renewing the Quest for the Good Life, by John Ivanko and Lisa Kivirist. 2004. New Society Publishers.

Eat Here – Reclaiming Homegrown Pleasures in a Global Supermarket, by Brian Halweil, 2004. W.W. Norton & Company.

Farming in the Dark

Ecoagriculture: Strategies to Feed the World and Save Wild Biodiversity, by Jeffrey McNeely and Sara Scherr. 2002. Island Press.

Farm as Natural Habitat, by Laura Jackson and Dana Jackson. 2002, Island Press.

New Agrarianism – Land, Culture, and the Community of Life, by Eric T. Freyfogle, 2001 Island Press.

French Fries and the Food System – A Year-round Curriculum Connecting Youth with Farming and Food, by Sara Coblyn and the Food Project Community, 2000, The Food Project.

Other books that could be used as text books in sustainable agriculture classes

Agroecology – The ecology of sustainable food systems, 2nd. Ed. By Stephen R. Gliessman, CRC Press, 2007

Developing and Extending Sustainable Agriculture – a new social contract. Charles A. Francis, Raymond P. Poincelot and George W. Bird, Editors. Haworth Food & Agriculture Products Press, 2006

Science-Based Organic Farming 2006: Toward Local and Secure Food Systems, Charles Francis, Katja Koehler-Cole, Twyla Hansen, Peter Skelton, Eds. University of Nebraska-Lincoln Extension Division, Center for Applied Rural Innovation. Published on-line at: www.cari.unl.edu/

Teaching Organic Farming & Gardening: Resources for Instructors, UC Santa Cruz Center for Agroecology and Sustainable Food Systems. 2003. http://casfs.ucsc.edu/education/instruction/tofg/index.html#order

The Next Green Revolution – Essential steps to a healthy, Sustainable Agriculture. By James E. Horne and Maura McDermott, Food Products Press, 2001

Agroecology; the Science of Sustainable Agriculture. Miguel Altieri. 1995. Westview Press, 2nd ed..

Introduction to Permaculture, By Bill Mollison, 1991, Tagari Publications, Australia.

Sustainable Agriculture in the Temperate Zones, Charles A. Francis, Cornelia Butler Flora and Larry D. King, Editors. John Wiley & Sons, Inc. 1990.

Acknowledgements:

I would like to offer special thanks to Terrie Becerra for her tireless efforts as a copy editor, and also as sounding board for ideas. In addition I would like to express gratitude and appreciation to my husband Raad, for his patience, companionship, and encouragement during this entire process.

Finally, thank you to all of those who have mentored me in the past, present, and into the future, including the farmers and others featured in this book.